# Ecology and Environmental Science

# Ecology and Environmental Science

Edited by Jeffery Clarke

New York

Published by Syrawood Publishing House,
750 Third Avenue, 9th Floor,
New York, NY 10017, USA
www.syrawoodpublishinghouse.com

**Ecology and Environmental Science**
Edited by Jeffery Clarke



International Standard Book Number: 978-1-68286-652-8 (Hardback)



**Cataloging-in-Publication Data**

Ecology and environmental science / edited by Jeffery Clarke.
p. cm.
Includes bibliographical references and index.
ISBN 978-1-68286-652-8
1. Ecology. 2. environmental sciences. I. Clarke, Jeffery.
QH541 .E26 2019
577--dc23

# TABLE OF CONTENTS

# PREFACE

The purpose of the book is to provide a glimpse into the dynamics and to present opinions and studies of some of the scientists engaged in the development of new ideas in the field from very different standpoints. This book will prove useful to students and researchers owing to its high content quality.

Environmental science is a vast field concerned with the study of numerous environmental phenomena. As a discipline, environmental science integrates theories and concepts of many other scientific fields like biology, chemistry, plant science, limnology, etc. Ecology, on the other hand, is an interdisciplinary field of biology that studies the relationships of organisms with their immediate surroundings. Ecology has a number of applications across various fields of study such as conservation biology, natural resource management, wetland management, etc. This book attempts to present the concepts and theories central to the field of ecology in detail and examine their relevance in the field of environmental science. It unfolds the innovative aspects of ecology and illustrates how the techniques of this field can be applied for the progress of environmental science in the future. The topics included in this book are of utmost significance and bound to provide incredible insights to readers. Those in search of information to further their knowledge will be greatly assisted by this book.

At the end, I would like to appreciate all the efforts made by the authors in completing their chapters professionally. I express my deepest gratitude to all of them for contributing to this book by sharing their valuable works. A special thanks to my family and friends for their constant support in this journey.

**Editor**

1

# A global perspective on decadal challenges and priorities in biodiversity informatics

A Townsend Peterson*, Jorge Soberón and Leonard Krishtalka

**Abstract**

Biodiversity informatics is a field that is growing rapidly in data infrastructure, tools, and participation by researchers worldwide from diverse disciplines and with diverse, innovative approaches. A recent 'decadal view' of the field laid out a vision that was nonetheless restricted and constrained by its European focus. Our alternative decadal view is global, i.e., it sees the worldwide scope and importance of biodiversity informatics as addressing five major, global goals: (1) mobilize existing knowledge; (2) share this knowledge and the experience of its myriad deployments globally; (3) avoid 'siloing' and reinventing the tools of knowledge deployment; (4) tackle biodiversity informatics challenges at appropriate scales; and (5) seek solutions to difficult challenges that are strategic.

**Keywords:** Biodiversity informatics, Data, Infrastructure, Training, Capacity-building

## Background

Biodiversity informatics (BI) is simultaneously an old field and a very young one. Its major sources of data are old: records associated with physical voucher specimens housed in museums and herbaria that, in many cases, are still in the form of cross-referenced card files, paper catalogs, and other pre-digital ledgers. As a new discipline, however, BI has a computer-aided history of only a few decades, evolving from simple databases of collections and observations to detailed, interactive, and flexible systems of information management, modeling, analysis, and interpretation. Indeed, BI as a research enterprise in terms of analytical and theoretical power, sophistication, and research output, has expanded enormously during the last two decades.

Several workers in the field, however, have expressed concern that this arena of research is not driven by conceptual inquiry and fundamental questions. For example, a recent analysis [1] concluded that developments in BI have been driven largely by availability of technologies and data, and rarely by important and exciting conceptual challenges and theoretical predictions. That is, BI's evolution to date has been driven by the kinds of inquiry that become tractable or feasible, rather than by grand challenge questions that seek to discover deep, underlying patterns and processes: e.g., how many species inhabit Earth and what processes govern their distributions? Such key questions have largely lain fallow.

Hardisty and Roberts [2] laid out a 'decadal view' of challenges and priorities in BI, with several goals that are sound and that we applaud. However, their viewpoint looks solidly northward, i.e., their BI world is explicitly and almost exclusively European. It is highly commendable that the European community advances its BI resources and capabilities. However, biodiversity, which is richest in the Tropics, is a global phenomenon: the majority of species are on other continents, as are the bulk of biodiversity scientists and users of the science. Finally, as many others have noted, northern institutions, including European museums and herbaria, hold much of the historical, legacy biodiversity information—voucher specimens and associated data—for many of the Tropical countries, owing to colonial-era explorations. Indeed, in this sense, the rest of the world requires and depends on advances in European BI, but ideally these efforts should be informed, designed, mediated, and implemented by a

*Correspondence: town@ku.edu
Biodiversity Institute, University of Kansas, 1345 Jayhawk Blvd., Lawrence, KS 66045, USA

global view, one framed in international and intercontinental contexts.

This communication offers an alternative decadal view for biodiversity informatics. Hardisty and Roberts [2] listed tasks that have largely already been initiated or, in some cases, resolved. A more profound and challenging set of tasks lies ahead: (a) capture data associated with the billions of biodiversity information records (i.e., scientific specimens) held in 'northern' museums and herbaria; (b) share those data efficiently and collaboratively, effectively repatriating the data to countries of origin; and (c) share investment in training new generations of scientists in the concepts, tools, and theory to model, analyze, and apply these vast new data resources. Accomplishing these three tasks will propel BI worldwide, and will create a potent force in the overriding goal of informing and advancing smart global environmental stewardship.

## Biodiversity-rich and (frequently) information-poor regions

The countries and regions of the Earth are characterized by marked differences in richness of biodiversity. Specifically, among well-known biodiversity gradients, the temperate-to-tropical one is dominant, with tropical regions holding biotas that are considerably more diverse. This imbalance links to the Linnaean and Wallacean shortfalls [3, 4], which, respectively, are the massive gaps in knowledge about the details of the diversity and distribution of units of biodiversity. These gaps are particularly acute in the developing world, where biodiversity tends to be understudied in spite of its richness, and for which the huge volume of existing biodiversity data is still not available.

In sharp contrast to this biodiversity gradient is the reverse pattern of the history and current status of the world's wealth, power, and education, and its collateral effect of much less access to information and education. Colonial history, among other factors, particularly in Tropical regions during the period of most intense biodiversity exploration (approximately 1850–1950), resulted in massive collections of animals and plants and associated data being extracted from these countries and deposited in institutions across Europe and North America (Figure 1). This bias is mirrored by the demographics of biodiversity specialists, who are similarly concentrated in North American and European institutions [5].

Now, however, the geography of the biodiversity science enterprise is in rapid flux, with strong growth in research, education, and infrastructure since the end of the twentieth century in many developing countries [6]. Indeed, many sectors of the developing world—most notably Mexico, Colombia, Brazil, and South Africa—have achieved such growth that they have 'flattened' the world of global biodiversity science; several other countries are not far behind. As such, this globalization of BI resources, expertise, and research is redefining and broadening the 'centers' of the biodiversity science universe to domains beyond North America and Europe; the process is far from complete, but the tendency is clear.

## The uneven state of biodiversity science in Europe

Biodiversity science in Europe is thriving. Numerous research groups are generating systematic revisions (e.g., [7]), molecular phylogenetic and phylogeographic studies (e.g., [8]), biogeographic and ecological models (e.g., [9]), and environmental syntheses [10]. Other initiatives are extending biodiversity science in Europe to related fields (e.g., B4Life, BEST, EBRI).

Simultaneously, however, the underlying promise of future European biodiversity science might be seriously constrained by institutional history and culture. Breakthrough advances in biodiversity science depend on harnessing and integrating two primary realms of evidence: one comprises legacy biodiversity data, such as those documented by existing biocollections in museums and herbaria; the second realm comprises data from new, rich, and geographically widespread biocollections that are focused by modern research questions. With some taxonomic and institutional exceptions [11], European biocollections appear to be failing both sides of this critical equation: the legacy collections, despite their overwhelming importance in documenting past global biodiversity [12], are not being digitized or shared at a rate that will bring them into currency for science and society in time to inform solutions to the planet's biodiversity crisis (note, e.g., that the Natural History Museum of the UK appears to serve no records via GBIF; the Royal Botanic Gardens of the UK serves only 728,527 records out of a total of 7 million specimens, or about 10%) [13]. At the same time, the impetus is modest, if not absent, for conducting new, collections-based surveys and inventories that document current global biodiversity with new methodologies and tools, even within Europe [14]. As such, and again with exceptions, European biocollections institutions are neither investing in the future of BI, nor evolving the BI potency of the enormous volume of data already resident in their museum cases and ledgers.

Instead, Europe appears to be a champion of biodiversity meetings, workshops, and conferences (e.g., the recent e-Biosphere and GBIC [15] congresses), the vast majority of which merely repeat the points and priorities from decades of previous meetings, and conclude, as action items, the need for more meetings. Of course, European institutions are not alone in this malaise, but the situation there appears to be more acute than in the Americas, Asia, Africa and Australia, where institutions are more actively grabbing the BI future.

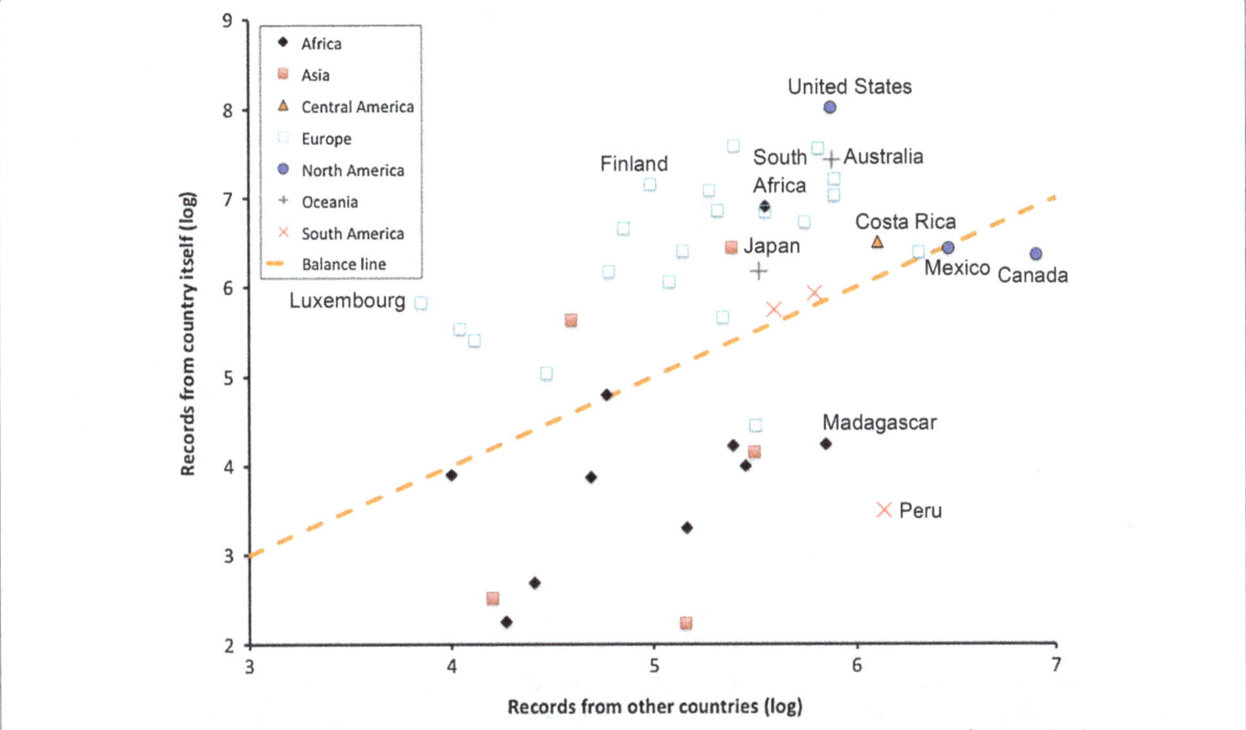


**Figure 1** Summary of Digital Accessible Knowledge for countries worldwide, drawn from the Global Biodiversity Information Facility in January 2014, showing $\log_{10}$ of numbers of records coming from the within the country versus those being provided by institutions in other countries. Countries (many, from all continents) that serve no data are omitted from the graphic. The *dashed line* indicates even balance between records from inside and outside of the country.

## Challenges and priorities

Of the detailed list of recommendations by Hardisty and Roberts [2], many appear to be post hoc and anticipatory of activities already begun. Essentially, their decadal view is of the past decade, not the next one, thus promising little new in the way of progress. For example:

1. Their challenge to assemble a comprehensive taxonomic summary of biodiversity does not seek a compendium of valid taxonomic names, but merely a list of names in use, a much more modest goal.
2. Their recommendation to develop persistent identifiers for biodiversity records has been a consistent topic of intense discussion and development [16–19] during the past decade; moreover, persistent identifiers are already in use in many institutions.
3. Their recommendation for mechanisms to evaluate data fitness for use in biodiversity studies misses major advances and effective solutions already in place [20, 21].
4. Their call to address the management and integration of observational data is apparently unaware of substantial accomplishments in this arena by the AudubonCore group [22].

These and other examples from their paper [2] illustrate a vision of slow, gradual, incremental change, often in areas in which significant change has already begun or occurred. Perhaps the most serious casualty of this incremental view are Europe's vast legacy collections and associated data records of past global biodiversity. With exceptions, digitization of this huge library of biodiversity information is either not occurring, or is occurring too slowly and haphazardly for global biodiversity science to progress. Indeed, digital mobilization of existing biodiversity knowledge is the first of five challenges we offer for the next decade. These challenges are designed to be fully global, applying equally across the international community of biodiversity institutions and infrastructures.

### Challenge #1: Mobilize existing knowledge

Biocollections of scientific specimens are, in effect, massive storehouses of irreplaceable biodiversity data. Although data aggregated from heterogeneous sources often have problems [23–28], such data are used extensively and increasingly by scientists in both developed and developing countries. For example, in 2013, the Global Biodiversity Information Facility (GBIF) data portal saw >130,000 visits from locations in the United States, but with many thousands of visits from Mexico, Colombia, Argentina, Brazil, and India, among many others; indeed, GBIF's new data portal (2013) served 3.67 billion records in its first 40 h [29]. As of its last report (August 2014), GBIF has compiled 886 scientific papers that used GBIF-mediated data in analyses ranging from basic research to applications of biodiversity policy (http://www.gbif.org/mendeley/usecases); although surely some of those papers do not actually use GBIF-mediated data in analyses, the utility of the resource is clear.

However, the enormous volume of biodiversity data that remains in analog format is nowhere near as easily accessed, shared, analyzed, or interpreted. Progress in accelerating and optimizing workflows and protocols for digitizing such data has demonstrated that meeting this challenge is increasingly feasible [30, 31]. Recent estimates are that museums and herbaria worldwide hold 1.5–2.0 billion specimens [32, 33], yet only about 10% of that total is currently accessible via GBIF, the largest aggregator of specimen records. Although smaller-scale initiatives provide access to additional specimen records (an excellent example is speciesLink, http://www.splink.org.br/), the bulk of the data associated with the world's biocollections remain inaccessible to biodiversity science.

The causes for this lack of progress in digitization and sharing of biocollections data are manifold, but common themes are budgetary and sociological [30]. Data 'owners' cite a spectrum of reasons: fear of activists or biopiracy; concern about insufficient data quality; a desire for economic return, or to control access and use of the scientific data; and the cost of digitizing collection data, on the order of US$1–10 per specimen [34], although initiatives have begun to reduce these costs significantly (http://beyondthebox.aibs.org/). Another possibility is that institutions may have assessed costs and benefits of such efforts, and decided that digitization is not worth the effort and investment. In many cases, however, the most serious hurdle is simply institutional inertia or strategic apathy—digital mobilization of their collections data is not a priority.

Even when collections data are in digital format, they often are not made available broadly and openly, despite major community technological initiatives to foster data access and sharing including development of the DarwinCore standard, information transfer protocols such as DiGIR and IPT, and implementation of large-scale biodiversity information portals (e.g., VertNet, GBIF, speciesLink, REMIB, UNIBIO, SEINet, iDigBio).

Whatever the reasons, in effect, by not moving ahead in digitizing data, institutions effectively quarantine and sequester biodiversity knowledge held in non-digital formats from modern research on biodiversity phenomena of considerable interest and currency. Rescuing these data digitally from stealth mode enables biodiversity informatics to transform a descriptive biodiversity enterprise into a powerfully predictive one [35–37]. A major challenge is, therefore, catalyzing the digital mobilization and sharing of the massive but dormant biocollections data in institutions across Europe, North America, Russia, Brazil, India, and China.

### Challenge #2: Share expertise globally

A corollary to 'flattening' [sensu 38] the biodiversity science world is the desperate thirst for more information, tools, knowledge, and conceptual frameworks. That is, as science communities develop and begin to thrive in the developing world, increasing numbers of students and researchers are eager to learn the newest techniques and frameworks. Despite these advances and growing opportunities, most expertise currently still resides in Europe and North America.

Therefore, global sharing of skills in systematics and biodiversity informatics is a requisite step for true globalization of the community and the science. Without such training and expansion of the user community, the science and policy potential of increasingly available data will go unexplored, particularly in the developing world—the geographic areas of greatest biodiversity and environmental concern.

Capacity-building and training opportunities, in the narrow sense, are only important in the shortest term. Rather, we contend that this new, 'flat' world of biodiversity science demands full educational opportunities for students from developing countries, equivalent to those in the developed world, i.e., the opportunity to complete a doctoral program in research and education. Where these opportunities have opened, developing countries have become leaders in biodiversity information management: South Africa with SANBI [39], Mexico with CONABIO [40], Colombia with Instituto von Humboldt [41], Costa Rica with INBio [42], Brazil with CRIA [43], and India with several initiatives (e.g., India Biodiversity Portal; http://indiabiodiversity.org/). Scientists at these institutions have tackled and solved complex problems of assembling, maintaining, and sharing large biodiversity databases, and routinely perform sophisticated analyses that provide the science underpinnings of policy. In turn, these institutions now have the capacity and capability to develop high-level training programs that formerly depended on North American or European leadership. This model is the good virus of biodiversity science: at the moment, programs that spread it are a cottage industry, when what is needed are industrial strength solutions.

**Challenge #3: Avoid silos and reinvented wheels**

A major challenge in biodiversity science is the degree to which information 'silos' are constraining integrative networks and deep insights. Quite simply, diverse data realms do not talk to one another very easily, as we pointed out in a recent review of the big questions in biodiversity informatics [1]. An excellent example is integrating the data that document and connect genome composition with the data that document species' geographic occurrences, which is critical to elucidating insights into drivers of speciation and diversification [44]. Formats and protocols for persistent individual record identifiers have been developed that would greatly facilitate such crosstalk and integration, but they are not broadly available in either the geographic-occurrence data world (e.g., GBIF) or the genomic-data world (e.g., GenBank). As a consequence, the two data realms remain as distinct islands of data. To be linked and related, data about individual organisms represented in both realms must often be analyzed by hand. Initiatives to connect the real biotic data realms of genomes and geographic occurrence [45, 46] require a massive boost.

More broadly, new initiatives in biodiversity science frequently wave the flag of innovation and synthesis, but in the competitive game of identity politics, turf, and science funding, each initiative is effectively siloed from other such projects, sometimes on purpose, and sometimes for lack of broader vision of the importance of cross-linkage. As a result, the wheels of biodiversity science—standards, tools, data schemas, and structures, etc.—are re-invented, without benefit to the advancement of the field (see, e.g., the discussion of PIDs by Hardisty and Roberts, when a major evaluation has been developed recently [47]). Indeed, in some instances, such re-invention has not resulted in competitive vigor but a growth in biodiversity's Babel—the non-interoperability of a plethora of data and analytical systems.

Most importantly, perhaps, biology lacks an underlying 'unified theory of biodiversity', and must rely on more local component frameworks, such as theories of natural selection and molecular evolution, ideas from island biogeography, etc. A broad, overarching theory would provide both the coherence and scaffolding on which to assemble and link the many entities of biodiversity information—molecular, physiological, morphological, systematic, ecological, phylogenetic, and spatial. Achieving this grand synthesis, however, is severely hampered by disciplinary and data silos; indeed, even exploration of component frameworks is hindered by lack of linkages among silos.

**Challenge #4: Deal with biodiversity science development challenges at the appropriate scale**

The challenge of understanding biodiversity is neither regional nor global, but highly multiscalar—a network of local challenges that sums to a global-scale enterprise that must be engaged on multiple levels [48]. As a corollary, all aspects of this enterprise—building data resources, protocols, and human resources in biodiversity science—should also be multiscalar.

This principle is precisely why building local capacity and institutions are indispensable components of the biodiversity enterprise. In biodiversity science, local questions, perspectives, values, and approaches are as critical to success at that scale as are regional or national issues at those scales. Indeed, the work of biodiversity scientists and the education and training of students occurs at everything from local to global scales. This multiscalar property of biodiversity science belies the more geographically narrow view of Hardisty and Roberts [2].

We contend that solutions to the challenges described above and in Hardisty and Roberts [2] can be found in multiscalar approaches. For instance, digital capture and mobilization of the world's biocollections data should be designed and implemented around resource partnerships between developing-world scientists, students, and institutions, whose biodiversity mandates often depend on acquisition of such data, and developed-world institutions equally eager to bring these data to the forefront of biodiversity research and synthesis. Such collaboration maximizes purpose, personnel, protocols, and institutional resources in meeting a daunting challenge.

Similarly, whereas GBIF has just passed the monumental mark of half of a billion records served via its data portal, too many of them are not adequately fit for use, particularly in lacking georeferencing. This challenge aches for a distributed global consortium of partners with expertise, tools, and experience in georeferencing data associated with biodiversity records, each partner being most knowledgeable about and committed to improving the data from its respective region. Global entities, perhaps even GBIF, could integrate and coordinate the effort, knowing that the geographic knowledge needed for broad and effective execution of this initiative is inherently regional or national.

In sum, although a truism, it bears repeating that solutions to challenges in biodiversity science require efforts at the appropriate scale, whether global, national or local, and often collaboration among entities at different levels. For example, a local issue will require coordination and implementation at that level, with funding and political will at national and regional levels [49]. This point is precisely the reason why the training of national and regional scientists, as well as local cadres (park managers, guides, rangers, etc.) is a *sine qua non* of biodiversity management.

### Challenge #5: Find strategic solutions

Goals cast so generally as to be unachievable are not particularly useful. In this sense, broad, overarching recommendations and targets that largely repeat initiatives already underway [2] are puzzling. Instead, setting limited, achievable goals with built-in rewards of accomplishment, significance, and impact will be much more strategic. As such, goals in biodiversity informatics, once reached, should bear near-term, exciting, and novel fruit.

For example, accumulating biodiversity information by convenience rather than explicit strategy will build the absolute number of records served, but at the severe expense of mere quantity over quality, i.e., fitness for use for biodiversity science [27, 28, 50]. An example was the goal set by GBIF some years ago of serving one billion biodiversity records by 2010. Rather, a different, multipronged strategy would begin with a comprehensive gap analysis of existing biodiversity data. One prong might be to complete the coverage of groups that are already well-represented and near-comprehensive (e.g., birds; Figure 2), which would provide a complete view of known diversity in a single group. In parallel, other prongs would address remaining taxa according to explicit criteria, protocols, and lessons learned.

## Conclusions

Our decadal view of biodiversity informatics stands in sharp contrast to that of Hardisty and Roberts [2]. To be fair, we acknowledge the political and funding realities of European science, and Hardisty and Roberts [2] are at least explicit in their exclusive focus on Europe's next decade. Nevertheless, the Hardisty and Roberts [2] paper is a useful cipher for the thinking and ills that pervade the field more broadly, which manifest in regional (not global) thinking and activity.

Unlike Hardisty and Roberts [2], our decadal view deliberately leapfrogs the well-worn points and priorities of the past decade or two, all of which were repeated as almost mantric recitations at innumerable meetings the three of us have attended. Instead, we focus our view on the fast-evolving global scientific and social landscape, which will govern the next generation of advances in biodiversity informatics. This landscape is increasingly being flattened and more evenly populated with scientists, students, institutions, initiatives, and data resources in countries that previously were considered scientifically underdeveloped. This flatter world is a powerful selective agent armed with big challenges and opportunities. Biodiversity science must adapt and adjust. Those sectors that won't, will see its world sweep on by.

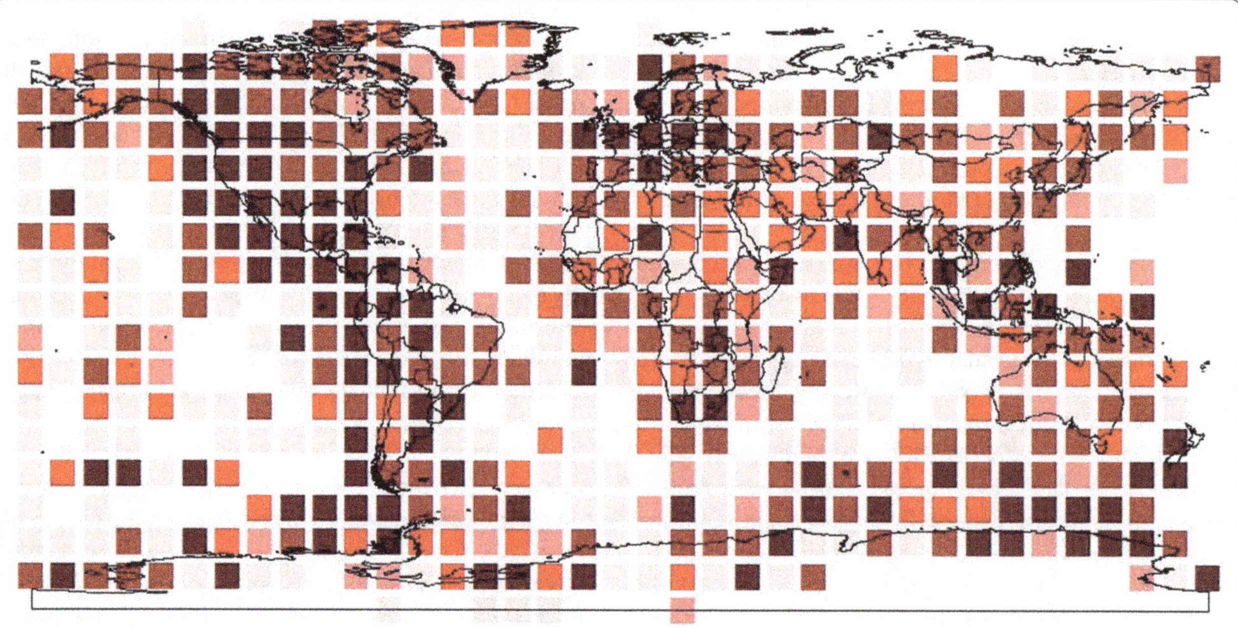

**Figure 2** Global summary of completeness of knowledge of birds of the world at 10° spatial resolution. *White* none of avifauna documented, *darkest red* avifauna completely documented. From Peterson et al. (in prep.).

**Authors' contributions**
All three authors contributed equally to the development of this paper. All authors read and approved the final manuscript.

**Acknowledgments**
We thank our many colleagues in biodiversity informatics for rich discussion and debate over many years, although they may not agree with much of what is said in this manuscript.

**Compliance with ethical guidelines**

**Competing interests**
The authors declare that they have no competing interests.

## Response

By Alex Hardisty

E-Mail: hardistyar@cardiff.ac.uk

Address: Cardiff University, School of Computer Science and Informatics, Queens Buildings, 5 The Parade, Cardiff CF24 3AA, UK

Our article [2] set out (for the first time, to our knowledge) a decadal view of challenges and priorities presently facing practitioners in biodiversity informatics. We presented a range of actions necessary to link the extensive array of available computerised resources and tools into a commonly-shared sustainable e-Infrastructure supporting all aspects of biodiversity and ecosystems science. We were explicit in saying we had considered the topic mainly from the European perspective. We provided a rallying point for community efforts, mainly in Europe it has to be said. We offered a baseline against which funding agencies could, if they choose assess new informatics proposals. However, we also said the vision is of global interest and relevance. The views were the result of a public consultation involving some 75+ contributing respondents, not all from Europe. On behalf of those contributors I'm grateful for the further correspondence by Peterson et al., which provides welcome additions to an important debate.

Biologists, ecologists, taxonomists, technologists and informaticians have to communicate and interact together. Only together as a global community can we achieve the right, interoperable, common informatics solutions to assist the science to generate the knowledge of how the biosphere works. Predicting the biosphere and providing sufficient evidence to manage it robustly is a greater challenge still. But, if we want to be able to do this in a scalable way, there are as Purves et al. [51] point out huge challenges to building useful models; not least in obtaining the appropriate types of data to validate the model predictions.

Data mobilisation, built on foundations of acquisition, whether by digitisation or other means; curation and preservation; discovery and open access; and ability to process; with inter-linkages and names playing their central roles is thus an essential strategic goal but one that has to be expressed as being for the explicit purpose. In this we can draw lessons from meteorology in the 1950s, 1960s and 1970s [52] where the purposes of geopolitics (nuclear arms race, and being first to put a man on the moon) were served with prioritised funding for meteorological models and supercomputing, and scientists collaborating together. This was not only to develop the models but also to identify and close data gaps and to re-work/invert the existing data. They "made global data and they made their data global". Today that modern data, collected almost continuously around the world and the models that rely on it have significant commercial as well as scientific value for all kinds of stakeholders.

Essential biodiversity variables (EBV) [53] or similar indicators are a parallel case and a core future business; potentially with high scientific and commercial value that demands removal of barriers to global interoperability [54]. Just like weather variables, EBVs imply the ability to measure and calculate for any geographic area, small or large, fine-grained or coarse; at a temporal scale determined by need and/or the frequency of available observations; at a point in time in the past, present day or in the future; at appropriate scale, for any species, assemblage, ecosystem, biome, etc.; using data for that area/topic that may be held by any and across multiple data resources; using a standardised and widely accepted workflow capable of executing in any research infrastructure; and by any person anywhere.

What we see today in biodiversity informatics is, to use terminology from the article, mainly a "cottage industry"; or worse a subsistence economy with pockets of cottage industry. In the era of global societal challenges, global cooperation and a flatter world we need to make the transition to industrial-grade solutions. We need to work collectively, engaging with industry such that biodiversity/ecology professionals and industry together improve the way computer systems share, utilise and process information for biodiversity science. We must promote the coordinated use of standards we already have, and identify and adopt or develop those new ones still needed. We must mobilise the data to serve the purpose, rather than mobilising for mobilising sake. This creates interoperability benefits for the sector overall and profit opportunities to stimulate industry interest. Lessons from other sectors (healthcare for example [55, 56]) can show us how to tackle the issue.

Responding to some of the specific points in the article:

1. Peterson et al. are concerned that we are northward looking and almost exclusively European. As noted, we were explicit about the European perspective but the main themes of the vision [integration of available resources; support for scientific synthesis; a shared

maintained multi-purpose network of computer-based data and processing services using a small set of (global) interchange standards] and the details needed to realise these themes are relevant in all corners of the world. This view is borne out by results from the international coordination project, CReATIVE-B working towards a global virtual environment for biodiversity research in its roadmap [54]. An international High Level Stakeholders Group comprising representatives of biodiversity and ecosystems research infrastructures from around the world serves to promote policy liaison and recommendations and coordinate towards that aim. The recently funded GLOBIS-B project to further coordinate informatics work to support EBVs, has support also from Australia, Brazil, China, South Africa, USA as well as Elixir, GBIF and GEO BON.

2. I see the alternative view and the five challenges offered by Peterson et al. not as a competing vision that "stands in sharp contrast" to our own but as a re-stating of or complement to what we propose. The issue is not that work remains to be started in all the areas we suggest nor that technical solutions still need to be found. Instead, it is that the works in progress need to become more widely known, to consolidate, to converge, and to embed in everyday practice right across the community. In this sense our vision is concerned much more with promoting infrastructure emergence and community consensus to achieve widespread buy-in, adoption and usage, than it is about solving any particular technical problem. We need to move more towards sustained funding anchored in pay-per-use or institutional commitments than to continue current hand-to-mouth dependencies on externally funded short-term projects.

3. Peterson et al. conclude with talk of leap-frogging, and I have some sympathy with that view. They evoke the fast-evolving, flatter more populous world of multiple stakeholders and encourage us to adapt to it or die. They ask for strategic solutions situated in this new world order and are right to do so but they do not offer the alternative scenarios that could play out in it. Without these we cannot yet find the best path to pursue for the most likely circumstances or more likely, for several different circumstances. We need to increase our depth of understanding by application of horizon scanning, scenario building and multi-path mapping techniques [57]. As should be clear by now, it is not the biodiversity informatics research that is the concern but the matter of how to translate results from that into everyday industrial-scale practice. Education and training curricula have an important role to play there as the authors have suggested but so does involvement of commerce/industry. I see with the hindsight of 2 years and from this perspective that our vision has not sufficiently addressed these and other sociological issues. Indeed in my own work establishing the Biodiversity Virtual e-Laboratory (BioVeL) infrastructure [58, 59] I see the new interest coming from eager young researchers outside of the established G8 and other western countries. However, I often ask myself whether we really sufficiently understand from the sociological and psychological perspectives how the complex technologies and methods we invent become effectively translated into practice. More work is needed.

In conclusion, I am happy that Peterson et al. have taken the time not only to read the original article but also to think about the issues and to write a response. I thank them for that and hope that such correspondence serves to further stimulate the debate and the consensus global action that has to follow. This is essential if modern biodiversity science, ecology and Earth stewardship are to fully benefit from the capabilities that informatics solutions offer.

**References**


1. Peterson AT, Knapp S, Guralnick R, Soberón J, Holder MT (2010) The big questions for biodiversity informatics. Syst Biodivers 8:159–168
2. Hardisty A, Roberts D (2013) A decadal view of biodiversity informatics: challenges and priorities. BMC Ecol 13:16
3. Bini LM, Diniz-Filho JAF, Rangel TF, Bastos RP, Pinto MP (2006) Challenging Wallacean and Linnean shortfalls: knowledge gradients and conservation planning in a biodiversity hotspot. Divers Distrib 12:475–482
4. Whittaker RJ, Riddle BR, Hawkins BA, Ladle RJ (2013) The geographical distribution of life and the problem of regionalization: 100 years after Alfred Russel Wallace. J Biogeogr 40:2209–2214
5. Gaston KJ, May RM (1992) Taxonomy of taxonomists. Nature 356:281–282
6. Soberón JM, Sarukhán JK (2009) A new mechanism for science-policy transfer and biodiversity governance? Environ Conserv 36:265–267
7. Buffetaut E (2014) Tertiary ground birds from Patagonia (Argentina) in the Tournouër collection of the Muséum National d'Histoire Naturelle, Paris. Bull Soc Geol Fr 185:207–214
8. Den Tex R-J, Leonard J (2014) The phylogeography of red and yellow coppersmith barbets (Aves: *Megalaima haemacephala*). Phylogenet Phylogenomics Syst 2:16
9. Boulangeat I, Gravel D, Thuiller W (2012) Accounting for dispersal and biotic interactions to disentangle the drivers of species distributions and their abundances. Ecol Lett 15(6):584–593
10. Garcia RA, Cabeza M, Rahbek C, Araújo MB (2014) Multiple dimensions of climate change and their implications for biodiversity. Science 344:1247579
11. Pignal M, Romaniuc-Neto S, Souza SD, Chagnoux S, Canhos DAL (2012) Saint-Hilaire virtual herbarium, a new upgradeable tool to study Brazilian botany. Adansonia 35:7–18
12. Navarro-Sigüenza AG, Peterson AT, Gordillo-Martínez A (2003) The big questions for biodiversity informatics. Bull Br Ornithol Club 123A:207–225
13. King N, Krishtalka L, Chavan V (2010) Thoughts on implementation of the recommendations of the GBIF Task Group on a Global Strategy and Action Plan for Mobilisation of Natural History Collections Data. Biodivers Inform 7:72–76
14. Fontaine B, van Achterberg K, Alonso-Zarazaga MA, Araujo R, Asche M, Aspöck H et al (2012) New species in the old world: Europe as a frontier in biodiversity exploration, a test bed for 21st century taxonomy. PLoS One 7:e36881

15. Hobern D, Apostolico A, Arnaud E, Bello JC, Canhos D, Dubois G et al (2013) Global biodiversity informatics outlook: delivering biodiversity knowledge in the information age. Global Biodiversity Information Facility, Copenhagen
16. Bafna S, Humphries J, Miranker DP (2008) Schema driven assignment and implementation of life science identifiers (LSIDs). J Biomed Inform 41:730–738
17. Page RDM (2008) LSID Tester, a tool for testing Life Science Identifier resolution services. Source Code Biol Med 3:2
18. Page RDM (2008) Biodiversity informatics: the challenge of linking data and the role of shared identifiers. Brief Bioinform 9:345–354
19. Roberts D, Chavan V (2008) Standard identifier could mobilize data and free time. Nature 453:449–450
20. Chapman AD (2005) Principles of data quality, version 1. Global Biodiversity Information Facility, Copenhagen
21. Chapman AD (2005) Principles and methods of data cleaning: primary species and species-occurrence data. Global Biodiversity Information Facility, Copenhagen
22. Morris RA, Barve V, Carausu M, Chavan V, Cuadra J, Freeland C et al (2013) Discovery and publishing of primary biodiversity data associated with multimedia resources: the Audubon core strategies and approaches. Biodivers Inform 8:185–197
23. Beck J, Ballesteros-Mejia L, Nagel P, Kitching IJ (2013) Online solutions and the 'Wallacean shortfall': what does GBIF contribute to our knowledge of species' ranges? Divers Distrib 19:1043–1050
24. Graham C, Ferrier S, Huettman F, Moritz C, Peterson AT (2004) New developments in museum-based informatics and applications in biodiversity analysis. Trends Ecol Evol 19:497–503
25. Soberón J, Arriaga L, Lara L (2002) Issues of quality control in large, mixed-origin entomological databases. In: Saarenmaa H, Nielsen E (eds) Towards a global biological information infrastructure, vol 70. European Environment Agency, Copenhagen, pp 15–22
26. Gaiji S, Chavan V, Ariño AH, Otegui J, Hobern D, Sood R et al (2013) Content assessment of the primary biodiversity data published through GBIF network: status, challenges and potentials. Biodivers Inform 8:94–172
27. Ballesteros-Mejia L, Kitching IJ, Jetz W, Nagel P, Beck J (2013) Mapping the biodiversity of tropical insects: species richness and inventory completeness of African sphingid moths. Glob Ecol Biogeogr 22:586–595
28. Yesson C, Brewer PW, Sutton T, Caithness N, Pahwa JS, Burgess M et al (2007) How global is the global biodiversity information facility? PLoS One 2:e1124
29. GBIF (2014) GBIF annual report 2013. Global Biodiversity Information Facility, Copenhagen
30. Vollmar A, Macklin JA, Ford L (2010) Natural history specimen digitization: challenges and concerns. Biodivers Inform 7:93–112
31. Beaman RS, Cellinese N (2012) Mass digitization of scientific collections: new opportunities to transform the use of biological specimens and underwrite biodiversity science. ZooKeys 209:7
32. Chalmers NR (1996) Monitoring and inventorying biodiversity: collections, data and training. In: Castri FD, Younes T (eds) Biodiversity, science and development: towards a new partnership. CAB International, Wallingford, pp 171–179
33. Ariño AH (2010) Approaches to estimating the universe of natural history collections data. Biodivers Inform 7:81–92
34. Smith GF, Steenkamp Y, Klopper RR, Siebert SJ, Arnold TH (2003) The price of collecting life. Nature 422:375–376
35. Krishtalka L, Humphrey PS (1998) Fiddling while the planet burns: the challenge for U.S. natural history museums. Mus News 77:29–35
36. Krishtalka L, Humphrey PS (2000) Can natural history museums capture the future? Bioscience 50:611–617
37. Krishtalka L (2009) Natural history museums as sentinel observatories of life on Earth: a public trust. In: Holo S, Alvarez M-T (eds) Beyond the turnstile: making the case for museums and sustainable values. AltaMira Press, London, pp 12–15
38. Freidman T (2005) The world is flat. Farrar, Straus and Giroux, New York
39. Crouch NR, Smith GF, Figuereido E (2013) From checklists to an e-flora for southern Africa: past experiences and future prospects for meeting target 1 of the 2020 global strategy for plant conservation. Ann Mo Bot Gard 99:153–160
40. CONABIO (2012) CONABIO: two decades of history, 1992–2012. In: Mexico DF (ed) Mexico: Comision Nacional para el Conocimiento y Uso de la Biodiversidad, pp 1–36
41. Samper C (1997) Linking science and policy: a research agenda for Colombian biodiversity. In: Press NA (ed) Nature and human society: the quest for a sustainable world. National Academy Press, Washington, pp 483–491
42. Sandlund OT (1991) Costa Rica's INBio: towards sustainable use of natural biodiversity. NINA Notat 7:1–25
43. Canhos DAL, Sousa-Baena MS, Souza S, Garcia LC, Giovanni RD, Maia LC et al (2014) Lacunas: a web interface to identify plant knowledge gaps to support informed decision-making. Biodivers Conserv 23:109–131
44. Chan LM, Brown JL, Yoder AD (2011) Integrating statistical genetic and geospatial methods brings new power to phylogeography. Mol Phylogenet Evol 59:523–537
45. Harmon LJ, Baumes J, Hughes C, Soberón J, Specht CD, Turner W et al (2013) Arbor: comparative analysis workflows for the tree of life. PLoS Curr 5. doi:10.1371/currents.tol.099161de5eabdee073fd3d21a44518dc
46. Miller JT, Jolley-Rogers G (2014) Correcting the disconnect between phylogenetics and biodiversity informatics. Zootaxa 3754:195–200
47. GBIF (2011) A Beginner's guide to persistent identifiers, version 1.0: http://links.gbif.org/persistent_identifiers_guide_en_v1.pdf. Global Biodiversity Information Facility, Copenhagen
48. Paavola J, Gouldson A, Kluvánková-Oravská T (2009) Interplay of actors, scales, frameworks and regimes in the governance of biodiversity. Environ Policy Gov 19:148–158
49. Folke C, Hahn T, Olsson P, Norberg J (2005) Adaptive governance of social–ecological systems. Annu Rev Environ Resour 30:441–473
50. Beck J, Böller M, Erhardt A, Schwanghart W (2014) Spatial bias in the GBIF database and its effect on modeling species' geographic distributions. Ecol Inform 19:10–15
51. Purves D, Scharlemann JPW, Harfoot M, Newbold T, Tittensor DP, Hutton J et al (2013) Ecosystems: time to model all life on Earth. Nature 493:295–297
52. Edwards P (2010) A vast machine: computer models, climate data, and the politics of global warming. MIT Press, Cambridge. ISBN 978-0-262-01392-5
53. Pereira HM, Ferrier S, Walters M, Geller GN, Jongman RHG, Scholes RJ et al (2013) Essential biodiversity variables. Science 339(6117):277–278
54. Flock Together with CReATIVE-B (2015) A roadmap of global research data infrastructures supporting biodiversity and ecosystem science. http://tinyurl.com/qcbx92q. Accessed 24 Feb 2015
55. Health Level 7 (2015) HL7 Web site. http://www.hl7.org. Accessed 24 Feb 2015
56. Integrating the Healthcare Enterprise (2015) IHE Web site. http://www.ihe.net. Accessed 24 Feb 2015
57. Hardisty AR, Peirce SC, Preece A, Bolton CE, Conley EC, Gray WA et al (2011) Bridging two translation gaps: a new informatics research agenda for telemonitoring of chronic disease. Int J Med Inform 80:734–744
58. Biodiversity Virtual e-Laboratory (BioVeL) infrastructure. http://www.biovel.eu/. Accessed 24 Feb 2015
59. Mathew C, Güntsch A, Obst M, Vicario S, Haines R, Williams AR et al (2014) A semi-automated workflow for biodiversity data retrieval, cleaning, and quality control. Biodivers Data J (2):e4221. doi:10.3897/bdj.2.e4221

2

# Propagule pressure increase and phylogenetic diversity decrease community's susceptibility to invasion

T. Ketola*, K. Saarinen and L. Lindström

**Abstract**

**Background:** Invasions pose a large threat to native species, but the question of why some species are more invasive, and some communities more prone to invasions than others, is far from solved. Using 10 different three-species bacterial communities, we tested experimentally if the phylogenetic relationships between an invader and a resident community and the propagule pressure affect invasion probability.

**Results:** We found that greater diversity in phylogenetic distances between the members of resident community and the invader lowered invasion success, and higher propagule pressure increased invasion success whereas phylogenetic distance had no clear effect. In the later stages of invasion, phylogenetic diversity had no effect on invasion success but community identity played a stronger role.

**Conclusions:** Taken together, our results emphasize that invasion success does not depend only on propagule pressure, but also on the properties of the community members. Our results thus indicate that invasion is a process where both invader and residing community characters act in concert.

**Keywords:** Bacteria, Competition, Invasion, Phylogenetic distance, Phylogenetic similarity and propagule pressure

## Background

Invasions of non-native species are considered a serious threat to native species across the globe. However, our ability to predict which species are invasive, and in what kind of communities invasions take place, is far from complete. Interestingly, understanding the reasons for the invasion success of certain species or populations seems to have gained more attention than those for why some communities are more prone to invasions [1, 2]. Perhaps the most renowned mechanistic explanation for invasions is the effect of propagule pressure. The amount of invading individuals, or the number of repeated introductions of invader species, have been found to positively affect the establishment of the invader, in both wild and laboratory experiments [1]. However, as the invasion process is a function of both the resident and invader species, not all communities are expected to be similar in their receptiveness to invasions. However, why some communities are more prone to invasions than others has received less attention than why some species are better invaders than others. The average and the variation in phylogenetic relatedness between the resident and invading species are likely drivers of community receptiveness to invasions.

The idea that community structure and, in particular, relatedness between an invader and the resident community could affect invasion probability has a long history [3, 4]. However, a consensus has not been reached on whether low or high phylogenetic relatedness could promote invasions [1, 5–10]. On the one hand competition-relatedness hypothesis, [9] suggests that the competitive exclusion of closely related species restricts invasion, for example because of overlapping resource use, as well as shared predators, herbivores, and pathogens. On the other hand, it has been proposed that invaders that are closely phylogenetically related to

*Correspondence: tketola@jyu.fi
Department of Biological and Environmental Science, Centre of Excellence in Biological Interactions, University of Jyvaskyla, P.O. Box 35, 40014 Jyvaskyla, Finland

the resident species could have better invasion success due to shared mutualists, facilitation, and an increased likelihood of tolerating similar conditions [1, 11]. Such opposite hypotheses could very well explain the lack of evidence for phylogenetic relatedness affecting invasion success in the wild, where these effects cannot be controlled. However, even in controlled experiments the evidence has been mixed. For example, a study of pairwise competition with protists and rotifers found support for the hypothesis that larger phylogenetic distances promote co-existence [6]. In contrast, in pairwise comparisons of algae [7, 8] and an extensive competitive experiment with 142 species of plants [9], coexistence was not affected by phylogenetic distance. Similarly, experimental evidence using several different kinds of two, three and four-species communities gave support to the idea that close average phylogenetic distances between the invader and resident community members resulted in reduced population sizes of the invader [12]. But again, evidence obtained from recent plant studies was equivocal (see [5]).

In addition to average phylogenetic distance, the likelihood of finding a free niche space could be affected by phylogenetic diversity within community. High phylogenetic diversity could affect the invaders chance of finding a free niche space, and thus the likelihood of invasion. The very limited existing evidence suggests that in plants large phylogenetic diversity decreases the receptiveness of communities to invasion [5, 13]. This mechanism is analogous to the mechanisms believed to be operating at species diversity level where invasion is expected to be smaller in more species diverse communities. This idea has gained support from small-scale experiments of species diversity [14, 15], but not from observations on a grander scale [16–18].

We designed, and ran an invasion experiment in which highly competitive bacteria *Serratia marcescens* invaded ten different kinds of three species bacterial communities over several bacterial generations. Our experimental setup allows us to test whether invasion success is positively or negatively linked to the phylogenetic distance between the invader and the community members (estimated from 16 sRNA based phylogeny). Under positive linkage, large average phylogenetic distance should facilitate *S. marcescens* invasion, whereas the opposite would be true if small phylogenetic distance facilitated invasion success. Moreover, we can test if greater variation in phylogenetic distances between the invader and resident community members potentiates invasion success, by lessening the competition. In addition, we tested the role of propagule pressure in invasion success by manipulating the amounts of individuals starting the invasion.

## Methods

### Experimental invasions

We initiated the experiment by creating ten different three-species bacterial communities out of five species: *Pseudomonas chlororaphis (ATCC 17418), Pseudomonas putida (ATCC 12633), Escherichia coli (ATCC 11775), Enterobacter aerogenes (ATCC 13048)* and *Leclercia adecarboxylata (ATCC 23216)* (Table 1). We selected easily culturable, laboratory-adapted, species to ease identification and culturing. All community species were propagated separately from frozen samples (1:1 high-density bacteria in nutrient broth and 80% glycerol at −80 °C) for 3 days at 30 °C prior to creating the communities [40 µl of thawed bacterial stock in 4 ml of Nutrient Broth (NB: 10 g Nutrient broth, 2.5 g Yeast extract in 1 l $ddH_2O$)]. The communities were initiated by pipetting 50 µl of each species (altogether 150 µl) into 6 ml of NB in 15 ml

**Table 1 List of species used in the communities (1–10) to which *S. marcescens* invaded**

| # | Resident species | Phylogenetic distance | Phylogenetic diversity |
|---|---|---|---|
| 1 | *Pseudomonas chlororaphis + Pseudomonas putida + Escherichia coli* | *0.109* | 4.60E−03 |
| 2 | *Pseudomonas chlororaphis + Pseudomonas putida + Enterobacter aerogenes* | *0.105* | 5.59E−03 |
| 3 | *Pseudomonas chlororaphis + Pseudomonas putida + Leclercia adecarboxylata* | *0.107* | 5.25E−03 |
| 4 | *Pseudomonas chlororaphis + Escherichia coli + Enterobacter aerogenes* | 0.066 | 5.08E−03 |
| 5 | *Pseudomonas chlororaphis + Escherichia coli + Leclercia adecarboxylata* | 0.067 | 4.90E−03 |
| 6 | *Pseudomonas chlororaphis + Enterobacter aerogenes + Leclercia adecarboxylata* | 0.063 | 5.38E−03 |
| 7 | *Pseudomonas putida + Escherichia coli + Enterobacter aerogenes* | 0.066 | 5.16E−03 |
| 8 | *Pseudomonas putida + Escherichia coli + Leclercia adecarboxylata* | 0.068 | 4.98E−03 |
| 9 | *Pseudomonas putida + Enterobacter aerogenes + Leclercia adecarboxylata* | 0.064 | 5.47E−03 |
| 10 | *Escherichia coli + Enterobacter aerogenes + Leclercia adecarboxylata* | 0.024 | *3.73E−05* |

Phylogenetic distances reflect average distances between resident community species and invader and phylogenetic diversity reflects the variance of distances between the invader and resident species (see Additional file 1 for phylogeny). Italic font in phylogenetic distance indicate communities belonging to group with large phylogenetic distance. In phylogenetic diversity italic font indicate a community with very low phylogenetic diversity centrifuge tubes. After 3 days of propagating the communities (at 30 °C in thermally controlled cabinet; ILP-12, Jeio Tech, Seoul, Korea), we performed invasions with *S. marcescens*, using four different propagule pressures (i.e. inoculum sizes): 12.5, 25, 50 and 75 µl, respectively. We did two replicates for each propagule pressure treatment, leading to a total of eight similar communities, which were invaded with four different amounts of the invader. In total we had 80 communities. The invading *S. marcescens* had been pre-grown for 3 days using the same procedure as with the community species.

On days 3 and 6 after invasion we transferred the communities to fresh medium. A well-shaken inoculum (500 µl) of each bacterial community was transferred to 5.4 ml of fresh NB medium. To mimic immigration, and to keep the community treatment effective throughout the experiment, we also added 33 µl of each pre-grown (3 days old stock) community species (according to community composition) into the communities. On 3, 6 and 9 days after invasion samples of communities were frozen (1:1 bacteria and 80% glycerol in cryotubes at −80 °C). The experiment continued at 30 °C for 9 days after the invasion.

### Invasion success

Invasion success was determined from frozen samples collected at days 3 and 9 after *S. marcescens* invasion. The population sizes of the invader and the rest of the community were determined using standard dilution series techniques (100 µl of thawed sample into 900 µl of sterile $dH_2O$ water. Dilution was repeated six times to achieve a $10^{-6}$ dilution). Diluted samples were plated on DNase test agar with methyl green (Becton and Dickinson and Company, Sparks, MD; premade at Tammer-tutkan maljat, Tampere, Finland), which allows growth of all used species, but distinguishes *S. marcescens* clones by the clear halo around colonies [19]. The plates were propagated at room temperature. 2 days after plating, both the total number of all bacterial colonies and colonies of *S. marcescens* were quantified to measure invasion success.

### Phylogenetic effects

The phylogenetic similarity among the used bacterial species was estimated from data obtained from NCBI Gene-Bank nucleotide sequences database using program Mega (v. 5.0). The phylogenetic tree includes the sequences FJ971882 (*E. aerogenes*), GQ856082 (*L. adecarboxylata*), NR_041980 (*S. marcescens* ssp. *marcescens*), NR_024570 (*E. coli*), AF094736 (*P. putida*), AB680102 (*P. chlororaphis*). The phylogenetic distances between species in Table 1 (invader, and resident community members) were calculated from a distance matrix based on the estimated phylogeny (see: Additional file 1). In the analyses we consider two variables. Average *phylogenetic distance* between the community members and the invader and *phylogenetic diversity*, which is the variance of phylogenetic distances between the community members and the invader (*S. marcescens* ssp. *marcescens*).

### Data analysis

Data analysis was conducted with glmer in R (Lme4), fitting generalized mixed model with a binomial error distribution and a logit link on colony counts of the invader *S. marcescens*. To control for the differences in the total colony count the amount of colonies of other species were used as a denominator (see [20]). By this way we modelled the odds of finding *S. marcescens* colonies from agar plates. We controlled also for the fact that the same community composition was used in four different propagule pressure levels and in two technical replicates by fitting community ID as a random factor in the model. Explanatory variables were propagule pressure, phylogenetic diversity and phylogenetic distance. Since phylogenetic distance was clearly either small or large (Table 1) we classified the phylogenetic distances to two groups (low < 0.09 < high). All other predictors were fitted as continuous fixed variables standardized to a mean of zero and standard deviation of one. Data analyses were done for both time points separately and testing two competing models; one with all linear effects, and one with both all linear and all quadratic effects. These two models were not simplified and their fit to the data were compared by AIC values. Since one community (number 10, Table 1) had a very low phylogenetic diversity compared to the other 9 communities the effect of phylogenetic diversity on invasion success was not fitted for this community (i.e. data were coded for this variable as missing values). Since one community (number 10, Table 1) had a very low phylogenetic diversity compared to the other 9 communities, this community has been removed from our analyses. We however provide the model with the 10th community as a Additional file 1.

## Results

In the beginning of the invasion (day 3), invasion success increased with increasing propagule pressure (est. = 0.245, s.e. = 0.052, $z = 4.740$, $p < 0.001$, Fig. 1a) and decreased with increasing phylogenetic diversity (est. = −0.612, s.e. = 0.167, $z = -3.676$, $p < 0.001$, Fig. 1b). In addition, large phylogenetic distance between invader and community did not favour invasion in comparison to small phylogenetic distance (small distance: 0.156, large distance: 0.252, $z = 1.720$, $p = 0.085$, Fig. 1c). Community ID had no effect on invasion (est. = 0.221, s.e. = 0.470, $z = 1.484$, $p = 0.172$). When we tested model with quadratic effects for phylogenetic diversity,

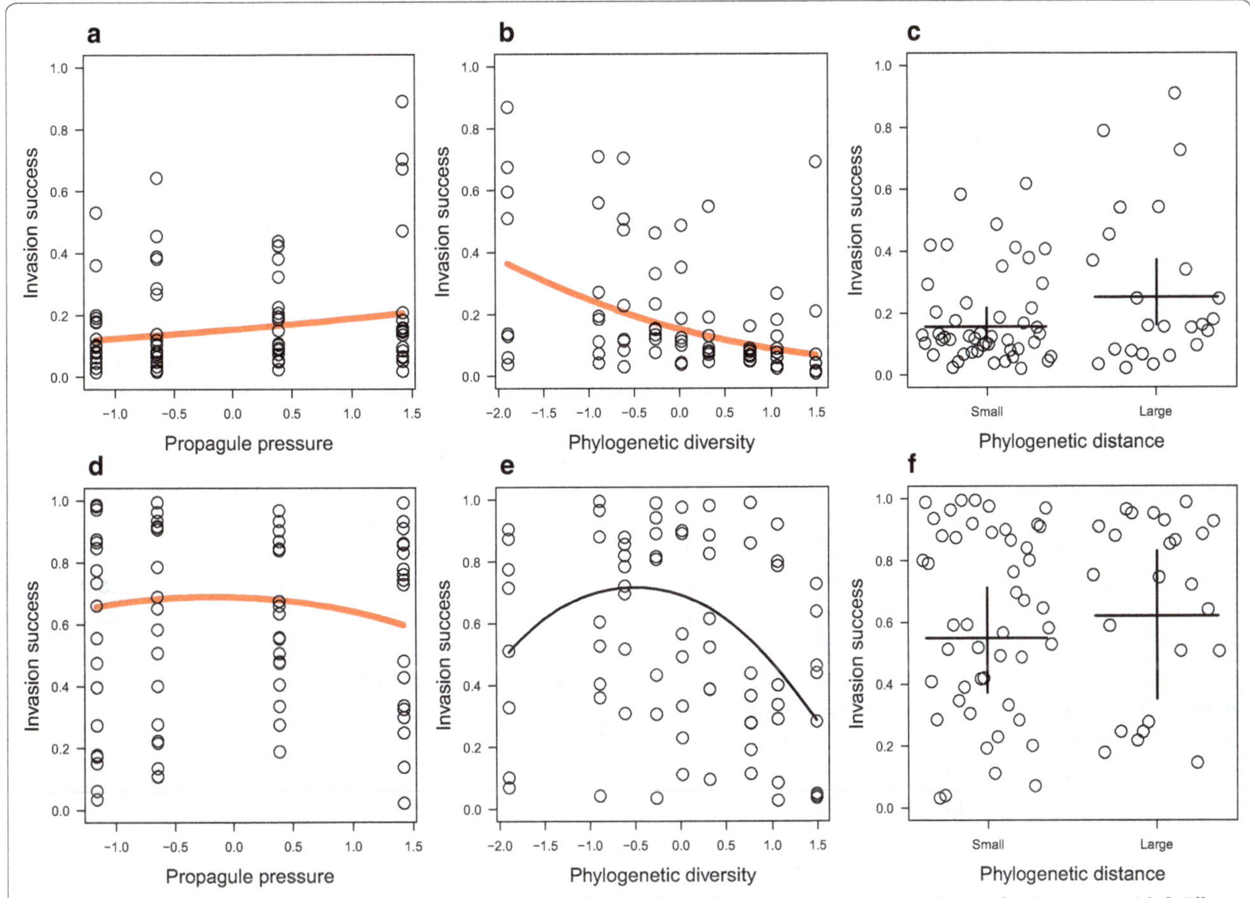


**Fig. 1** Invasion success of invader *Serratia marcescens* (proportion of invader from all colonies) three (**a**–**c**) and 9 days after the invasion (**d**–**f**). Effect of propagule pressure (**a**, **d**), phylogenetic diversity (**b**, **e**) and phylogenetic distance (**c**, **f**) on invasion success. *Panels* containing significant effects are highlighted with *red fit line* (see "Results" for detailed statistics). *Whiskers* denote ± 1.96 × standard errors of the estimate

and for propagule pressure, both quadratic coefficients were clearly non-significant and model fit decreased (AIC = 383.0) compared with the model with linear coefficients (AIC = 379.1, ΔAIC = 3.9).

After 9 days from invasion we found no evidence for linear effect of propagule pressure on invasion success (est. = −0.057, s.e. = 0.043, z = −1.331, p = 0.183). However, intermediate propagule pressures facilitated higher invasion success (est. = −0.158, s.e. = 0.056, z = −2.818, p = 0.005, Fig. 1d). Compared to the beginning of invasion community ID affected invasion success (est = 0.606, s.e. = 0.779, z = 2.462, p = 0.036) 9 days after invasion. No other explanatory variable affected invasion success (phylogenetic diversity: est. = −0.466, s.e. = 0.282, z = −1.651, p = 0.099; quadratic effect of phylogenetic diversity: est. = −0.463, s.e. = 0.314, z = −1.475, p = 0.140; small phylogenetic distance: 0.548, large distance: 0.620, z = 0.410, p = 0.682, Fig. 1f). The quadratic model (AIC = 588.0) with quadratic effects for phylogenetic diversity, and for propagule pressure, explained invasion better than model with only linear effects (AIC 593.8, ΔAIC = 5.8).

## Discussion

Despite the long history of invasion biology, fewer studies have focused on the properties of receiving communities which interact with the invader, than on the properties of the invaders themselves [1]. Most of the experimental studies on community properties affecting invasion have so far concentrated on exploring the effects of competition and mean phylogenetic relatedness between the invader and resident species on invasion success. However, when experiments are taken to the community level, the mean phylogenetic relatedness between species might not be the only denominator, as invasions could also be affected by phylogenetic diversity.

We found out that large phylogenetic distance was not linked strongly to invasion success (Fig. 1c). Although our results points to a direction that phylogenetically less related communities could be invaded more easily

(Fig. 1c, p = 0.085), it is noteworthy that phylogenetic distance is a proxy of resource/niche use complementarity, and is sensitive to the phylogenetic history of a community [1]. For example, related species that evolved in sympatry could be less likely to compete with each other, whereas related species that evolved in allopatry could share ecological similarities, leading to strong competition (but see [21] for opposite prediction). Such effects could obscure the relation between competition and phylogenetic relatedness and perhaps be one additional reason for why evidence for the role of phylogenetic distance in invasions is mixed (see "Background") and sometimes weak (as here). It is also evident that our dataset is relatively small (10 communities), which is suboptimal in estimating perhaps small effect of phylogenetic similarity on invasion.

In addition to phylogenetic similarity the diversity in communities could play an important role in dictating the invasion success. We found that smaller phylogenetic diversity between the invader and the invaded community promoted *S. marcescens* invasion success (Fig. 1b, e). These results could be explained if large phylogenetic diversity lowers the invader's chances of finding a free niche space. Our results, thus, resemble those few studies done with plants, where higher phylogenetic diversity within community was found to hinder invasion [5, 13]. Moreover, our results are also in line with what has been expected to occur in invasions to species rich and poor communities.

We show that, during the early stage of invasion, the strongest explanatory variable for invasion success was the propagule pressure, as has been found in many cases before [22, 23]. Interestingly, intermediate propagule pressure at the beginning of the experiment was associated with slightly higher overall proportion of invaders at the latter stage of invasion. This finding could be linked with intrinsic population dynamics of the invader. For example those microcosms where propagule pressure has been larger could have attained maximum yield earlier than in those microcosms where propagule pressure was lower. This could lead to lowered survival after the resources have been used up (see [24]) and consequently this would cause lower invader densities, especially at latter stages of invasion when smaller differences at the beginning of the experiment would have grown larger in each renewal.

All models fitted on invasion success, measured after 9 days, indicated significant effect of community ID, whereas none of the analyses done on first time step indicated strong effects of community. This result suggest that proxies of invasion, like phylogenetic diversity could be more accurate descriptors of invasion propensity at the early stages of invasions. At latter stages individual communities could act more idiosyncratic manner, masking proxies of invasion success and emphasizing properties of particular species assemblages. We had two technical replicates, and four repeated measurements of the same community, which is relatively low sample to fully disentangle the effects of community ID on invasion success. Moreover, in fast growing bacteria, 9 days could already be considered a late stage of invasion and these results could also reflect "equilibrium" densities of these systems. Moreover, 9 days is long enough to observe evolution of invader and communities [19], which could also conceal the role of phylogenetic distance.

To summarize we found out that high propagule pressure, high phylogenetic distance and low phylogenetic diversity between the invader and the community species facilitated invasions, at early stage of invasions. At latter stage the invader numbers were more affected by the identity of the community, and to a lesser extent propagule pressure. Our results thus indicate that invasion is a process where both invader and residing community characters act in concert.

## Additional file

**Additional file 1: Figure S1.** Phylogeny of the study species, based on 16S rRNA. The tree includes the sequences FJ971882 (*E. aerogenes*), GQ856082 (*L. adecarboxylata*), NR_041980 (*S. marcescens* ssp. *marcescens*), NR_024570 (*E. coli*), AF094736 (*P. putida*), and AB680102 (*P. chlororaphis*). The sequence accession numbers were obtained from the NCBI nucleotide sequences database. Metrics for mean phylogenetic distances and variance of distances were calculated based on standardized distances between species. Results from data analysis of full data containing all communities. **Figure S2.** Effects of propagule pressure (A), phylogenetic diversity (B) and phylogenetic distance (C) on invasion success after 3 days from invasion of depicted from quadratic model. Effects of propagule pressure (D), phylogenetic diversity (E) and phylogenetic distance (F) on invasion success after 9 days from invasion of depicted from quadratic model. Effects of propagule pressure (G), phylogenetic diversity (G) and phylogenetic distance (I) on invasion success after 9 days from invasion of depicted from model containing only linear effects (panels G-I). Both linear and quadratic models are represented due to model selection uncertainty indicated by very similar AIC values. Panels containing significant effects are highlighted with red fit line. Analysis was performed on the whole dataset. In dataset used in the paper the outlier of the phylogenetic diversity (the left most observations) were omitted from the analysis. Whiskers denote $\pm$ 1.96 $\times$ standard error of the mean.

### Authors' contributions

TK, LL and KS designed the study, KS conducted the experiments, TK analysed the data, TK, KS and LL wrote the manuscript. All authors read and approved the final manuscript.

### Acknowledgements

We thank the Biological Interactions Doctoral Programme and the University of Jyväskylä Doctoral Programme in Biological and Environmental Science (KS), Academy of Finland Projects 278751 (TK), 250248 (LL), and the Centre of Excellence in Biological Interactions for funding and facilities. We thank also Elina Aho and Jari Mantere for helping in the lab, and Emily Burdfield-Steel for editing the language.

**Competing interests**
The authors declare that they have no competing interests.

**Funding**
Academy of Finland Projects 278751 (TK), 250248 (LL), and the Centre of Excellence in Biological Interactions, and Biological Interactions Doctoral Programme and the University of Jyväskylä Doctoral Programme in Biological and Environmental Science (KS). Funding bodies had no effect on the design of the study and collection, analysis, and interpretation of data and in writing the manuscript.

**References**

1. Davis MA. Invasion biology. Oxford: Oxford University Press; 2009.
2. Capellini I, Baker J, Allen WL, Street SE, Venditti C. The role of life history traits in mammalian invasion success. Ecol Lett. 2015;18:1099–107.
3. Darwin C. On the origin of species. London: Murray; 1859.
4. Elton C. The ecology of invasions by animals and plants. London: Methuen; 1958.
5. Whitfeld T, Lodge AG, Roth AM. Community phylogenetic diversity and abiotic site characteristics influence abundance of the invasive plant *Rhamnus cathartica* L. J Plant Ecol. 2014;7:202–9.
6. Violle C, Nemergut DR, Pu Z, Jiang L. Phylogenetic limiting similarity and competitive exclusion. Ecol Lett. 2011;14:782–7.
7. Narwani A, Alexandrou MA, Oakley TH, Carroll IT, Cardinale BJ. Experimental evidence that evolutionary relatedness does not affect the ecological mechanisms of coexistence in freshwater green algae. Ecol Lett. 2013;16:1373–81.
8. Venail PA, Narwani A, Fritschie K, Alexandrou MA, Oakley TH, Cardinale BJ. The influence of phylogenetic relatedness on species interactions among freshwater green algae in a mesocosm experiment. J Ecol. 2014;102:1288–99.
9. Cahill JF, Kembel SW, Lamb EG, Keddy PA. Does phylogenetic relatedness influence the strength of competition among vascular plants? Perspect Plant Ecol. 2008;10:41–50.
10. Alexandrou MA, Cardinale BJ, Hall JD, Delwiche CF, Fritschie K, Narwani A, et al. Evolutionary relatedness does not predict competition and co-occurrence in natural or experimental communities of green algae. Proc R Soc B Biol Sci. 2015;282:20141745.
11. Mitchell CE, Agrawal AA, Bever JD, Gilbert GS, Hufbauer RA, Klironomos JN, et al. Biotic interactions and plant invasions. Ecol Lett. 2006;9:726–40.
12. Jiang L, Tan J, Pu Z. An experimental test of Darwin's naturalization hypothesis. Am Nat. 2010;175:415–23.
13. Gerhold P, Pärtel M, Tackenberg O, Hennekens SM, Bartish I, Schaminée JHJ, et al. Phylogenetically poor plant communities receive more alien species, which more easily coexist with natives. Am Nat. 2011;177:668–80.
14. Levine JM. Species diversity and biological invasions: relating local process to community pattern. Science. 2000;288:852–4.
15. Tilman D. The ecological consequences of changes in biodiversity: a search for general principles. Ecology. 1999;80:1455–74.
16. Jiang L, Morin PJ. Productivity gradients cause positive diversity—invasibility relationships in microbial communities. Ecol Lett. 2004;7:1047–57.
17. Stohlgren TJ, Binkley D, Chong GW, Kalkhan MA, Schell LD, Bull KA, et al. Exotic plant species invade hot spots of native plant diversity. Ecol Monogr. 1999;69:25–46.
18. Stohlgren TJ, Barnett DT, Kartesz JT. The rich get richer: patterns of plant invasions in the United States. Front Ecol Environ. 2003;1:11.
19. Ketola T, Mikonranta L, Mappes J. Evolution of bacterial life-history traits is sensitive to community structure. Evolution. 2016;70:1334–41.
20. Warton DI, Hui FKC. The arcsine is asinine: the analysis of proportions in ecology. Ecology. 2011;92:3–10.
21. Zelezniak A, Andrejev S, Ponomarova O, Mende DR, Bork P, Patil KR. Metabolic dependencies drive species co-occurrence in diverse microbial communities. PNAS Natl Acad Sci. 2015;112:6449–54.
22. Jeschke JM, Strayer DL. Determinants of vertebrate invasion success in Europe and North America. Glob Change Biol. 2006;12:1608–19.
23. Wittmann MJ, Metzler D, Gabriel W, Jeschke JM. Decomposing propagule pressure: the effects of propagule size and propagule frequency on invasion success. Oikos. 2014;123:441–50.
24. Pekkonen M, Korhonen J, Laakso JT. Increased survival during famine improves fitness of bacteria in a pulsed-resource environment. Evol Ecol Res. 2011;13:1–18.

3

# Quality control in public participation assessments of water quality: the OPAL Water Survey

N L. Rose[1*], S. D. Turner[1], B. Goldsmith[1], L. Gosling[2] and T. A. Davidson[3]

**Abstract**

**Background:** Public participation in scientific data collection is a rapidly expanding field. In water quality surveys, the involvement of the public, usually as trained volunteers, generally includes the identification of aquatic invertebrates to a broad taxonomic level. However, quality assurance is often not addressed and remains a key concern for the acceptance of publicly-generated water quality data. The Open Air Laboratories (OPAL) Water Survey, launched in May 2010, aimed to encourage interest and participation in water science by developing a 'low-barrier-to-entry' water quality survey. During 2010, over 3000 participant-selected lakes and ponds were surveyed making this the largest public participation lake and pond survey undertaken to date in the UK. But the OPAL approach of using untrained volunteers and largely anonymous data submission exacerbates quality control concerns. A number of approaches were used in order to address data quality issues including: sensitivity analysis to determine differences due to operator, sampling effort and duration; direct comparisons of identification between participants and experienced scientists; the use of a self-assessment identification quiz; the use of multiple participant surveys to assess data variability at single sites over short periods of time; comparison of survey techniques with other measurement variables and with other metrics generally considered more accurate. These quality control approaches were then used to screen the OPAL Water Survey data to generate a more robust dataset.

**Results:** The OPAL Water Survey results provide a regional and national assessment of water quality as well as a first national picture of water clarity (as suspended solids concentrations). Less than 10 % of lakes and ponds surveyed were 'poor' quality while 26.8 % were in the highest water quality band.

**Conclusions:** It is likely that there will always be a question mark over untrained volunteer generated data simply because quality assurance is uncertain, regardless of any post hoc data analyses. Quality control at all stages, from survey design, identification tests, data submission and interpretation can all increase confidence such that useful data can be generated by public participants.

**Keywords:** Citizen science, Open Air Laboratories, Quality assurance, Water quality, Water survey

## Background

In aquatic science, and especially in water quality assessment, volunteer monitoring has been used for nearly 50 years. Lee [1] provides a history of volunteer water quality monitoring for the United States from its beginnings in the 1960s and the initiation of volunteer water clarity monitoring in Minnesota lakes in 1973 leading to the present annual 'Secchi Dip-In' where more than 2000 lakes nationally are monitored (http://www.secchidipin.org/index.html). In the UK, almost all public participation projects relating to freshwaters have been concerned with lotic water quality. In 1971, the Advisory Centre for Education (ACE), supported by a national newspaper, *The Sunday Times*, organised a river water quality survey for school children [2]. Nearly 5000 participants, mainly aged between 10 and 13, used a series of simple

*Correspondence: n.rose@ucl.ac.uk
[1] Environmental Change Research Centre, Department of Geography, University College London, Gower St, London WC1E 6BT, UK
Full list of author information is available at the end of the article

metrics including chemical tests and the identification of 'indicator' benthic macroinvertebrates to estimate the extent of water "pollution" across England and Wales. The received data was found to provide good agreement with that previously collected by professional biologists, but covered a greater geographical area. A similar exercise was undertaken between 1991 and 1993 with three annual surveys organised by Riverwatch and sponsored by National Power, the National Rivers Authority and The Wildlife Trusts. The first of these surveys asked participants to provide a description of the site, an assessment of the aquatic biota within the river (benthic invertebrate survey; fish information from anglers; aquatic plants) and simple chemical tests for nitrate, pH and carbonate. Data from the 500 responses were compared directly with the ACE survey from 20 years previously to show how water quality had improved or deteriorated in rivers at a regional scale [3]. Since then, such large-scale river and stream surveys have not been repeated although the Riverfly Partnership's Riverfly Monitoring Initiative launched nationally in 2007, uses trained volunteer anglers to assess water quality on a monthly basis using estimates of caddisfly, mayfly, stonefly and *Gammarus* abundance (http://www.riverflies.org/rp-riverfly-monitoring-initiative). By contrast, no similar large-scale water quality surveys of standing waters have been undertaken in the UK. The 'National Pond Survey' in 1989 surveyed 200 minimally-impacted ponds while the 'Impacted Ponds Survey' and the 'Lowland Ponds Survey', both in 1996 surveyed 350 and 150 sites respectively [4]. None of these surveys employed the public to generate data. Pond Conservation's (now Freshwater Habitat Trust) annual "Big Pond Dip" launched in 2009 focuses on garden ponds and received data from 250 participants in its first year (J. Biggs, Freshwater Habitat Trust, pers. comm.). Hence, at the start of the OPAL project there was scope for a national lake and pond surveying programme.

The Open Air Laboratories (OPAL) programme was launched across England in 2007 with the aim of bringing scientists and communities together to observe and record the natural world in local neighbourhoods, and is now being expanded across the whole of the UK [5]. Participation is principally via national surveys used to assess changes to biodiversity, environmental degradation and climate change [6]. The programme, funded by the UK Big Lottery Fund, provides educational materials to aid these investigations [6]. One of OPAL's primary objectives is to encourage and facilitate participation in science among people who might not otherwise have the opportunity, so while OPAL survey participation is national and for all ages and abilities, the focus is on urban areas and in particular, deprived communities [7]. This principal of inclusion requires that all OPAL activities are 'low barrier to entry' with no requirement for training except that included within the survey materials themselves. However, nine regional Community Scientists were available during 2010 to offer training and advice on the water survey to groups and individuals when requested. The OPAL Water Survey was launched in May 2010. As with all OPAL surveys (e.g., [8–10]), and some other public participation water surveys (e.g., 'Waterwatch Victoria') [11] there were dual objectives of education and generating useful data, here, specifically an assessment of lake and pond water quality.

Public participation has been widely used in monitoring water quality [12] but there is a widespread concern over quality assurance of volunteer generated data (e.g., [13–16]) and participant objectivity [17–20]. This seems to be poorly addressed in many surveys using trained volunteers [12, 21] but is exacerbated by the OPAL approach of using untrained volunteers and largely anonymous data submission. However, in essence, the problems associated with either professional or volunteer generated data are the same. Both are of little value if monitoring or surveying is undertaken the wrong way [22] and without quality assurance and quality control measures, only a proportion is likely to be useable [23]. Appropriate tools therefore need to be in place to allow participants to produce data of known quality as well as helping users to extract useful information [15].

A number of recent papers describe the stages and requirements for constructing a successful public participation programme (e.g., [24–27]) but the scientific value of using simplified methods within these has been little studied [10]. The aim of this paper is to consider how quality assurance can be addressed in a public participation water quality survey especially where that involves untrained volunteers. We then apply these approaches to the OPAL Water Survey responses from 2010 to assess the extent to which data generated may be useable as a scientific dataset.

## The OPAL Water Survey

The main objective of the OPAL Water Survey was to gain a national 'snap-shot' assessment of water quality for as many lakes and ponds across England as possible. The use of public participation allowed access to many more lakes and ponds than would be possible by traditional monitoring programmes [24, 28], including some in private grounds that had never been surveyed before. To this end, a water survey 'pack' was compiled that included a series of activities and materials with the aim of providing something of interest to as many people as possible whilst generating useful data [5]. 40,000 packs were printed and freely distributed. All materials were (and remain) freely available to be downloaded from the OPAL website (http://www.opalexplorenature.org/WaterSurvey). Both biological and non-biological assessments were included in order to stress the importance of considering lakes and ponds in an holistic way. As with all OPAL survey materials, the OPAL Water Survey activities were pilot tested with 'naive audiences' [25] in order to ensure clarity of the step-by-step approaches and survey forms. Such an approach is crucial for untrained volunteer surveys. All instructions and protocols for undertaking the survey activities and submitting information (either directly online or Freepost return of paper copies for participants who had no internet access) were present within the pack. Once data had been entered onto the OPAL website, anyone with internet access was able to interrogate and explore all submitted data using a variety of online tools, data mapping and visualization techniques.

The OPAL Water Survey comprised four activities. Participants could take part in as many or few of these as they wished:

#### An assessment of water quality using the presence and absence of broad, and easily identifiable, classes of aquatic invertebrate

The use of indicator species and freshwater communities to assess water quality has been in use for over 100 years ([29, 30] and references therein). Benthic macroinvertebrates are the most commonly used organisms for biomonitoring [31] with over 50 different macroinvertebrate-based assessment methods currently in use [32]. Aquatic macroinvertebrates are generally localised so their response to any stress is related to local conditions, they live for a period sufficient to identify impacts and display a wide range of sensitivity to water quality. They are also found in even the smallest water bodies, and are relatively easy to sample and identify to a broad classification level. These latter qualities in particular make the use of these organisms well suited to public involvement studies [33] and especially with school children [2, 34] while their use also avoids the need for the equipment required for equivalent chemical determinations [35].

Generating indices of water quality or scales of pollution from macroinvertebrate assemblages in streams has also been used for many years, from indices of general pollution or disturbance such as the Trent Biotic Index [36] and the Chandler Biotic Score [37] to more specific indices such as Hilsenhoff's 'Family level Biotic Index' (FBI) for organic pollution [38] and the multi-metric 'Benthic Index of Biotic Integrity' (B-IBI) [39]. Many of these indices use a three-category tiered system for classifying degradation either as indicator tolerance classes of the invertebrate groups that are compiled to create the index (e.g., [21, 31, 40, 41]) or as a means to classify the scale of degradation of the stream itself [39].

The OPAL Water Survey used a similar approach, classifying aquatic invertebrates into 13 broad taxonomic classes, to each of which was allocated a 'health score' using a three-tiered system based on the invertebrate group's tolerance to a broad range of stressors (Table 1). This system was based on the classification used by Pond Conservation (now Freshwater Habitats Trust) in their 2009 Big Pond Dip, itself developed from methods used by the National Pond Survey and the Predictive System for Multimetrics (PSYM) [42]. Invertebrates collected were identified into broad taxonomic groups using a four-page fold-out guide which included photographs, size guides and bullet-point key identification features for each. Health scores relating to the presence of each identified group (Table 1) were then summed to obtain a total 'Pond health score' for the lake or pond. A maximum score of 78 was therefore obtainable if all 13 taxonomic groups were found. 'Pond health scores' were then allocated into three classes also using the 'Big Pond Dip' system: Very healthy (score $\geq$31); Quite healthy (6–30) and Poor (or 'Could be improved') (0–5). A classification of a pond to 'quite healthy' therefore required the presence of at least one medium-sensitivity (5 score) invertebrate class while a 'very healthy' classification required the presence of at least one high-sensitivity (10 score) class (Table 1).

The use of simple sampling protocols that are not too demanding [28] and identification to broad taxonomic groups, which avoid taxonomic jargon [13] and do not require significant training, are widely used in volunteer water surveys (e.g., [2, 31, 34, 40, 41]) in order to maximise participant numbers [25]. Simple, scientifically tested methodologies available to the public have been found to obtain unbiased samples and accurate classification into taxonomic classes [41]. These can provide comparable patterns to officially accepted modern monitoring methods [43] and agree well with professionally collected data [2]. While there are certain key taxonomic groups which volunteers should be able to identify (e.g., Ephemeroptera, Plecoptera, Trichoptera, chironmids, oligochaetes and isopods) [21], attempting greater taxonomic identification in the field with volunteers is considered likely to introduce excessive errors [31, 33]. Classification to the broad taxonomic groups used within the OPAL Water Survey may therefore be the most appropriate especially for un- or self-trained volunteers [31]. A balance is clearly required between the simplicity that allows untrained citizens to undertake water quality monitoring and the level of sophistication required to make the resulting data useful [44].

**Table 1 OPAL Water Survey invertebrate classification and 'Invertebrate group health' score based on the tolerance of the group to a range of stressors. Below is a comparison of descriptors for the derived pond health score ranges for the OPAL Water Survey (2010) and the 2014 Big Pond Dip [54]**

| Tolerance class | Groups | Group health score |
|---|---|---|
| High sensitivity | Cased caddisfly larvae; Caseless caddisfly larvae; Dragonfly larvae; Damselfly larvae; Alderfly larvae | 10 |
| Medium sensitivity | Mayfly larvae; Water beetles (adults and larvae); Water bugs (including water boatmen, water scorpions, water stick insects etc.,); Pond skaters; Water shrimps | 5 |
| Low sensitivity | Water slaters (water hoglice); Worm-like animals (including chironomid larvae; flatworms; leeches; worms etc.,); Water snails (spired; limpets; planorbids) | 1 |
| **Pond health score** | **OPAL Water Survey (2010) description** | **Big Pond Dip (2014) description** |
| 0–5 | Poor or 'could be improved' | Not yet great |
| 6–30 | Quite healthy | Good |
| 31 and above | Very healthy | Brilliant |

Engel and Voshel [31] suggest that common protocols based on the presence and absence of benthic invertebrates, identified to broad classification levels and divided into three pollution categories are likely to overrate ecological condition. This is in contrast to other volunteer program data (e.g., [41]) (and our own—see below) which indicate that volunteers preferentially tend to miss smaller types of invertebrate thereby underestimating 'pond health' or water quality. While underestimation may be as 'dangerous' as over-estimation, especially where such an assessment may act as a trigger for possibly expensive further work, for a broad-scale national snap-shot such as the OPAL Water Survey, under-estimation is probably preferable as it provides a worst case scenario. Whatever the approach, an assessment of potential volunteer sampling bias, accuracy of invertebrate identification and sensitivity of the protocols to sampling location and individual effort are required in order to determine the value of generated data. This can only be achieved through quality assurance programmes applied to each public participation study.

### An assessment of water clarity

Water clarity is a fundamental determinand in limnology as it provides an assessment of turbidity or suspended material in the water column. This determines the extent of light penetration and hence the potential for primary productivity and macrophyte growth on the lake bed. As a result there are indirect effects on habitat availability for aquatic fauna, stability of littoral areas, nutrient availability and pollutant transport within lakes [45, 46]. In the OPAL Water Survey, water clarity was measured using a bespoke device, termed the "OPALometer", which comprised a white disc with 12 OPAL symbols arranged around the edge shaded in specific percentile gradations from light grey (5 %) to black (100 %) [5]. Participants weighted the disc by taping a small coin to the reverse and then pushed it through the neck of an empty 2 L clear plastic drinks bottle. The bottle was filled with pond water to a pre-determined level to provide a standard depth of water column. The participant then looked through the neck of the bottle, counted and recorded the number of OPAL symbols they could see. This device was calibrated against standard empirical measures of water clarity such as the Secchi Disc [47] and laboratory measures of suspended matter concentration (see below). Such a simple, visual approach to the assessment of turbidity allowed mass participation in this activity and is considered to be of greater educative value than, for example, taking a reading from an electronic meter [11].

### The measurement of lake water pH

pH, the measure of hydrogen ion ($H^+$) concentration, is also fundamental in limnology. It drives many chemical processes in both catchment soils and freshwaters and can also determine which biological organisms may exist or thrive within a water body. Because of this, it is a parameter that is widely included in public participation surveys, for example the World Water Monitoring Challenge (http://www.worldwatermonitoringday.org/); the International Year of Chemistry's 'pH of the Planet' (http://my.rsc.org/globalexperiment); and the OPAL Soil Survey [8].

There are many ways to assess pH and all have their advantages and disadvantages for large-scale surveys. For the OPAL Water Survey, two cheap, commercially-produced pH strips were included in each of the 40,000 packs and additional strips were sent free of charge to participants upon request. These were Pehanon® pH 4.5–9.0 test strips manufactured by Macherey–Nagel (Germany). They are considered especially appropriate for a range of coloured waters and so are particularly useful for a national survey where lakes with high concentrations of dissolved organic carbon (DOC) may also be included. The test strips have the same accuracy over the whole measurement range (±0.25 pH unit) as tested against standard acid and basic solutions. However, as natural freshwaters tend to be weakly buffered, colour development of the test strip can take several minutes. Therefore, reading the strip too soon could result in the under-reading of water pH (A.Herzig and J. Tomatzky; Macherey–Nagel, pers. comms.).

## Methods: quality control and calibration approaches

Many studies have shown that volunteer-based schemes can provide reliable data and unbiased results [8, 13, 14] but Schmeller et al. [48] suggest that the quality of data collected by volunteers is more likely determined by survey design, methodology and communication skills than by the involvement of the volunteers per se. It is therefore as important to investigate quality assurance within the OPAL Water Survey methodologies as that of the participant-generated data itself.

### Invertebrate sampling method: sensitivity analysis

The OPAL Water Survey protocols provided simple instructions on how to sample for aquatic invertebrates. It was suggested that participants should look in as many different habitats around the lake and pond as possible and that, in each sampling location, a "vigorous sweep" of the pond net for 15–20 s amongst plants or other habitats should be undertaken. It is not possible to quantify how closely a volunteer (or indeed a professional scientist) adheres to a stated method from the reported data [11] but it is possible to determine how changes in sampling approach can affect those results. To provide such an assessment we undertook a multiple sampling exercise at ten lakes.

The lakes were selected to provide a broad cross-section of English standing water bodies for which the British Geological Survey already had archived data on catchment soil and inflow stream sediment chemistry to provide a calibration for the concurrent OPAL metals survey (http://www.opalexplorenature.org/metalssurvey). These lakes included three upland tarns with moorland catchments mostly used only for rough grazing and with largely homogenous, stony littoral zones, and six lowland lakes where catchments were more impacted (e.g., agriculture, dwellings) and where littoral habitats varied more widely including areas of emergent and floating macrophytes, and areas shaded by overhanging trees. Loweswater, in the Lake District, fell somewhere between these two site-types possessing a more agricultural catchment than the moorland Blea, Stickle and Burnmoor Tarns. Locations of these lakes are shown in Fig. 1 (sites 1–10) with site details provided in Table 2.

At each of these ten lakes, ten locations were selected around the perimeter and at each of these locations, three experienced surveyors each undertook a 10-, a 20- and a 30-s net sweep. This resulted in 90 invertebrate records for each lake except at Blea Tarn where only two surveyors were available (N = 60). As the surveyors worked together, surveys at each location were undertaken at the same time of day, precluding the influence of any diurnal changes. These surveys allowed an assessment of the effect on pond health score by individual sampling effort; on the variability of data generated by different individuals at the same location and time; and the effect of sampling at multiple locations around a lake compared to (and between) any one single location.

#### *Effect of sampling effort*

For the 870 surveys undertaken (all sampling sites; all surveyors; all sweep times), the highest 'Pond health' scores increased with sampling duration. 24.2 % of the highest scores were obtained by the 10 s sweep, 31.2 % by the 20 s sweep and 44.6 % by the 30 s sweep. Only the difference between 10 and 30 s sweep was significant at the $p < 0.01$ level (N = 290). Of the six individual

**Fig. 1** Site location map. Location map of multiple survey ponds: Marney's Pond and Weston Green Pond (M); the participant experiment at the Little Wittenham Wood Pond (W) and the 10 calibration experiment lakes (*1–10*). *1* Loweswater, *2* Burnmoor Tarn, *3* Stickle Tarn, *4* Blea Tarn, *5* Combe Pool, *6* Compton Verney Lake, *7* Hydelane Lake, *8* Bonningtons Lake, *9* Preston's Lake, *10* Scampston Park Lake

**Table 2 Site details for the ten calibration experiment lakes**

| Site | Latitude | Longitude | Altitude (m a.s.l) | Lake area (ha) | Max. recorded depth (m) |
|---|---|---|---|---|---|
| Loweswater | 54°34′52″N | 03°21′19″W | 125 | 60.3 | 16.5 |
| Blea Tarn | 54°31′02″N | 03°05′44″W | 478 | 7.4 | 12.0 |
| Stickle Tarn | 54°27′28″N | 03°06′16″W | 473 | 7.4 | 12.5 |
| Burnmoor Tarn | 54°25′46″N | 03°15′28″W | 253 | 23.9 | 13.0 |
| Scampston Park Lake | 54°09′53″N | 00°40′27″W | 32 | 4.9 | 1.2 |
| Coombe Pool | 52°24′43″N | 01°25′15″W | 73 | 30.6 | 2.0 |
| Compton Verney Lake | 52°10′07″N | 01°33′04″W | 78 | 13.1 | 2.0 |
| Hydelane Lake | 52°00′37″N | 00°56′40″W | 74 | 11.4 | 2.5 |
| Preston's Lake | 51°57′22″N | 00°41′51″E | 51 | 7.7 | 3.5 |
| Bonnington's Lake | 51°47′59″N | 00°42′37″E | 57 | 2.8 | 1.8 |

surveyors, four had similar results to these overall scores (i.e., increased score with effort) while the frequency of the scores from the other two (who undertook the fewest surveys) varied more widely between the different sweep times (Fig. 2a). Only one surveyor returned any significant difference ($p < 0.01$) between sweep times, again between the 10 and 30 s sweeps ($N = 30$).

The same was also observed on a site-by-site basis (Fig. 2b). For seven of the ten lakes, highest scores were observed with the 30 s sweep (all sampling locations around the lake; all surveyors). The 20 s sweep produced the greatest frequency of high scores at Hydelane Reservoir, while at Bonnington's Lake and Burnmoor Tarn, all three sweep times produced very similar frequencies of highest score. The greatest differences between sweep times appear to be at the lowland sites Compton Verney, Preston's Lake and Combe Pool but also at Loweswater, and this may be due to greater habitat diversity around these sites. However, only the differences between 10 and 30 s sweeps (all surveyors) at Preston's Lake and Compton Verney were significant ($p < 0.01$).

### *Effect of individual surveyors*

This dataset may also be used to determine the differences between individual surveyors at the same sampling locations using the same sweep times. Considering the three surveyors who undertook the most simultaneous surveys (1–3 in Fig. 2a), significant differences ($p < 0.01$) exist between two of the three combinations of pairs. Given this level of variability between experienced surveyors, at least a similar level of variation might be expected between untrained OPAL Water Survey participants (see 'multiple participant surveys' below). However, while a certain level of variability between individuals is to be expected it is important to consider how these data are submitted and used within OPAL. As

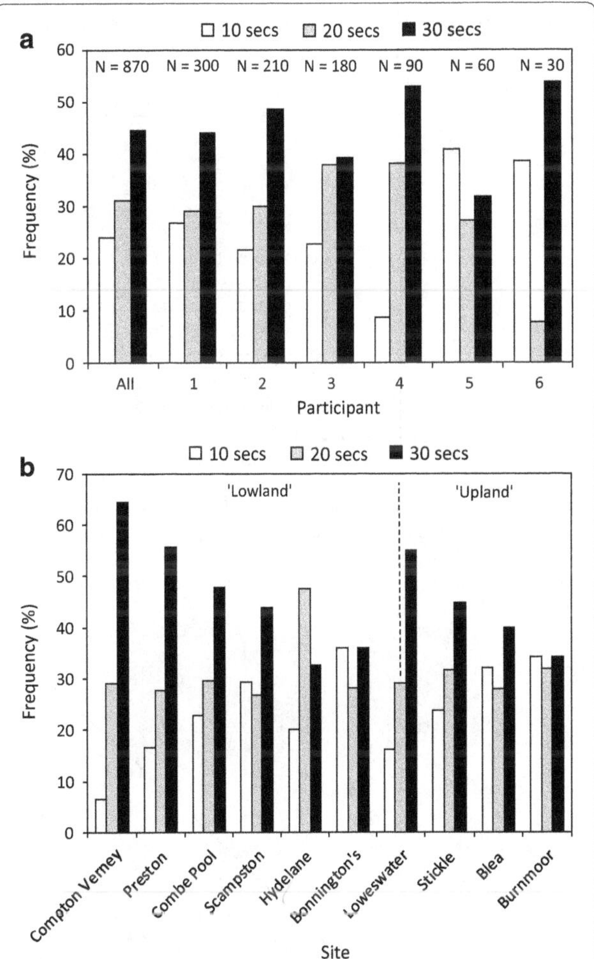

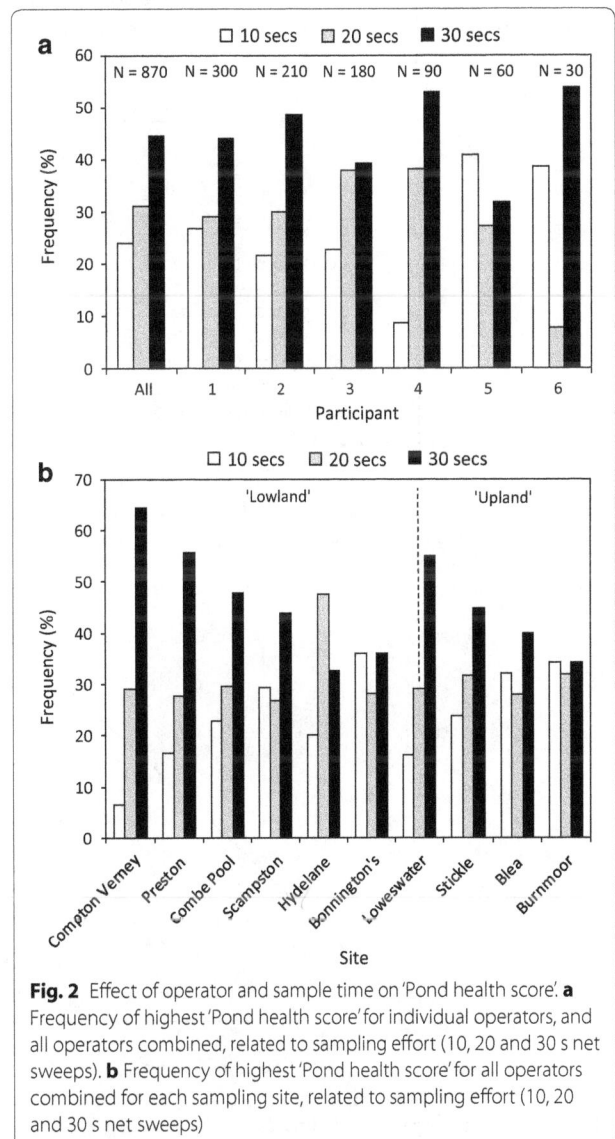

**Fig. 2** Effect of operator and sample time on 'Pond health score'. **a** Frequency of highest 'Pond health score' for individual operators, and all operators combined, related to sampling effort (10, 20 and 30 s net sweeps). **b** Frequency of highest 'Pond health score' for all operators combined for each sampling site, related to sampling effort (10, 20 and 30 s net sweeps)

described above, the OPAL Water Survey protocol states that sampling should be in as many different habitats as possible around a pond and that, in each sampling location, a 15–20 s sweep should be undertaken. These data are submitted to the OPAL website at the 'whole pond' level and allocated to one of three tiers of water quality. Figure 3 shows how this approach affects individual survey scores. Figure 3a shows all the individual samples (all lakes; all sampling locations; all times) for one of our pairs of experienced surveyors. While there is obviously a considerable amount of scatter, the agreement between the two surveyors is reasonable ($r^2 = 0.39$; $p < 0.001$; N = 210) and when water quality tiers are compared there is a 74 % agreement. If these individual samples are then amalgamated within a sampling location around a lake (i.e., 10, 20 and 30 s sweeps amalgamated for each location) then while the significance of the relationship decreases ($r^2 = 0.46$; $p = 0.015$; N = 70) the agreement within the water quality tier increases to 81 % (Fig. 3b). Furthermore, if these are then amalgamated to the whole pond level, (the level at which they would be reported to OPAL) then this agreement increases to 100 % even though the relationship between the scores from the individual surveyors is no longer significant at any level ($r^2 = 0.25$; $p = 0.86$; N = 7) (Fig. 3c). While the variability between individual samples even for experienced participants is quite high, once amalgamated to the 'whole pond' level and allocated to a water quality band, agreement is very good.

### Invertebrate identification: participant vs. scientist comparison

While previous studies have determined that field sampling of invertebrates for water quality assessments by

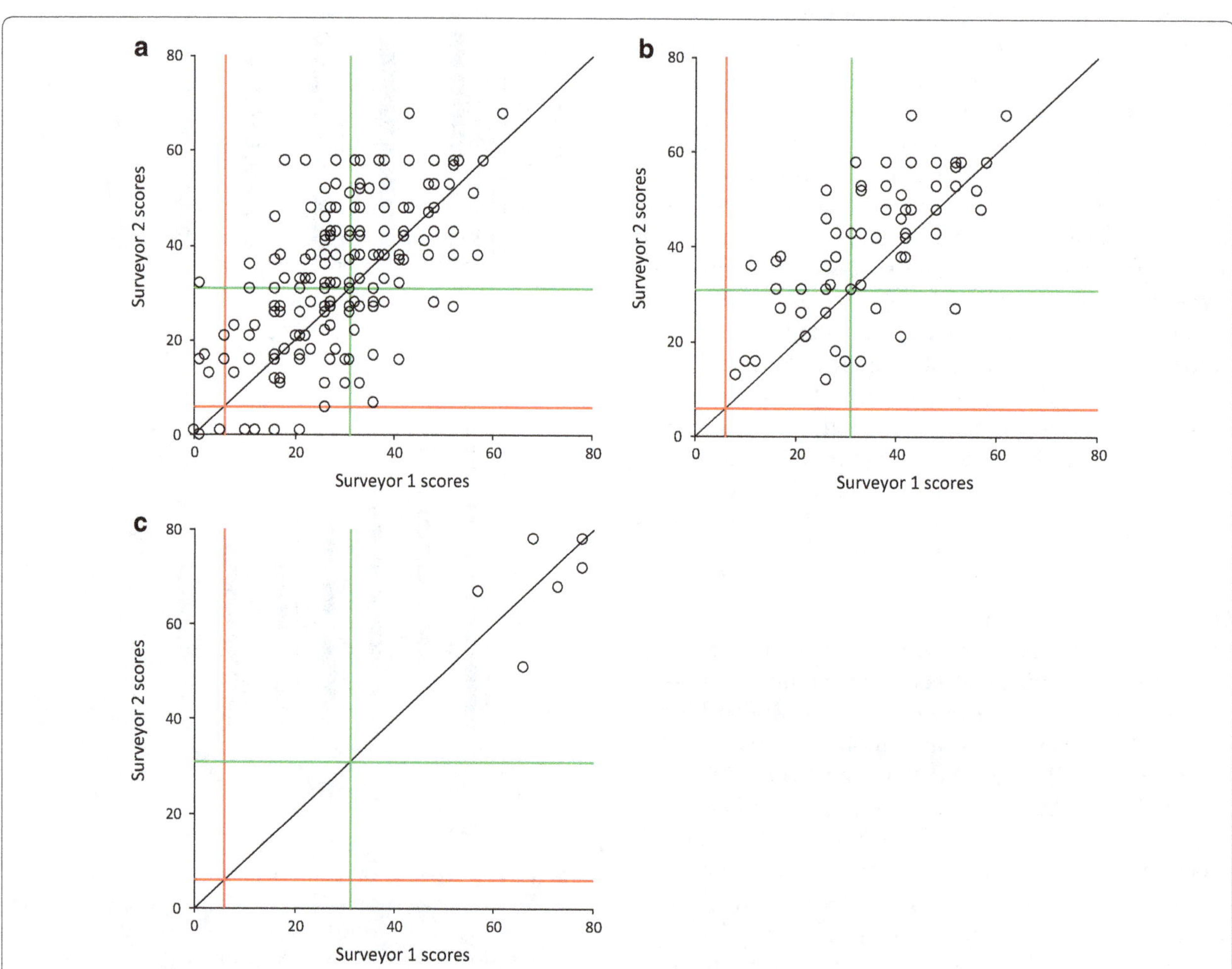

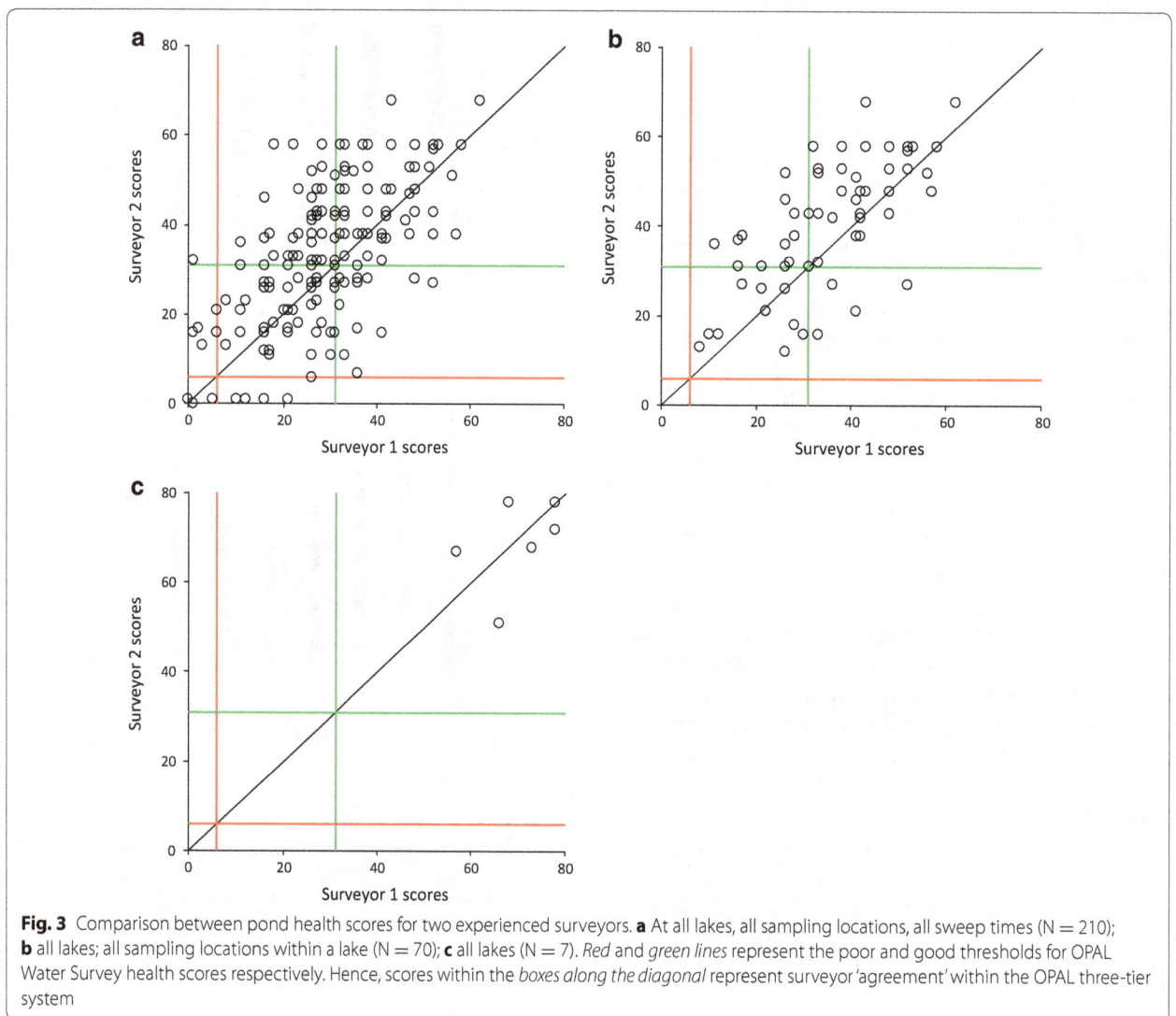

**Fig. 3** Comparison between pond health scores for two experienced surveyors. **a** At all lakes, all sampling locations, all sweep times (N = 210); **b** all lakes; all sampling locations within a lake (N = 70); **c** all lakes (N = 7). *Red* and *green lines* represent the poor and good thresholds for OPAL Water Survey health scores respectively. Hence, scores within the *boxes along the diagonal* represent surveyor 'agreement' within the OPAL three-tier system

trained volunteers and professional scientists can result in statistically indistinguishable data, especially where taxonomic identification is at Family level or above [13, 14], the emphasis of these studies has been on the adequate training of the participant volunteers. Little work has been undertaken on equivalent assessments with untrained participants. Although Heiman [49] suggests that professionals should sample side-by-side with volunteers in order to make a comparison, the results from our own 10-lake calibration dataset shows considerable variability occurs even between experienced people. Therefore, in order to assess the identification skills of untrained participants (rather than that individual's sampling ability) the scientist needs to work on the same collected samples.

Using an approach similar to that of Au et al. [43] to make this assessment, a group of eight untrained participants aged 17–18 years undertook the OPAL Water Survey at a pond in Little Wittenham Wood, Oxfordshire (Fig. 1; Site W). Each participant sampled the same pond following the OPAL Water Survey guide but without any other specific instructions. They collected and identified their individual samples but were allowed to work in pairs as OPAL safety protocols indicate water survey activities should never be undertaken alone. They used only the OPAL Water Survey guide for identification but were allowed to confer with each other as would undoubtedly happen during normal survey participation. Three participants had time to undertake a second sample resulting in 11 individual analyses. Each individual sample was then also studied by a member of the OPAL Water Survey team to make a direct comparison.

In general, 'water bugs' (mainly corixids), mayfly larvae, snails and worm-like animals were all correctly identified. However, very small damselfly larvae were also present in a number of samples and these were often mis-identified as mayflies. This had the effect of reducing the derived pond health scores (Fig. 4) such that all except one sample lie below or on the 1:1 line. The participant group, although small, reflected a good cross-section of interest in aquatic environments and this appeared to be reflected in performance. The most enthusiastic participant undertook two samples, correctly identified all the invertebrates present in both and hence produced identical scores to the member of the OPAL team. By contrast, another less engaged participant only identified a single invertebrate group even though a number of others were present. This resulted in the largest discrepancy in the dataset. Without this latter score, the $r^2$ value for this comparison was 0.83 ($p < 0.01$; N = 10) but including this score, the $r^2$ dropped to 0.35 ($p = 0.054$; N = 11). This would suggest that in general invertebrate identification among untrained volunteers is reasonably good,

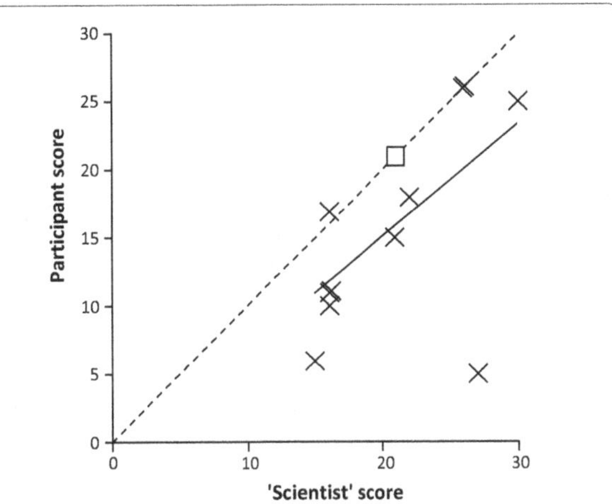

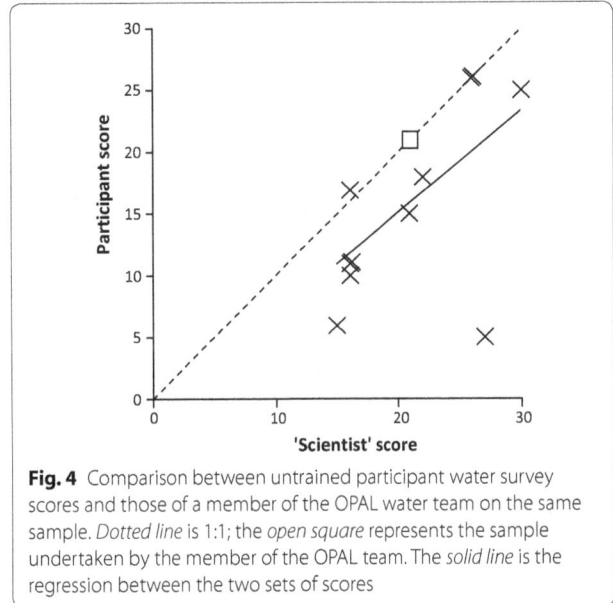

**Fig. 4** Comparison between untrained participant water survey scores and those of a member of the OPAL water team on the same sample. *Dotted line* is 1:1; the *open square* represents the sample undertaken by the member of the OPAL team. The *solid line* is the regression between the two sets of scores

especially amongst those motivated to participate by enthusiasm or interest in the activity.

These results agree very well with those of other studies (e.g., [34, 41]) who also reported that volunteers tended to miss smaller invertebrates in collected samples. This had the effect of reducing the numerical value of the derived metric (a biological integrity score) although these remained strongly correlated with professionally-derived data. Other studies have also indicated a lower taxonomic resolution in volunteer samples although performance is improved where additional aids (e.g., software keys; identification guides) are available [40, 50]. Greater experience in participation by undertaking further surveys would undoubtedly lead to more accurate identification (i.e., higher 'observer quality') [28].

#### Invertebrate identification: self-assessment

The use of quizzes and games within participatory activities provides a tool by which to evaluate observer skill and determine a criterion for data inclusion [28]. Within the OPAL Water Survey, a short identification quiz was included at the end of the online data submission procedure. This multiple-choice quiz involved six pictures of aquatic invertebrates each with a number of possible identifications. Participants selected the name they considered correct and received a score out of six. It is to be assumed that as the participants had concluded the OPAL Water Survey they would be familiar with the provided identification guide and would probably have used this in undertaking the quiz. As this is presumably how the animals collected in their surveys were also identified, this was not considered a problem, but the quiz scores should be considered a 'best available score' as a result (cf. [40, 50]).

Figure 5 shows the results from 2239 participants who attempted the identification quiz while inputting OPAL Water Survey data online in 2010. These data show a sharp decline in success rate with 56.8 % of participants getting all six identifications correct, 16.7 % getting five correct, and declining through to 1.1 % who identified none of the invertebrate pictures correctly. Hence, by accepting (for example) only those data for participants who correctly identified five or six pictures, over 73 % of the data (1644 surveys) would remain for further analysis. Interestingly, the participants in the Wittenham Pond experiment, described above, all scored five or six correct identifications in this quiz. So, while using this criterion may lead to greater confidence in the dataset, it does not necessarily guarantee accurate identification within collected samples.

One further part of the online submission process allowed participants to add feedback about the survey, or additional information. These comments regularly showed participants' own concerns over their identification ability, for example, not being "good enough for a scientific study". Other feedback suggested that while they had enjoyed taking part in the activity, they didn't submit their data for these same reasons. This anecdotal evidence indicates a general willingness for participants to try to identify the invertebrates to the best of their ability while the quiz provides a measure of how well they succeeded.

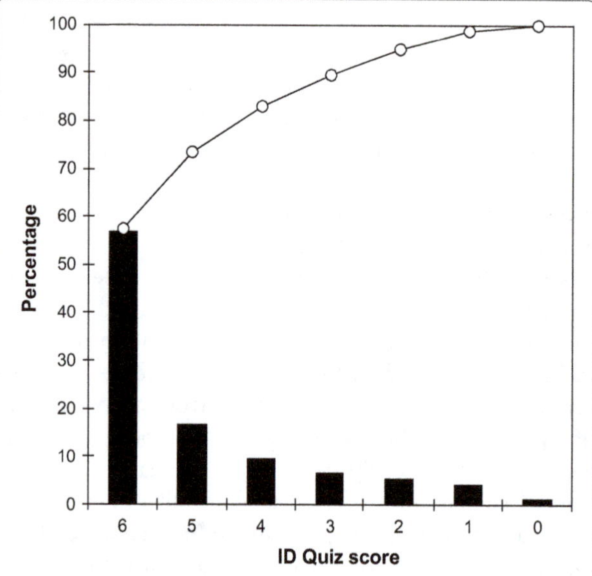


**Fig. 5** Results of the OPAL Water Survey invertebrate identification quiz undertaken during on-line data submission. Histogram shows the scores for the 2239 people who undertook the quiz as a percentage of the total. *Open symbols* show the cumulative percentage for each subsequent score

**Use of multiple participant surveys**

On 6 July 2010 over 80 school children aged 10–11 each undertook the OPAL Water Survey on two ponds, Marney's Pond and Weston Green Pond, both in Esher, Surrey (Fig. 1, site M). The surveys were undertaken within a short period of time and from a number of access points around each pond, although these were limited as the ponds are quite small (Fig. 6). Such 'class activities' allow an assessment of the variability in estimates of the different survey parameters. Here, these were recorded by a large number of individuals all with little experience either in invertebrate identification or in pH and water clarity measurement.

Figures 7 and 8 show the pond health scores, and pH and OPALometer data respectively, recorded for Marney's Pond and Weston Green Pond (a and b). For both sites there is a broad distribution of pond health scores indicating considerable variability in the invertebrates recorded by the participants in their individual sample. Indeed, for both ponds all 13 invertebrate classes were recorded at least once providing a 'theoretical health score' for each pond of a maximum 78. However, no individual score got close to this maximum. For Marney's Pond, individual scores tended to be lower and less distributed ( $\bar{x}$= 14.1; $\sigma$ = 6.6; max = 31) than those at Weston Green Pond ($\bar{x}$ = 24.3; $\sigma$ = 10.8; max = 47). Our sensitivity analysis data (above) showed that experienced surveyors can also obtain quite variable scores for the same sampling location at the same time (Fig. 3a) and therefore this broad within-site variability may not be a 'fault' in participant sampling or identification, but rather sample variability, noise and, in this case, the possibility that the same location was sampled many times within a short period. To compensate for the first of these, the

**Fig. 6** Aerial photograph of the two multiple-survey ponds. Google Earth image of Marney's Pond (*left*) and Weston Green Pond (*right*) used for multiple OPAL Water Surveys by over 80 school children in July 2010. Image date 24th May 2009

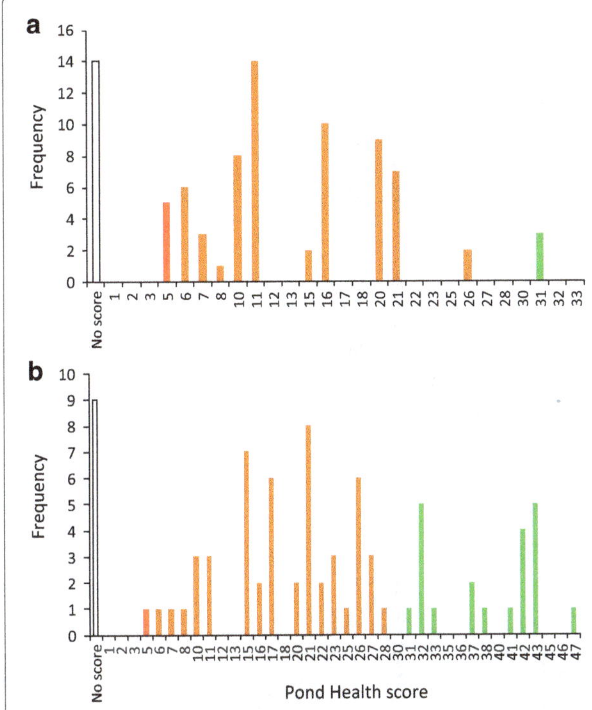

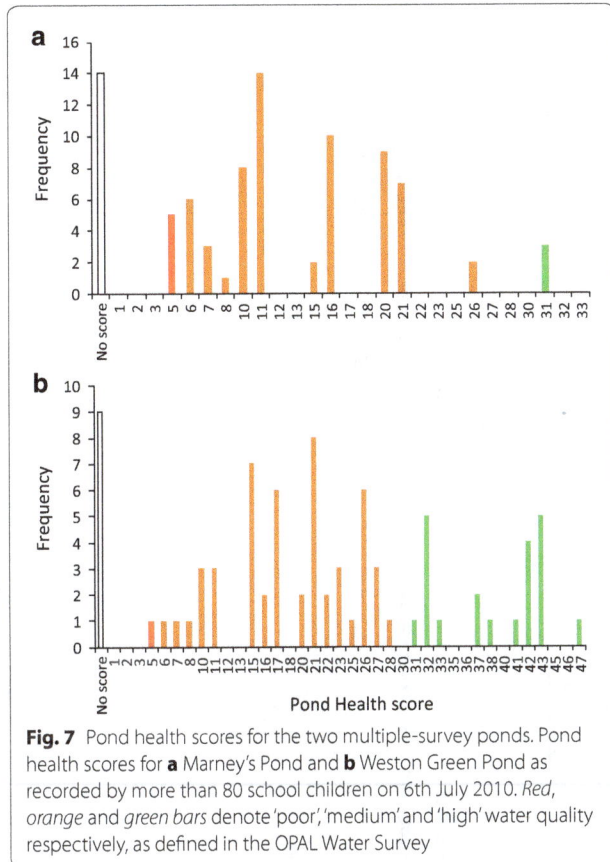

**Fig. 7** Pond health scores for the two multiple-survey ponds. Pond health scores for **a** Marney's Pond and **b** Weston Green Pond as recorded by more than 80 school children on 6th July 2010. *Red*, *orange* and *green bars* denote 'poor', 'medium' and 'high' water quality respectively, as defined in the OPAL Water Survey

OPAL Water Survey allocates pond health scores to three 'tiers', reducing the emphasis placed on an individual score. Using this approach, of those participants providing invertebrate data for Marney's Pond, 90 % produced values in the 'quite healthy' range of 6–30 (orange bars in Fig. 7a). Only eight participants produced other values and these were all close to the threshold (i.e., 5 and 31). For Weston Green Pond, of those participants recording invertebrate data, 70.8 % generated 'quite healthy' scores while a further 27.8 % derived higher 'very healthy' scores (Fig. 7b). Only one (1.4 %) recorded a 'poor' score of five. Hence, despite both ponds recording the same theoretical maximum these data would indicate both ponds are quite healthy, while the multiple participant data suggest Weston Green Pond may have a better water quality than that of Marney's Pond.

The non-biological data generated by the children appear to show less variability. At Marney's Pond over 92 % of the participants who returned water clarity data gave an OPALometer score of zero, while at Weston Green Pond over 91 % gave scores of 10 or 11 (Fig. 8). Verification of this distinction between the two ponds is clear from satellite images (Fig. 6) as are other catchment and pond characteristics recorded by the participants. By contrast, the pH data are similar as would be expected from two very closely located ponds (similar rainfall; geology; catchment characteristics). For Marney's Pond over 94 % of participants recording pH data gave a value of 6.0–6.5, while for Weston Green Pond over 93 % recorded values 5.5–6.0, but predominantly (48.1 %) 6.0 (Fig. 8). Although these pH data are not independently verifiable they do show that the OPAL Water Survey approach does provide consistent results even among participants new to the techniques.

### Comparison with other variables

While participant data show good replicability for the non-biological parameters, there is a further question regarding how these data compare with more standard means of measurement. This is particularly important for the water clarity data as it can be used to provide a national picture of suspended solids in the water column.

Figure 9 shows OPALometer data compared against both empirically measured Secchi disc depths and suspended solids measurements undertaken at nine lake sites monitored every 3 months over the course of the OPAL project [51]. There are good relationships between all these variables ($r^2 = 0.56$ and 0.47 for OPALometer vs. Secchi depth and suspended solids respectively) as observed in previous water clarity comparisons between turbidity meter and turbidity tubes [11]. One limitation of the OPALometer is that suspended solids can increase beyond the point at which no OPAL logos are visible resulting in a broad range at this lowest value (Fig. 9b). Similarly, Secchi disc depth can also increase beyond the point at which all 12 OPAL logos are visible, again resulting in high variability at this point on the scale (Fig. 9a). Comparison between these measurement approaches therefore needs to be interpreted with caution at highest and lowest water clarities and this is in agreement with previous community-generated water clarity data which was found to be most inaccurate at highest and lowest clarity levels [11]. Consequently, we again used a three-tier approach to the intermediate OPALometer scores and calculated a relationship between these, and with both suspended solids concentrations and Secchi Disc depth (Fig. 9c, d), which could be applied to the national OPAL Water Survey data.

### Use of other metrics

The pH dip-strips employed in the OPAL Water Survey may only realistically be expected to provide an indication of water pH but it is of interest to assess their performance against other, more standard, means of measurement in order to determine how these data may best be interpreted. Estimates of pH made using the dip-strips were compared simultaneously in the field with a

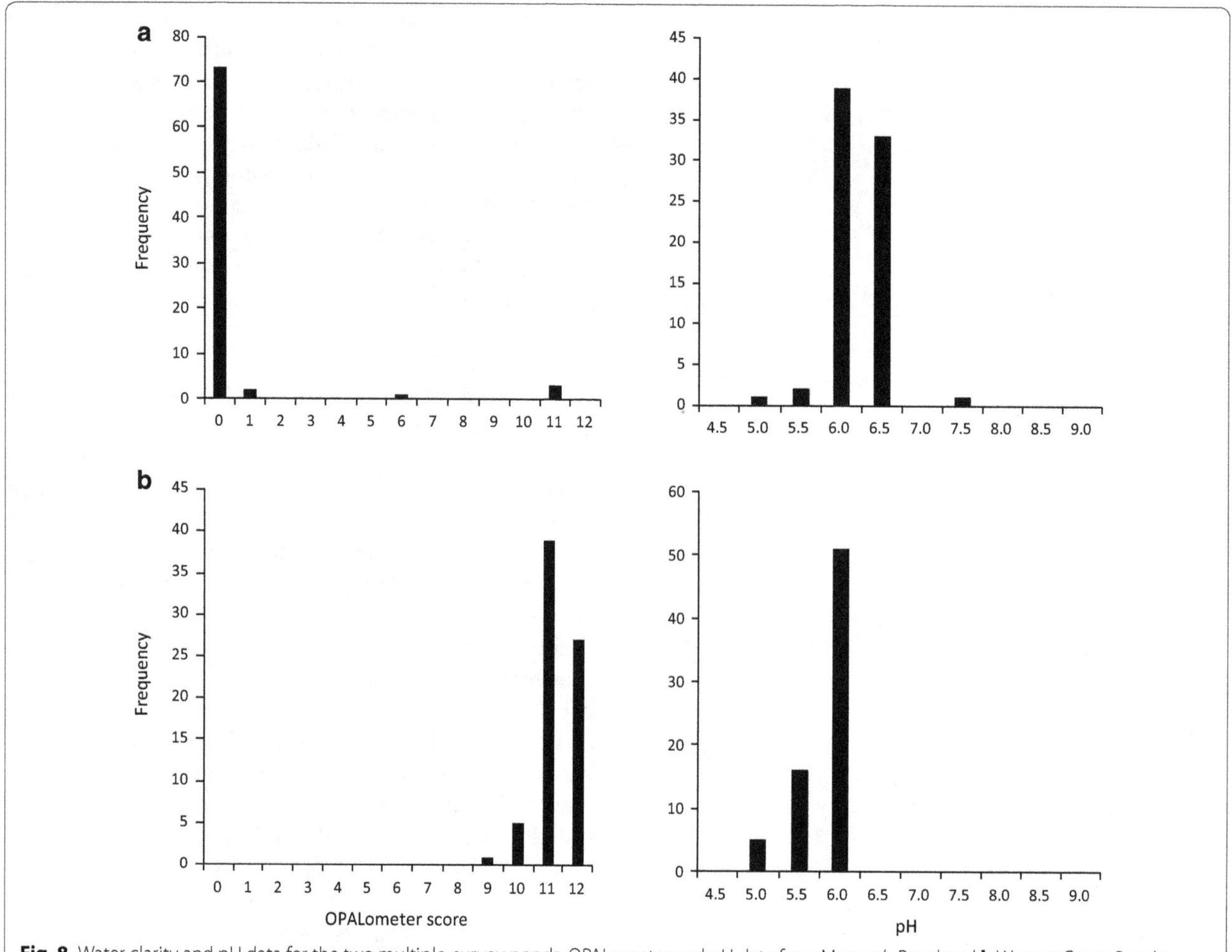


**Fig. 8** Water clarity and pH data for the two multiple-survey ponds. OPALometer and pH data for **a** Marney's Pond and **b** Weston Green Pond as recorded by more than 80 school children on 6th July 2010

calibrated pH probe (Hach HQ30d with PHC101 probe) and also against laboratory measurements on water samples collected at the same time (National Laboratory Service; automated electrode probe; RSD of 0.2, 0.5 % accuracy measured against pH 7.6 quality control standard, N = 398; NLS pers comm.). Figure 10 shows a comparison of these three methods. As expected, the field pH probe and laboratory measurements show very good agreement ($r^2 = 0.82$; $p < 0.01$; N = 80; Fig. 10b) and largely plot along the 1:1 line although the field probe may slightly under-read with respect to the laboratory analysis. By contrast, the dip-strips under-read and appear to show a considerable range against both probe and laboratory measurements (Fig. 10a, c respectively). This may be at least partly due to insufficient time being given for dip-strip colour development in weakly-buffered natural waters. Including a further set of data from upland lakes with lower pHs in the comparison between dip-strips and laboratory measurements (Fig. 10c; no probe data available) appears to improve the relationship ($r^2 = 0.58$; $p < 0.01$; N = 80) but this is undoubtedly driven by the lower pH values. While the dip-strips give an unchanged value of pH 5.0 there is considerable variability in laboratory pH measurement for the equivalent water samples (pH 4.45–6.0). Therefore, while the pH dip-strips employed in the OPAL Water Survey may provide a rudimentary field assessment of pH and undoubtedly allow participants to consider important non-biological parameters in lakes and ponds, this approach is unlikely to generate robust data.

### Summary of quality assurance

It is evident that there is considerable variation between individual samples taken from around a lake or pond even when sampling is undertaken by experienced surveyors. However, the approach used by the OPAL Water Survey, of using broad taxonomic classes, amalgamating multiple habitat surveys into a single 'Pond health score'

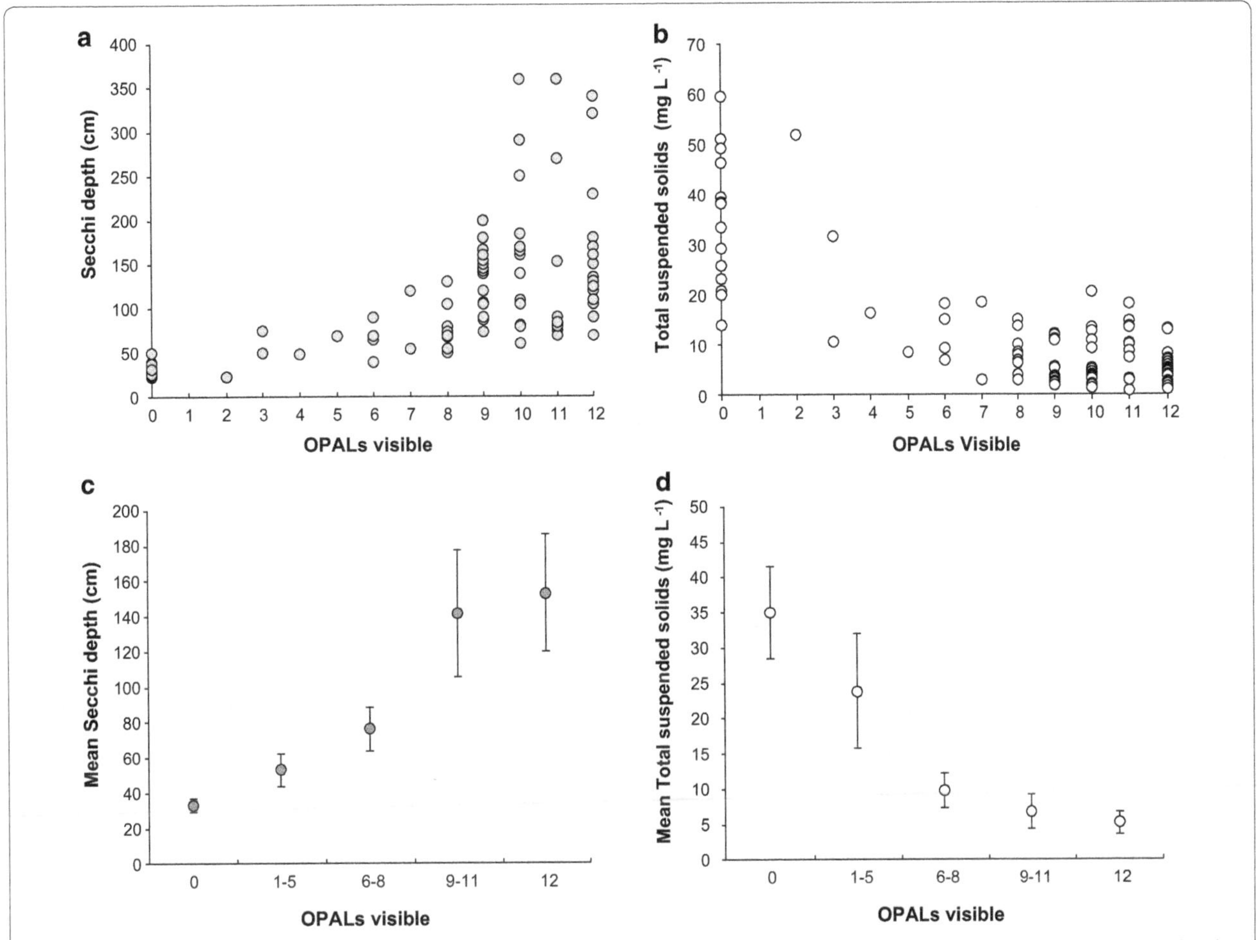

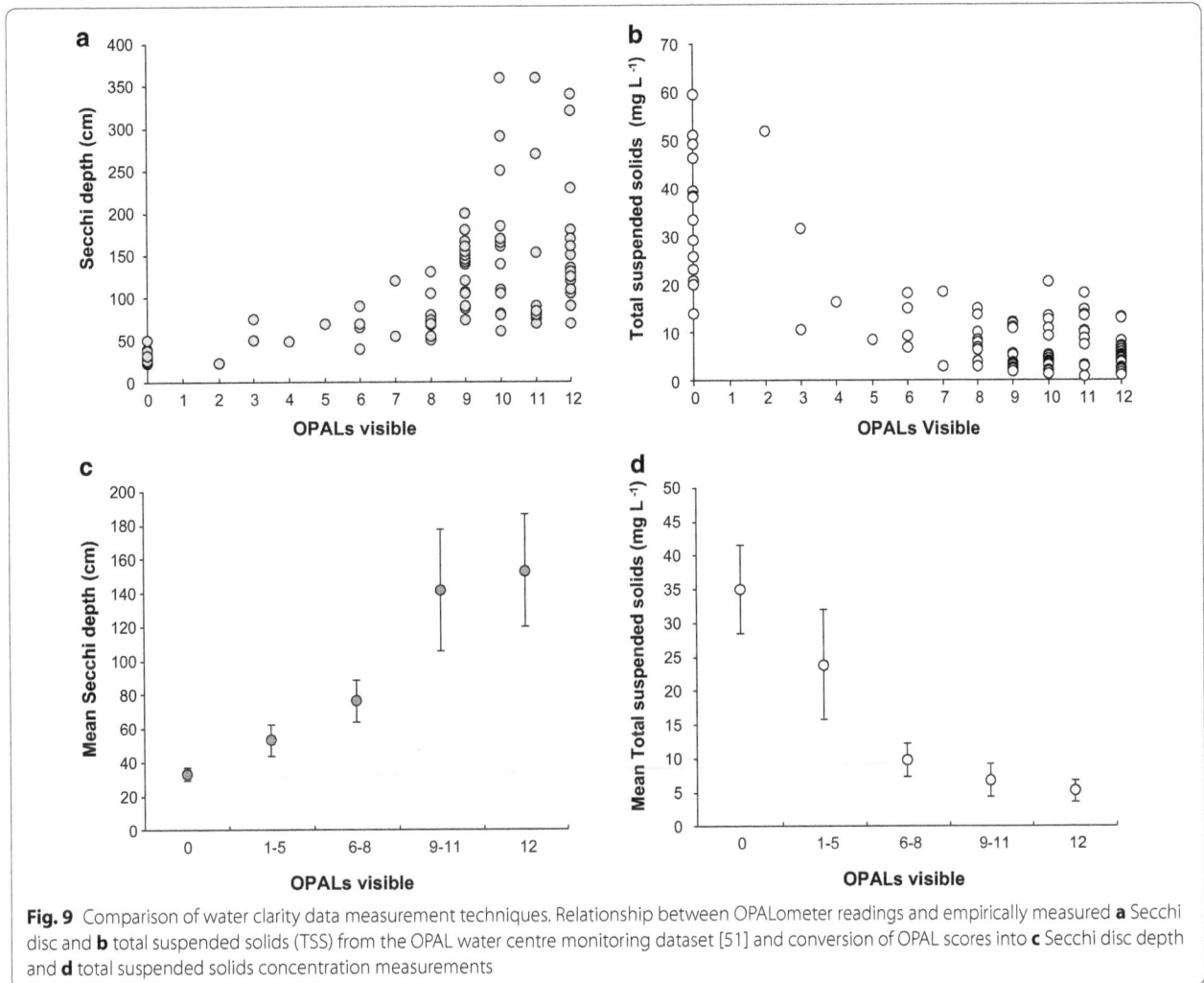

**Fig. 9** Comparison of water clarity data measurement techniques. Relationship between OPALometer readings and empirically measured **a** Secchi disc and **b** total suspended solids (TSS) from the OPAL water centre monitoring dataset [51] and conversion of OPAL scores into **c** Secchi disc depth and **d** total suspended solids concentration measurements

and allocating that score into three tiers of water quality provides sufficient latitude for variations in individual sampling technique (sampling time; number of locations; individual effort) and, to some extent, inexperience in invertebrate identification. However, increasing discrepancy is to be expected moving through the continuum of sampling possibilities from a single, short sweep in one location at a lake through to a multi-habitat, multi-sample strategy. This will also be influenced by lake size and the availability of differing habitats at the site.

It would appear that most participants will try to undertake the survey to the best of their ability, and untrained volunteers, motivated to take part by enthusiasm and interest, generally appear to be concerned about data quality and some even decline to submit their data as a result. Untrained or inexperienced volunteers are most likely to miss, or mis-identify, smaller invertebrates resulting in lower pond health scores and providing an under-estimate for any generated data. They may also sample less efficiently than experienced participants thereby also potentially reducing their sample scores. Performance would undoubtedly improve with experience leading to greater 'observer quality' and hence more reliable data. Use of a self-assessment invertebrate identification quiz provides a means to make a broad judgement of a participant's taxonomic skills and could be used to remove data associated with the lowest identification scores and increase confidence in the remaining dataset. However, this approach does not help with the possible reduced effectiveness of an inexperienced participant's sampling technique.

Of the activities within the OPAL Water Survey, the invertebrate sampling is able to provide useful water quality scores while the water clarity results, when calibrated to empirical measurements, can generate broad-scale suspended solids data. By contrast, and despite

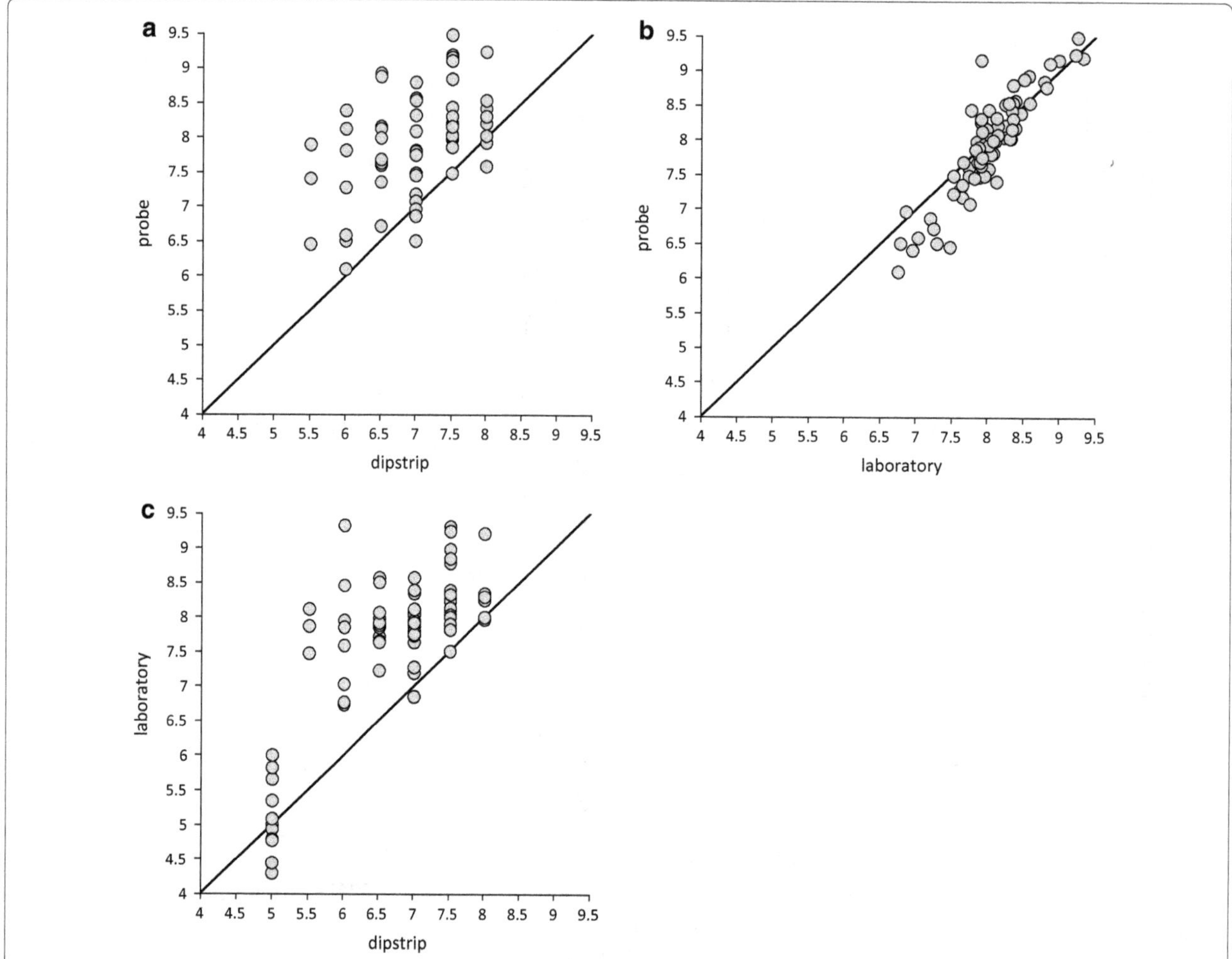


**Fig. 10** Comparison of pH data measurement techniques. Comparison of pH measurements between field probe and **a** dipstrip and **b** laboratory measurement. Also **c** the comparison between laboratory and dipstrip measurements. 1:1 lines are also shown

the consistent data generated by the school children at Marney's Pond and Weston Green Pond (Fig. 8) there appears to be considerable variation in the pH measurements in some waters depending on the time allowed for the colour to develop. This is likely due to the response of the pH strips in natural, low ionic-strength waters but also due to the survey instructions which did not provide sufficient emphasis on allowing time for the colour to develop. Although improving these instructions would help, it is likely that, while cheap and indicative of acid/base status to a broad level, within the context of public participation surveys, these pH strips may not be providing data of sufficient quality to interpret further. Here, we simply present the pH data as submitted (Fig. 11). In summary, with careful consideration of the data and some simple quality assessment we believe that untrained volunteers can provide useful water quality data.

## OPAL Water Survey results and discussion

### Participation

The OPAL Water Survey was launched 4 May 2010. Data continue to be submitted to the OPAL website and more than 4600 data entries had been entered by the end of 2013. A few surveys were submitted prior to the official launch date as a result of local training days for OPAL Community Scientists. Here, we focus only on the data submitted between April and November 2010 to remove any issues relating to inter-annual variability. During this period 3090 surveys were submitted online or returned as hard copies. Peak submissions occurred at the end of June and in July (Fig. 12) probably because this represents the last few weeks of the academic year when it was reported that many school groups took part.

For the first phase of OPAL from 2007 to 2013, England was divided into nine regions [6]. 27.9 % of the total surveys submitted in 2010 were undertaken within the West

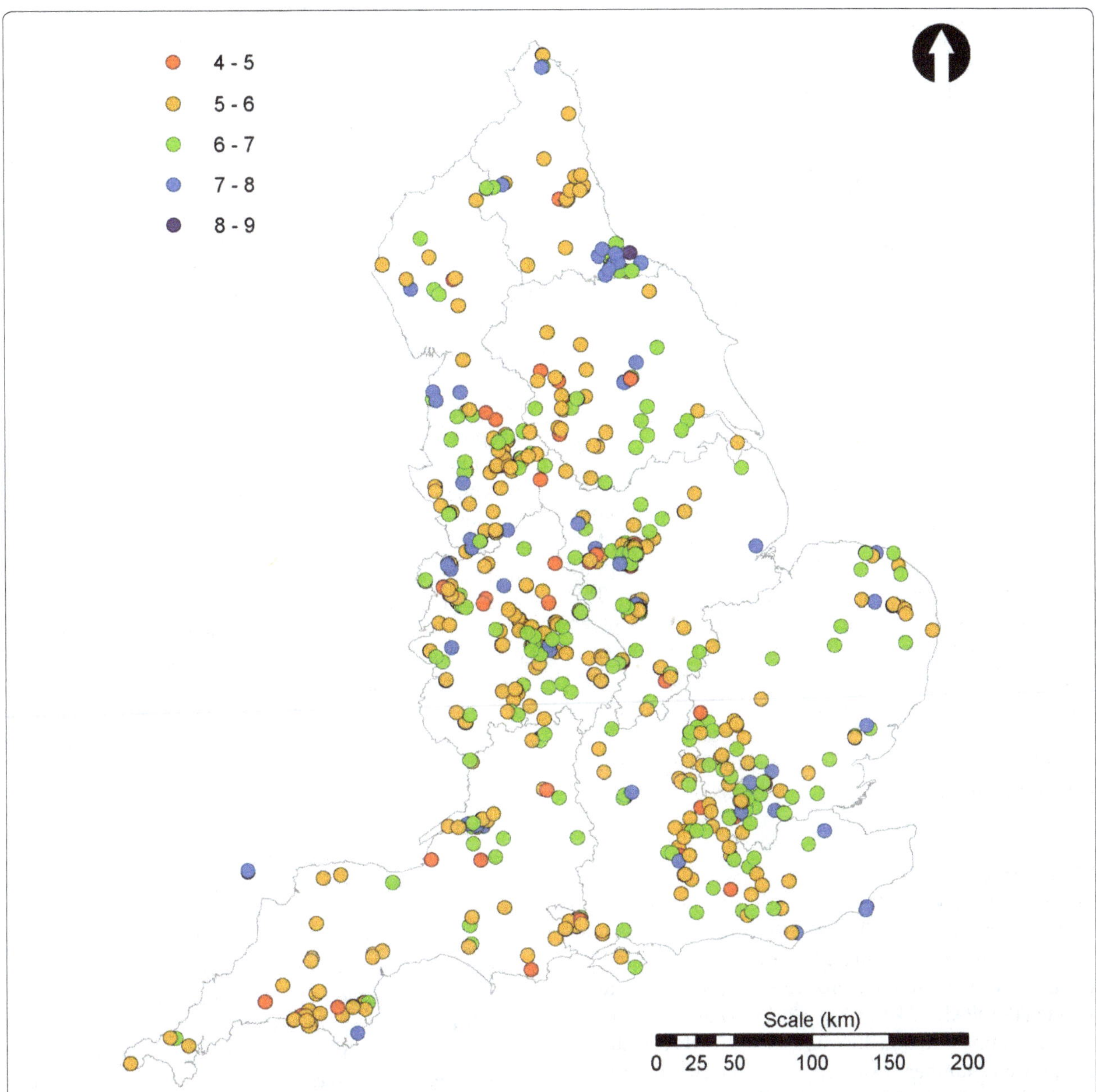


**Fig. 11** pH data for the 2010 OPAL Water Survey final dataset. The 1609 OPAL Water Survey sites included in the final dataset showing the reported pH data. The data points apparently located in the sea in the south-west are from ponds on Lundy island

Midlands, 14.3, 12.5 and 10.9 % were undertaken in the southeast, northwest and southwest regions of England respectively. The remainder of the regions all returned <10 % of the total with London, the smallest region by area, submitting just 3.9 %. It is difficult to estimate the total number of people who took part as respondents did not include the number of participants for each survey. However, in addition to independent data returns, OPAL Community Scientists reported that they had worked with over 4900 people on the water survey in 2010. Of these, 15 % could be classified as 'hard to reach' and these included people from areas of deprivation, black and ethnic minority groups and people with disabilities [7]. In terms of educative value, 94.7 % of survey questionnaire respondents said they had learned something new from the OPAL Water Survey compared with 89.8 % for OPAL overall [7]. Four per cent of water surveys were carried out in areas in the top 10 % of most deprived areas in England [52].

#### Initial data screening

The 3090 data submissions were initially 'cleaned' by removing sites outside England; sites providing locations

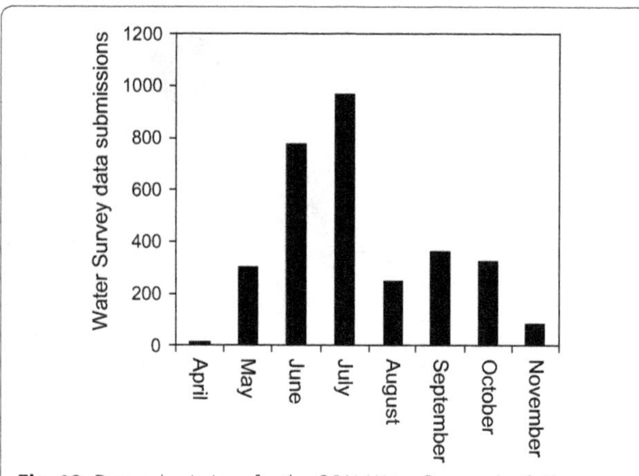


**Fig. 12** Data submissions for the OPAL Water Survey: April–November 2010

which were in the sea; and those where no lake or pond was identifiable from a satellite image and where that location indicated a very low likelihood of a lake or pond being present or close by (e.g., roads, buildings). This resulted in 2897 data entries. Submissions without an invertebrate ID quiz score and those with quiz scores of less than five were then removed. This resulted in a final dataset of 1609 sites distributed across England (Figs. 11, 13, 14) and represented 52 % of the total surveys submitted in 2010. These 1609 sites included large lakes to garden ponds, and, from within each region, urban, suburban and rural sites.

#### Water clarity

Of the 1609 surveys, 63 (3.9 %) were submitted without an OPALometer score. The results from the remaining 1546 surveys were dominated by the two end members of visibility; none visible (zero OPAL logos, 21 %) and all visible (12 OPALs, 24 %) (Figs. 13, 15). High water clarity values dominate nationally (median = 10 OPALs) due to a significant number of surveys recording 10 (10.7 %) and 11 (11.6 %). This largely bi-modal distribution of OPALometer scores was also observed in the quarterly monitoring programme of nine lakes and ponds during the OPAL Water Centre monitoring project (n = 142) [51].

The mid-range OPALometer scores (2–8) indicate a Secchi depth <1.5 m (Fig. 9c). The range of Secchi depths between nine and 12 OPALs can be explained by data from upland lakes with higher dissolved organic carbon (DOC) content which are 'transparent' in the short water column depth of the OPALometer but give a shallow (<1.5 m) Secchi depth. The same effect was observed when algal blooms reduced the Secchi depth but had little effect on the number of OPALs observable.

Total Suspended Solids (TSS) vs. OPALometer scores were similarly affected. Between 1 and 5 OPALs, TSS values reduced significantly with improved water clarity (Fig. 9b) while an OPALometer score greater than 6 provides a TSS estimation of <20 mg $L^{-1}$ (Fig. 9d). Viewed nationally, good and very good water clarity in ponds and lakes dominate (Fig. 13) but at a smaller scale the pattern becomes random and site specific. The spatial autocorrelation of water clarity is highly clustered (Moran's I = 0.26, Z-score = 19.53) and this is caused by two factors. First, multiple measurements were taken at the same site(s) (e.g., school class exercises) and second, that the sampling of ponds and lakes did not occur systematically.

#### Invertebrate-derived estimates of pond health

The pond health scores from the 1609 sites are presented in Fig. 14. Overall, 8.4 % of all sites showed 'poor' (or 'could be improved') water quality; 64.8 % were 'quite healthy' and 26.8 % were 'very healthy'. All regions showed similar distributions to this national picture except East of England where the 'quite healthy' and 'very healthy' categories scored approximately equally and in East Midlands where the frequency of 'very healthy' lakes and ponds exceeded the 'quite healthy' ones (54.9: 41.5 %). In all regions the 'poor' category included only 1.4 % (northeast) to 9.8 % (West Midlands) of the total number of sites. Similar to the measurements of water clarity, the spatial autocorrelation of health scores is clustered (Moran's I index = 0.3, Z-score = 22.9).

Health scores from all categories were present in both urban and rural sites in each region. Furthermore, in each region, some sites scored very highly indeed, despite concerns over the possibility for volunteer underscoring. In order to identify these highest quality sites an additional 'excellent' category was added where pond health score exceeded 52, requiring the presence of at least three classes of invertebrate from the highest sensitivity band. Sites in this category were present in every region (Fig. 14) but nationally included just the highest 4.0 % (* on Fig. 16b). However, no observed criteria were able to distinguish these from other lakes or ponds in the dataset.

Although the addition of this new 'excellent' category simply split the upper group into two (Fig. 16b), the alteration of classes during monitoring programmes should be avoided. In order to make meaningful and reliable comparisons there is a need for consistency not only by participants in sampling and identification but also in data assessment. The strength of monitoring lies in longevity and consistency of method, and public participation monitoring is no exception. Participants should be able to observe how, over time, water quality is improving

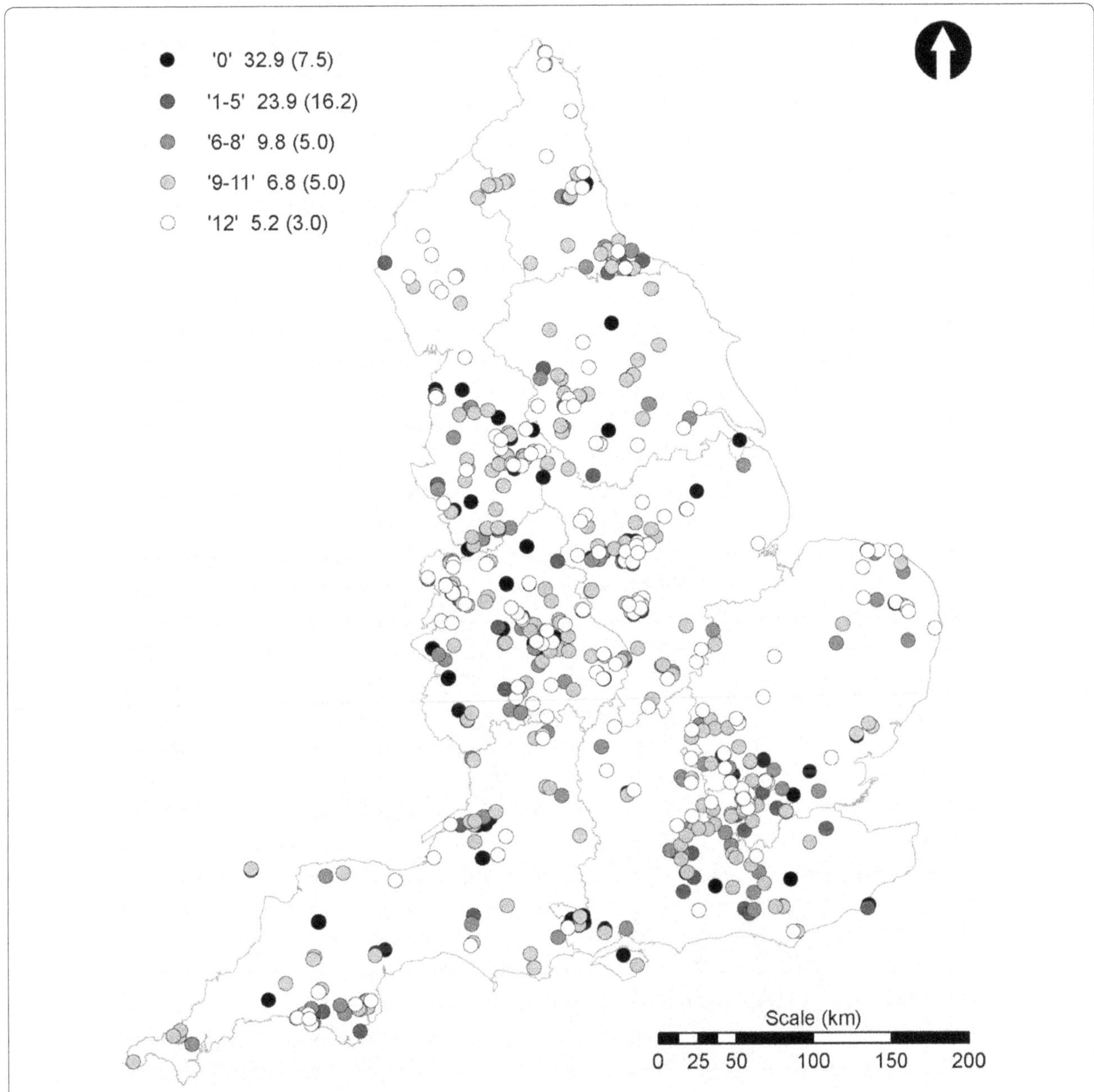


**Fig. 13** Distribution of total suspended solid (TSS) concentrations. *Scale* shows the OPALometer score range (e.g., '1–5'), and the mean and standard deviation (in *parentheses*) of the TSS concentration for each (mg $L^{-1}$)

or deteriorating, the extent to which any attempts at improvement have been successful, or conversely, how any new impacts have had deleterious effects. An example of how changing data assessment criteria could influence interpretation is presented in Fig. 16. This divides the 1609 pond health scores into water quality categories using approaches from recent UK pond surveys. Pond Conservation's (now Freshwater Habitat Trust) 'Big Pond Dip' (BPD) launched in 2009 used a three-tier scheme with classifications of 0–4 ("could be improved"); 5–25 ("good") and 26 and above ("really good") [42]. The OPAL Water Survey classification was based upon this, with minor modification, and so the distributions between the first BPD and OPAL classifications are very similar especially when the OPAL 'excellent' category is included in the 'very healthy' class (Fig. 16). However, in 2010, Pond Conservation changed their classification to "four bands of equal width.. [to].. assist in interpretation" [42]. If this classification is applied to the 1609 scores a vast increase in the number of lakes and ponds allocated to the two

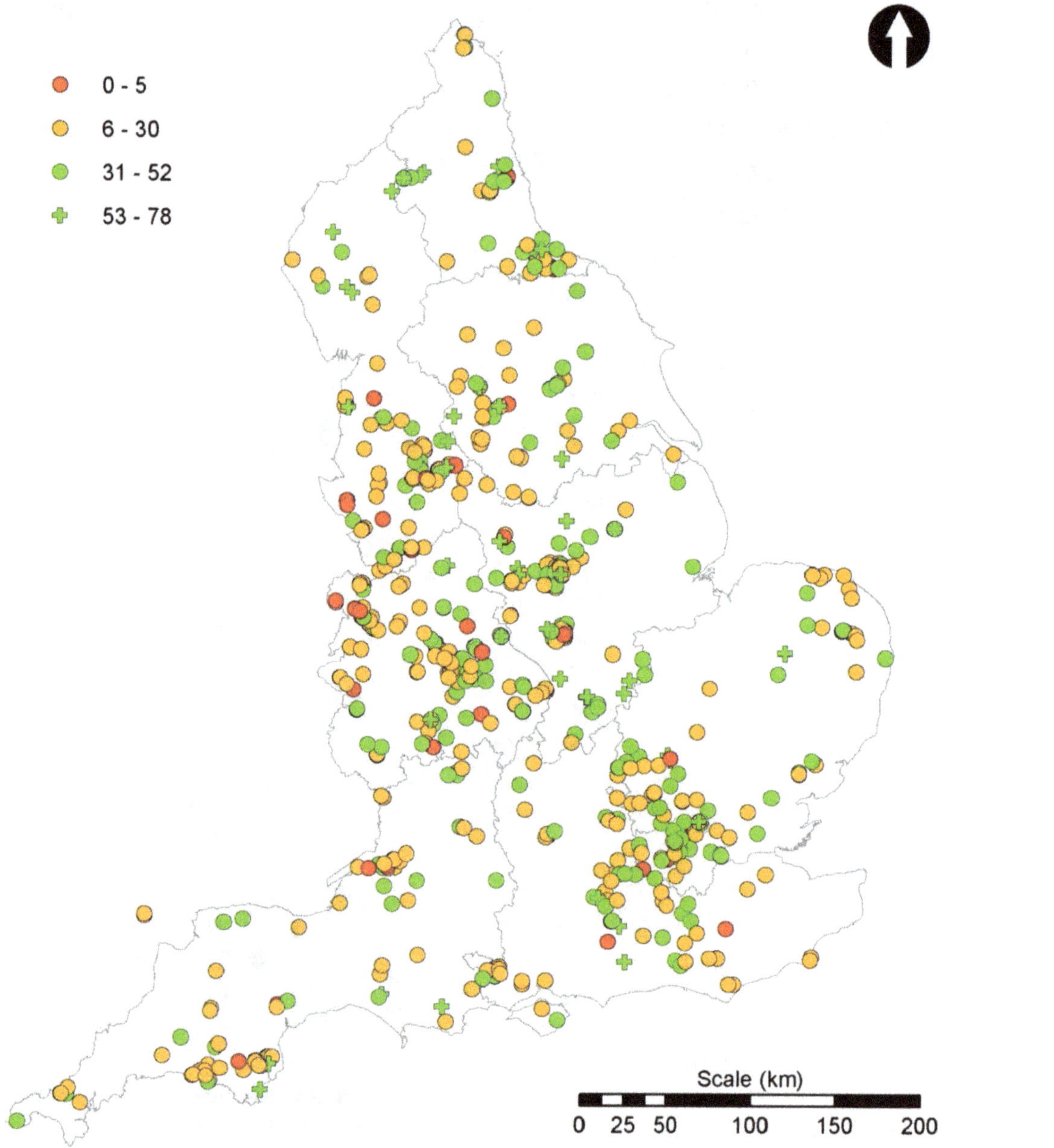

**Fig. 14** Water quality data for the 2010 OPAL Water Survey final dataset. Pond health scores derived from the invertebrate data are divided into the three water quality tiers and an additional 'excellent category' (53–78) (see text). The nine OPAL regions of England are also shown

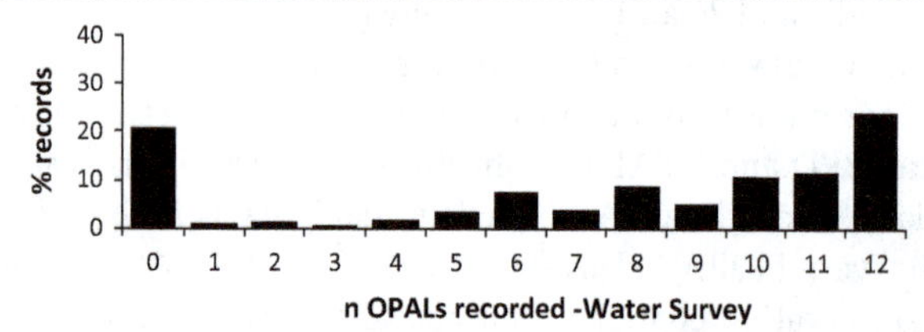


**Fig. 15** Breakdown of individual OPALometer scores recorded as a percentage of the 1546 participant submissions

poorest health classes results (i.e., 0–17 "low"; 18–34 "moderate"; Fig. 16c). This alters the frequency distribution of pond health scores and hence any interpretation that would stem from it. This may explain the apparently contradictory conclusions of BPD and OPAL whereby BPD 2009 concluded "about half of all ponds were in poorer condition" [53] whereas the OPAL data from 2010 show over 60 % were 'quite healthy' and a further 26 % were 'very healthy'. In 2014, the BPD reverted to using

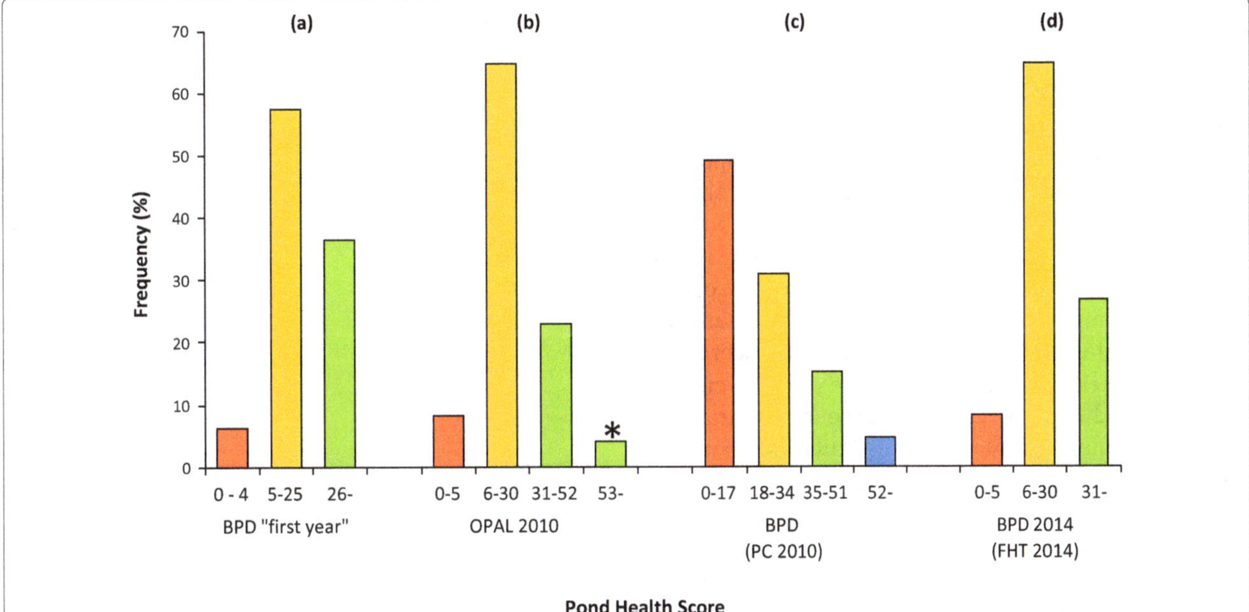


**Fig. 16** The effect of changing water quality tiers on the 2010 OPAL Water Survey final dataset. OPAL Water Survey 2010 pond health score dataset classified into water quality bands using **a** the original Big Pond Dip scheme [42]; **b** the OPAL Water Survey scheme with the additional "excellent" classification (marked as *asterisk*; see text); **c** the Big Pond Dip 'equal bands' scheme "for detailed interpretation" [42] and **d** the Big Pond Dip 2014 classification [54]

the same classification scheme as the OPAL Water Survey with three tiers of 0–5 ("not yet great"), 6–30 ("good") and 31 and above ("brilliant") (Table 1) [54]. Applying these to the OPAL 2010 dataset (Fig. 16d) provides a similar frequency distribution to the original BPD and, of course, the OPAL Water Survey.

### Using volunteers for water quality monitoring

The OPAL Water Survey generated a wealth of data on a variety of ponds and lakes across England (and beyond) many of which had not been surveyed before. The high pond health scores reported for these sites show that both natural and artificial ponds, in rural and urban settings, can have a high aquatic diversity [55]. In particular, the value of these ponds lies in the varied habitats they can provide [56], highlighting the need to sample in as many habitats as possible for a more reliable and repeatable pond health score, particularly in lowland sites where lakes may have a greater number of habitat types.

While the requirement for multiple habitat sampling was stressed in the OPAL protocol it is not possible to tell from how far along the sampling continuum any particular datum was derived but it must be assumed that participants attempted the sampling programme to the best of their abilities. Anecdotal evidence from within OPAL would certainly suggest that this was the case. The data presented above shows that the OPAL approach allows a certain amount of latitude in sampling and that the simple identification and classification allows the generation of repeatable results especially where information from multiple habitats around a pond are amalgamated to a single pond health score. While such considerations make the 'worst case' presumption that no participants had undertaken similar exercises before, many will undoubtedly have been enthused to take part by having done surveys previously while the OPAL Water Survey experience will hopefully encourage others to undertake more in the future. Hence, over a longer term where participants undertake the monitoring of 'their' lake or pond on multiple occasions it would be expected that sampling reliability and identification will improve and therefore the data produced more robust. However, it is important to note that as a participant's experience and skill improves they may also sample more efficiently and therefore find and identify more invertebrate groups than they would previously have done. Hence, scores might increase due an increase in sampling skill rather than because the pond water quality has improved.

The detection of broad-scale trends at multiple sites is a particular strength of volunteer monitoring. While the identification of invertebrates to broad classes may never be sensitive enough to detect subtle changes in water quality [21], a lower taxonomic level is sufficient to detect the impact of a perturbation on an aquatic community [33, 57]. Detection of trends is strengthened by the three-tier classification approach where at least one mid-sensitivity invertebrate class is required for the 'quite healthy' score while a 'very healthy' classification requires the presence of at least one high sensitivity invertebrate group. This is particularly important in surveys such as the OPAL Water Survey and the Big Pond Dip where the same pond health score could be generated by different combinations of invertebrates. This 3-tier approach therefore avoids potential confusions, for example where combinations of mid-sensitivity invertebrates could raise the pond health score to a higher health tier without the presence of high-sensitivity classes and may also help take account of variations due to sampling technique or dissimilarities between volunteer observations.

## Conclusions

To conclude, we return to the question of whether untrained volunteer participants can provide scientifically useful water quality data. It is likely that there will always be a question mark over such data simply because quality assurance is uncertain, regardless of any number of post hoc data analyses. This is exacerbated by the approach used by OPAL where data submissions can also be anonymous even though this was designed to increase participation number. In undertaking such surveys, there is a need to assume that participants have undertaken the activities using the described protocols to the best of their abilities. If this is the case, and the questions and techniques are simple and clearly explained, then there is no reason why these data should not be useful and available on a much greater spatial scale than would otherwise be possible.

There are means by which quality assurance can be improved in public participation water quality surveys. Training volunteers where possible (e.g., Freshwater Habitat Trust's PondNet survey [58]); the use of repeat surveys to gain experience; re-surveying by experienced personnel; and the ability to provide feedback (although this requires non-anonymity) would all provide more confidence in data collection. Further, the inclusion of quality control at all stages, from survey design, identification tests, data submission and interpretation can also increase the confidence in a final dataset. As with all monitoring and survey work, either professional or voluntary, consistency of approach in sampling, interpretation and assessment of data are key, while experience through undertaking more surveys would also undoubtedly improve data quality even for initially untrained and anonymous volunteers. However, for projects such as OPAL, data collation is only one of the aims. A consideration of the benefits to education, raising environmental awareness and ecological literacy, and an appreciation of the natural world are also important.

### Abbreviations

ACE: Advisory Centre for Education; ARK: action for the river Kennet; B-IBI: Benthic Index of Biotic Integrity; BPD: Big Pond Dip; DOC: dissolved organic carbon; EPA: Environmental Protection Agency; FBI: Family-level Biotic Index; FHT: Freshwater Habitats Trust; ha: hectare; m: metre; m.a.s.l: metres above sea-level; NLS: National Laboratory Service; OPAL: Open Air Laboratories; PSYM: Predictive System for Multimetrics; TSS: total suspended solids; UK: United Kingdom.

### Authors' contributions

NR, ST, LG and TD designed and developed the OPAL Water Survey and NR, ST and BG coordinated it. NR, ST and BG devised and undertook the quality assurance programme. ST and LG undertook data analysis. NR drafted the manuscript with contributions from ST and LG. All authors read and approved the final manuscript.

### Author details

[1] Environmental Change Research Centre, Department of Geography, University College London, Gower St, London WC1E 6BT, UK. [2] Centre for Environmental Policy, Imperial College London, 13-15 Prince's Gardens, London SW7 1NA, UK. [3] Department of Bioscience, Aarhus University, Vejlsøvej 25, Silkeborg, Denmark.

### Acknowledgements

We are grateful to the Big Lottery Fund for supporting the OPAL project and to OPAL staff at Imperial College (in particular Linda Davies and Roger Fradera) and the Natural History Museum (Chris Howard, Kate Martin, Katherine Bouton, Lucy Robinson, and John Tweddle) for project coordination, website development and help with producing the OPAL Water Survey. We would also like to thank:

Pond Conservation (now Freshwater Habitats Trust); Buglife; Botanical Society of the British Isles; British Dragonfly Society; Amphibian and Reptile Conservation for their help in developing the OPAL Water Survey activities;

The Field Studies Council (especially Simon Norman); for designing and producing the OPAL Water Survey guide and workbook;

Mark Astley, Alyson Rawnsley and Penny Webb (National Trust); Martin Sleath (Lake District National Park); Joe Taylor (Coombe Abbey Country Park); Gary Webb (Compton Verney); Simon Bumstead (Buckingham and District Angling Association); Paul Dixon (Bonnington's Estate); John Garnham (Preston's Lake); Chris Legard (Scampston Hall) and Steven Keyworth for permission to work at the calibration sites;

Jacqui Lundquist; Alice Milner; Ewan Shilland for their help with the calibration exercise;

Megan Boyt; Charlotte Cantwell, Philippa Gledhill; Kerri McBride; Mary McGowan; Ellen Rose; Kirsty Ross and Abigail Wright for their help with the Wittenham Pond exercise and Chris Parker at Earthtrust for permission to work at that site;

Staff and pupils at Thames Ditton primary school, Surrey, for their surveys of Marney's Pond and Weston Green Pond;

Alison Johnston and Cath Rose for survey data entry;

Two anonymous reviewers for their constructive comments on improving the manuscript;

All participants in the OPAL Water Survey.

### Declarations

This article has been published as part of *BMC Ecology* Volume 16 Supplement 1, 2016: Citizen science through the OPAL lens. The full contents of the supplement are available online at http://bmcecol.biomedcentral.com/articles/supplements/volume-16-supplement-1. Publication of this supplement was supported by Defra.

### Competing interests

The authors declare that they have no competing interests.

**References**


1. Lee V. Volunteer monitoring: a brief history. Volunt Monit. 1994;6:29–33.
2. Mellanby K. A water pollution survey, mainly by British school children. Environ Pollut. 1974;6:161–73.
3. National Riverwatch. The River report. A three year project review. 1994.
4. Biggs J, Williams P, Whitfield M, Nicolet P, Weatherby A. 15 years of pond assessment in Britain: results and lessons learned from the work of Pond Conservation. Aq Cons Mar Freshwat Ecosys. 2005;15:693–714.
5. Davies L, Fradera R, Riesch H, Lakeman Fraser P. Surveying the citizen science landscape: an exploration of the design, delivery and impact of citizen science through the lens of the Open Air Laboratories (OPAL) programme. 2016;16(s1). doi:10.1186/s12898-016-0066-z.
6. Davies L, Bell JNB, Bone J, Head M, Hill L, Howard C, Hobbs SJ, Jones DT, Power SA, Rose N, Ryder C, Seed L, Stevens G, Toumi R, Voulvoulis N. White PCL Open Air Laboratories (OPAL): a community-driven research programme. Environ Pollut. 2011;159:2203–10.
7. Davies L, Gosling L, Bachariou C, Eastwood J, Fradera R, Manomaiudom N, Robins S, editors. OPAL Community Environment Report. London: OPAL; 2013.
8. Bone J, Archer M, Barraclough D, Eggleton P, Flight D, Head M, Jones DT, Scheib C, Voulvoulis N. Public participation in soil surveys: lessons from a pilot study in England. Environ Sci Technol. 2012;46:3687–96.
9. Seed L, Wolseley P, Gosling L, Davies L, Power SA. Modelling relationships between lichen bioindicators, air quality and climate on a national scale: results from the UK OPAL Air Survey. Environ Pollut. 2013;182:437–47.
10. Tregidgo DJ, West SE, Ashmore MR. Can citizen science produce good science? Testing the OPAL Air Survey methodology, using lichens as indicators of nitrogenous pollution. Environ Pollut. 2013;182:448–51.
11. Nicholson E, Ryan J, Hodgkins D. Community data—where does the value lie? Assessing confidence limits of community collected water quality data. Wat Sci Technol. 2002;45:193–200.
12. Conrad CC, Hilchey KG. A review of citizen science and community-based environmental monitoring: issues and opportunities. Environ Monitor Assess. 2011;176:273–91.
13. Fore LS, Paulsen K, O'Laughlin K. Assessing the performance of volunteers in monitoring streams. Freshwat Biol. 2001;46:109–23.
14. Foster-Smith J, Evans S. M: the value of marine ecological data collected by volunteers. Biol Cons. 2003;113:199–213.
15. Gouveia C, Fonseca A, Câmara A, Ferreira F. Promoting the use of environmental data collected by concerned citizens through information and communication technologies. J Environ Manage. 2004;71:135–54.
16. Sharpe A, Conrad C. Community based ecological monitoring in Nova Scotia: challenges and opportunities. Environ Monitor Assess. 2006;113:395–409.
17. Stokes P, Havas M, Brydges T. Public participation and volunteer help in monitoring programs: an assessment. Environ Monitor Assess. 1990;15:225–9.
18. Root TL, Alpert P. Volunteers and the NBS. Science. 1994;263:1205.
19. Whitelaw G, Vaughan H, Craig B, Atkinson D. Establishing the Canadian community monitoring network. Environ Monitor Assess. 2003;88:409–18.
20. Pollock RM, Whitelaw GS. Community-based monitoring in support of local sustainability. Local Environ. 2005;10:211–28.
21. Penrose D, Call SM. Volunteer monitoring of benthic macroinvertebrates: regulatory biologists' perspectives. J North Am Benthol Soc. 1995;14:203–9.
22. Legg CJ, Nagy L. Why most conservation monitoring is, but need not be, a waste of time. J Environ Manage. 2006;78:194–9.
23. Mackechnie C, Maskell L, Norton L, Roy D. The role of 'Big Society' in monitoring the state of the natural environment. J Environ Monitor. 2011;13:2687–91.
24. Cooper CB, Dickinson JL, Phillips T, Bonney R. Citizen science as a tool for conservation in residential ecosystems. Ecol Soc. 2007;12:1–11.
25. Bonney R, Cooper CB, Dickinson J, Kelling S, Phillips T, Rosenberg KV, Shirk J. Citizen science: a developing tool for expanding science knowledge and scientific literacy. BioScience. 2009;59:977–84.
26. Silvertown J. A new dawn for citizen science. TREE. 2009;24:467–71.
27. Tweddle JC, Robinson LD, Pocock MJ, Roy HE. Guide to citizen science: developing, implementing and evaluating citizen science to study biodiversity and the environment in the UK. Natural History Museum and NERC Centre for Ecology & Hydrology; 2012.
28. Dickinson JL, Zuckerberg B, Bonter DN. Citizen science as an ecological tool: challenges and benefits. Ann Rev Ecol Evol Syst. 2010;41:149–72.
29. Patrick R. A proposed biological measure of stream conditions, based on a survey of the Conestoga basin, Lancaster County, Pennsylvania. Proc Acad Nat Sci Philadelphia. 1949;101:277–341.
30. Cairns J Jr. Indicator species vs. the concept of community structure as an index of pollution. Wat Res Bull. 1974;10:338–47.
31. Engel SR, Voshell JR. Volunteer biological monitoring: can it accurately assess the ecological condition of streams? Am Entomol. 2002;48:164–77.
32. Friberg N, Bonada N, Bradley DC, Dunbar MJ, Edwards FK, Grey J, Hayes RB, Hildrew AG, Lamouroux N, Trimmer M, Woodward G. Biomonitoring of human impacts in freshwater ecosystems: the good, the bad and the ugly. Adv Ecol Res. 2011;44:1–68.
33. O'Leary N, Vawter T, Plummer Wagenet L, Pfeffer M. Assessing water quality using two taxonomic levels of benthic macroinvertebrate analysis: Implications for volunteer monitors. J Freshwat Ecol. 2004;19:581–6.
34. Reynoldson T, Hampel L, Martin J. Biomonitoring networks operated by schoolchildren. Environ Pollut (Series A). 1986;41:363–80.
35. Savan B, Morgan AJ, Gore C. Volunteer environmental monitoring and the role of universities: the case of Citizens' Environment Watch. Environ Manage. 2003;31:561–8.
36. Woodiwiss FS. The biological system of stream classification used by the Trent River Board. Chem Ind. 1964;11:443–7.
37. Chandler JR. A biological approach to water quality management. Water Pollut Cont. 1970;69:415–22.
38. Hilsenhoff WL. Rapid field assessment of organic pollution with a family-level biotic index. J North Am Benthol Soc. 1988;7:65–8.
39. Kerans BL, Karr JR. A benthic index of biotic integrity (B-IBI) for rivers of the Tennessee valley. Ecol Appl. 1994;4:768–85.
40. FrostNerbonne J, Vondracek B. Volunteer macroinvertebrate monitoring: assessing training needs through examining error and bias in untrained volunteers. J North Am Benthol Soc. 2003;22:152–63.
41. Gowan C, Ruby M, Knisley R, Grimmer L. Stream monitoring methods suitable for citizen volunteers working in the coastal plain and lower Piedmont regions of Virginia. Am Entomol. 2007;53:48–57.
42. Pond Conservation. The development of the Big Pond Dip invertebrate survey method [http://www.freshwaterhabitats.org.uk/wordpress/wp-content/uploads/2013/09/Pond-Conservation-Invertebrate-Survey-Method.pdf].
43. Au J, Bagchi P, Chen J, Martinez R, Dudley SA, Sorger GJ. Methodology for public monitoring of total coliforms, *Escherichia coli* and toxicity in waterways by Canadian high school students. J Environ Manage. 2000;58:213–30.
44. Environmental Protection Agency [http://water.epa.gov/type/watersheds/monitoring/vol.cfm].
45. Wood PJ, Armitage PD. Biological effects of fine sediment in the lotic environment. Environ Manage. 1997;21:203–17.
46. Henley WF, Patterson MA, Neves RJ, Lemly AD. Effects of sedimentation and turbidity on lotic food webs: a concise review for natural resource managers. Rev Fish Sci. 2000;8:125–39.
47. Tyler FE. The Secchi disc. Limnol Oceanogr. 1968;13:1–6.
48. Schmeller DS, Henry P-Y, Julliard R, Gruber B, Clobert J, Dziock F, Lengyel S, Nowicki P, Déri E, Budrys E, Kull T, Tali K, Bauch B, Settele J, van Swaay C, Kobler A, Babij V, Papastergiadou E, Henle K. Advantages of volunteer-based biodiversity monitoring in Europe. Cons Biol. 2008;23:307–16.
49. Heiman MK. Science by the people: grassroots environmental monitoring and the debate over scientific expertise. J Plan Ed Res. 2013;16:291–9.
50. Lepczyk CA, Boyle OD, Vargo TL, Gould P, Jordan R, Liebenberg L, Masi S, Mueller WP, Prysby MD, Vaughan H. Citizen science in ecology: the intersection of research and education. Bull Ecol Soc Am. 2009;2009:308–17.
51. Turner SD, Rose NL, Goldsmith B, Harrad S, Davidson TA. OPAL Water Centre monitoring report 2008–2012. London: OPAL; 2013.
52. Department for Communities and Local Government. English indices of deprivation 2010. [https://www.gov.uk/government/publications/english-indices-of-deprivation-2010].
53. Pond Conservation. The Big Pond Dip: summary of findings from the 2009 survey. [http://www.biodiversitysouthwest.org.uk/docs/2009_Big_Pond_Dip_Summary_Final.pdf].
54. Freshwater Habitats Trust. Big Pond Dip 2014 [http://www.freshwater-habitats.org.uk/projects/big-pond-dip/score-means/].

55. Céréghino R, Boix D, Cauchie H-M, Martens K, Oertli B. The ecological role of ponds in a changing world. Hydrobiologia. 2014;723:1–6.
56. Gaston KJ, Smith RM, Thompson K, Warren PH. Urban domestic gardens (II): experimental tests of methods for increasing biodiversity. Biodivers Cons. 2005;14:395–413.
57. Guerold F. Influence of taxonomic determination level on several community indices. Wat Res. 2000;34:487–92.
58. Freshwater Habitat Trust. PondNet. [http://www.freshwaterhabitats.org.uk/projects/pondnet/].

# 4

# Fire-severity effects on plant–fungal interactions after a novel tundra wildfire disturbance: implications for arctic shrub and tree migration

Rebecca E. Hewitt[1,2*], Teresa N. Hollingsworth[3], F. Stuart Chapin III[1] and D. Lee Taylor[1,4]

**Abstract**

**Background:** Vegetation change in high latitude tundra ecosystems is expected to accelerate due to increased wildfire activity. High-severity fires increase the availability of mineral soil seedbeds, which facilitates recruitment, yet fire also alters soil microbial composition, which could significantly impact seedling establishment.

**Results:** We investigated the effects of fire severity on soil biota and associated effects on plant performance for two plant species predicted to expand into Arctic tundra. We inoculated seedlings in a growth chamber experiment with soils collected from the largest tundra fire recorded in the Arctic and used molecular tools to characterize root-associated fungal communities. Seedling biomass was significantly related to the composition of fungal inoculum. Biomass decreased as fire severity increased and the proportion of pathogenic fungi increased.

**Conclusions:** Our results suggest that effects of fire severity on soil biota reduces seedling performance and thus we hypothesize that in certain ecological contexts fire-severity effects on plant–fungal interactions may dampen the expected increases in tree and shrub establishment after tundra fire.

**Keywords:** *Alnus viridis*, Arctic tundra, ARISA, Climate change, Fire severity, Fungal internal transcribed spacer (ITS), *Picea mariana*, Shrub expansion, Treeline

## Background

In the last half century, warming in the Arctic and Subarctic has been correlated with the expansion of tundra shrubs into graminoid tundra [1] and the migration of forest into tundra in some locations [2]. These changes in vegetation could have strong positive feedbacks to the climate system, accentuating warming, through decreases in albedo, carbon storage, and increases in landscape flammability [3, 4]. Evidence suggests that factors influencing seedling establishment are the most critical determinants of global treeline [5] and shrubline advances [6]. However, the suite of ecological factors that influence seedling establishment in novel environments beyond current range limits are still not well understood.

Soil biota may influence the capacity of boreal trees to migrate into tundra and tundra shrubs to expand into non-shrubby tundra. Vegetation establishment can be influenced by soil biota, both mutualists and pathogens, which can affect both individual performance and plant species interactions. Although microbial symbionts can strongly influence plant performance and community structure [7], their impact on landscape-scale vegetation change is often overlooked. For example ectomycorrhizal (EM) fungi, are essential to seedling establishment and growth both inside [8] and outside [9] the native range of the host plant. Compared with non-mycorrhizal seedlings, EM-seedlings may display greater nutrient acquisition, lower levels of disease, and lower drought stress [10–12]. Although less well-studied, dark septate

*Correspondence: rebecca.hewitt@nau.edu
[2] Center for Ecosystem Science and Society, Northern Arizona University, PO Box 5620, Flagstaff, AZ 86011, USA
Full list of author information is available at the end of the article

endophytes (DSE), may also influence seedling establishment due to their suggested mutualistic influences, which largely overlap with those of EM fungi [13–15]. On the other hand, the establishment of seedlings can also be limited by the presence of species-specific enemies, including fungal pathogens [16–18]. The net outcome of negative interactions with pathogens and positive interactions with mutualists can influence the relative abundance and migration capacity of a plant species [11, 18].

The fire regime directly affects seedling recruitment success and migration [19] in the boreal forest, shrub growth and reproduction in the Subarctic [20], and shrub expansion in the Arctic [21] primarily through the effects of the severity of fire on the availability of high-quality, mineral soil seedbeds and time since fire on successional dynamics. Although fire disturbance has been relatively rare in the Arctic tundra for the last 11,000 years [22], in the last half-century the extent of tundra fires has increased due to warm and dry weather [23]. Increased fire frequency and severity in Arctic tundra is therefore expected to facilitate both tree migration and shrub expansion.

Despite the expected acceleration of tree migration and shrub expansions associated with warming in the Arctic [24], the effects of fire disturbance on fungal mutualists and pathogens may exert positive and negative indirect effects. For example, the same severe burns that increase the availability of high-quality establishment sites on mineral soil [19] can also alter the community structure of soil-dwelling fungal symbionts [25] and thus the fungal taxon-specific provision of soil resources to host plants. Recent research describes the broad geographic patterns of fungi across Arctic Alaska [26–28], yet the response of these fungal communities to wildfire disturbance is unknown. In temperate and boreal regions, severe fires decrease EM richness [29] and root colonization [30], and post-fire EM colonization of seedlings can be dependent on spores, sclerotia, or other components of the post-fire resistant propagule community (RPC) [31] as an inoculum source. Whether fire-severity affects DSE abundance and colonization in a similar manner to EM is not known. In addition to fire-effects on potentially beneficial mycobionts, burns can induce infection in fire-damaged roots of vegetation that survives fire [32], thus affecting the prevalence of pathogens that may associate with establishing seedlings. In tundra where fire has been relatively rare, the effect of fire severity on the prevalence of pathogens and availability of beneficial mycobionts in tundra is largely unknown.

Due to the historic rarity of tundra fires, current predictions of vegetation change after fire in Arctic Alaska are based on assumptions derived from the richer body of boreal forest research, which suggest that seedbed quality, i.e. exposure of moist, mineral seedbeds, is the primary ecological filter that drives seedling establishment [33]. The importance of post-fire, mineral seedbeds to seedling establishment in the boreal forest implies that in Arctic Alaska higher severity fires will produce a better seedbed and thus facilitate treeline advance and shrub expansion. As a first step towards understanding the potentially important biotic effects of root-associated fungi on tree and shrub establishment in tundra, we used a growth chamber experiment to investigate the role of post-fire soil microbes on seedling performance for two plant species predicted to migrate into tundra under future scenarios of warming and fire. In a companion field study, we observed that fire severity influenced the fungal communities associated with the dominant tundra shrub *Betula nana* and that these shrubs appear to provide a source of inoculum resilience by maintaining some mycorrhizal fungi after wildfire [34]. However, whether there is adequate inoculum in soils immediately after fire and how the inoculum varies with fire severity is unknown in tundra ecosystems. Specifically, we tested the hypothesis that increasing fire severity decreases performance in establishing seedlings due to fire-severity effects on root-associated fungal mutualists and pathogens. We examined the relationships between seedling biomass, fungal community composition, and fire severity. To our knowledge this study is the first investigation of fire-severity effects on plant–fungal interactions after tundra fire and thus provides an opportunity for hypothesis development regarding the importance of post-fire soil biota to seedling establishment, a process key to vegetation change in treeline and tundra ecosystems.

## Methods

### Study species and field sampling

In Alaska, extensive woody expansion into tundra has been documented for alder shrubs, *Alnus viridis* (Chaix) DC. [1], and its growth and reproduction are greater in burned sites [20]. Black spruce, *Picea mariana* (Mill.) Britton, Sterns & Poggenb., is the dominant latitudinal treeline species throughout most of North America. In boreal Alaska, occurrence of black spruce on the landscape is in large part determined by fire history [35] and, in addition, is often found in the coldest, wettest areas of the boreal forest. These black spruce communities share many plant species with the Alaska Arctic moist acidic tundra communities [36]. Both alder and spruce species are obligately EM [37]. Seeds of both species were collected in Interior Alaska at Washington Creek and Fairbanks and stored at −20 °C until used in our growth chamber experiment.

Between July and October 2007 the Anaktuvuk River fire (ARF), the largest tundra fire ever recorded on the North Slope of Alaska, burned 1039 $km^2$ of upland shrubby tussock tundra underlain by continuous permafrost [38] approximately 100 km north of present-day latitudinal treeline. The dominant vegetation before the fire was moist acidic tundra (54 %) with moist nonacidic tundra (15 %) and shrubland (30 %) covering smaller areas [39]. We focused our study on moist acidic tundra that is dominated by sedges (*Eriophorum vaginatum* L., *Carex bigelowii* Torr. Ex Schwein), evergreen shrubs (*Ledum palustre* L. and *Vaccinium vitis-idea* L.) that associate with ericoid mycorrhizal fungi (ERM), deciduous shrubs (*B. nana* L. and *Salix pulchra* Cham.) that are in symbiosis with EM fungi, mosses, and lichens [40] and thus are very similar to the understory composition of some boreal black spruce forests [36]. There is overlap in the composition of EM fungi that associate with EM tundra shrubs in this region and EM host trees and shrubs in the boreal forest [27, 41, 42].

In July 2008, the first growing season after the fire, we visited eight burned sites within the ARF burn scar corresponding with different fire severities via a one-time opportunity to access the burn scar with helicopter-based logistical support and approval from the Bureau of Land Management (Fig. 1; Additional file 1). Fire severity was measured in the field as the composite burn index (CBI) and with site descriptions of post-fire vegetation composition and structure and combustion of vegetation and soil [39]. At 20 points along a 50 m transect, we collected approximately 15 ml of organic soil and 15 ml mineral soil from the top 5 cm of the soil horizon. We then pooled and homogenized our 20 samples per site by soil horizon, and stored them at 4 °C at the University of Alaska Fairbanks for 3 weeks until we inoculated the host plants.

### Experimental design

We conducted a growth chamber experiment using a randomized block design with 18 treatments and 10 replicate blocks and two host plant species. There were 16 treatments with inoculum from field soil [8 sites × 2 soil horizons (organic and mineral)] and two additional treatments to test for unintentional inoculation of seedlings in the growth chamber with sterile inoculum (twice-autoclaved mineral and autoclaved organic soils). In July 2008 we surface-sterilized black spruce seeds with a solution of 5 % household bleach, 5 % ethanol, and liquinox soap for 5 min and the smaller alder seeds for 1 min followed by ten rinses with ultrapure water. Seeds were placed in sterile petri dishes on autoclaved filter paper, and RO water was used to keep the seeds moist.

We transplanted seedlings from petri dishes into standard 150 ml cone-tainers (Stuewe and Sons, Inc., Tangent, Oregon, USA) filled with sterilized silt soil 2 weeks after germination and then inoculated them with 1 of the 18 treatments. Twelve milli-liters of treatment soil was added to the top of each cone-tainer and watered into the autoclaved substrate. Seedlings received one 5 ml application (14 ppm N) of a 9:20:9 NPK fertilizer solution one and half months after inoculation and one 5 ml application (50 ppm N) 4 months after inoculation due to the chlorotic appearance of the seedlings. Seedlings were grown in a controlled-environment chamber (Conviron CMP 3246, Winnipeg, Manitoba, Canada) for 7 months at 25/10 °C day/night with 16-h photoperiod at 300 $\mu mol\ m^{-2}\ s^{-1}$ irradiance at ambient humidity (26–96 % RH, mean 73.75 RH ± 0.16 SE, NOAA National Climate Data Center http://www.ncdc.noaa.gov/). We watered seedlings to excess with RO water as needed. Height and survival were measured monthly. At the time of harvest, roots were rinsed gently with RO water and separated from shoots. We dried shoots and roots after root tip sampling (described below) at 60 °C for 48 h in a drying oven (VWR Scientific Products Forced Air Oven, Radnor, Pennsylvania, USA) and determined the dry weight of roots, stems, and leaves.

### Characterization of fungal communities

Harvested root systems were cut into 4 cm segments and floated in ultrapure water in a petri dish. Using a dissecting microscope (40× magnification) we randomly selected ten live root tips that exhibited signs of fungal colonization, e.g., no root hairs, from each root system. For each seedling, root tips were pooled for automated ribosomal intergenic spacer analysis (ARISA) and DNA sequence analysis of root-associated fungal community structure. ARISA community profiles provide information on the number and relative abundance of taxa, ribotypes, within a sample; however, ARISA does not provide taxonomic identities of the taxa. Therefore, we used Sanger sequencing to assign taxonomic identities to the dominant ARISA ribotypes (see Additional file 2). Pooled root tip samples were placed in a single 0.6 ml Eppendorf tube, frozen in a small amount of ultrapure water, and stored at −80 °C.

In September 2010 we extracted DNA from lyophilized pooled root tip samples of each seedling. From these pooled genomic DNA samples, the fungal ITS gene region was amplified using the primers ITSF and ITS4 following the protocol of Bent and Taylor [43]. We obtained ARISA fungal community profiles and ITS sequences following Bent et al. [44]. Fungal taxa were inferred from ITS sequences and matched to dominant ARISA ribotypes based on sequence and ribotype fragment lengths ([45]; Additional file 3). Fungal identities were assigned though comparison of our ITS sequences

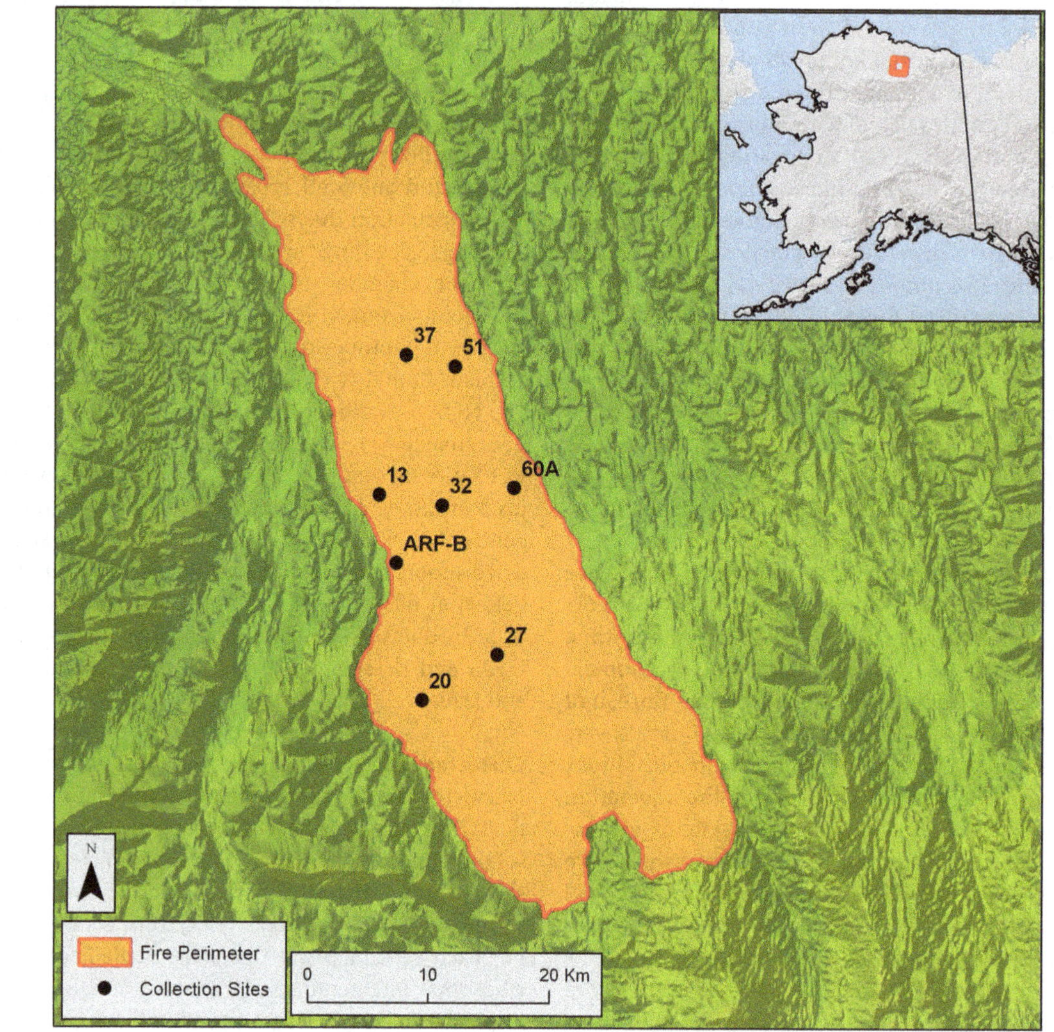


**Fig. 1** Map of soil collection sites that represent a fire-severity gradient within the Anaktuvuk river fire burn scar: low severity = sites *27* and *51*; moderate severity = sites *20*, *32*, and ARF-B; high severity = sites *13*, *37*, and *60A*

to those from GenBank using a curated fungal-ITS BLAST search (http://www.borealfungi.uaf.edu/) that excludes environmental and uncultured sequences ([46]; Additional file 2). In addition, we constructed maximum likelihood trees to infer identities for sequences with inconsistent identities resulting from the BLAST search (Additional file 2). Nomenclature for our sequences follows Timling et al. [27]. Functional groups were assigned based on Operational Taxonomic Unit (OTU) identities. For example, for species and genera we could categorize OTUs as pathogens, endophytes, DSE, saprotrophs, or ERM. OTUs that were assigned identities at family or higher levels of taxonomic resolution were assigned multiple functional groups (Additional file 4). In a few cases the closest sequence identities matched with a ribotype were from different functional groups, in which case we described ribotypes as either pathogenic or nonpathogenic (i.e. ERM, DSE).

### Statistical analysis

We used nonmetric multidimensional scaling (NMDS) [47] ordinations to interpret the variability in fungal composition across seedlings. Ordinations were based on ribotype abundance data, and we used Beal's smoothing to relieve the "zero truncation problem" [48]. We observed some fungal colonization of roots when we ran ARISA on seedlings that received the sterile inoculum (23 ribotypes on 31 of the 40 seedlings that received the sterile treatment; five of the ribotypes occurred more than once). We examined the relative abundance of the five ribotypes that occurred more than once and found that their abundance did not vary significantly between treatments. Therefore, we made the assumption that any variability in biomass detected is due to treatment effect and not the effects of the five ribotypes. In order to adjust for this contamination we used the most conservative approach, and all ribotypes observed on these seedlings were excluded from multivariate analysis of fungal composition. Following McCune & Grace [49] we eliminated all rare ribotypes, i.e. those that occurred in less than 5 % of the samples. We used the Sorensen distance measure and a random starting configuration with a final solution generated using 500 iterations [49]. First, to determine whether fungal composition was related to seedling biomass, we ordinated the fungal communities associated with all seedlings regardless of treatment or host species. Secondly, to investigate whether treatment differences in fungal composition were related to seedling biomass, we pooled treatment replicates into one fungal community profile per treatment (i.e. combined soil horizon and burn severity) for each host plant. In both cases, axis scores were produced and then used in regression analysis as a measure of fungal composition. We also used multiple-response permutation procedures (MRPP) [49] with the Euclidean distance measure [49] to investigate whether there were differences between fungal communities in burned sites grouped by low, moderate, and high burn-severity categories (low = 2 treatments, moderate = 3 treatments, high = 3 treatments) and between soil horizons (organic = 8 treatments, mineral = 8 treatments). Fire-severity categories were defined by CBI and site severity descriptions.

We assessed normality graphically for all response variables (total biomass, shoot biomass, root biomass, root:shoot, stem weight, leaf weight, and life span,) and considered skewness, kurtosis, and Shapiro–Wilk's W values before log-transforming data. We evaluated correlations between response variables using Spearman correlations. All response variables were significantly correlated to total seedling biomass (all correlations $P < 0.05$ and Spearman's $\rho > 0.14$), so we used log-transformed total biomass as the response variable for subsequent analyses. We used ANOVA to investigate inoculation effects on log-transformed seedling biomass with host plant species and inoculation status as factors, and ANCOVA to investigate the relationship between log-transformed seedling biomass and treatment, which includes a continuous fire severity factor and a categorical soil horizon factor for each host species. To evaluate the relationship between fungal composition (NMDS axis scores) and log-transformed seedling biomass, we used stepwise regression to evaluate the best-fit model comprised of all or a subset of NMDS axes. We used regression to test for relationships between the proportion of functional groups of fungi and both fire severity and log-transformed seedling biomass. Stepwise regression was then used to determine whether the relative abundance of particular taxa within defined functional groups were related to log-transformed seedling biomass. We used T test to compare the survivorship between host plants. Survivorship was expressed as the percentage of seedlings that survived the 7-month experiment compared to the beginning of the experiment.

Each inoculum treatment reflects both fire severity and soil horizon, so we tested for differences in treatment means of seedling biomass (ten seedlings/inoculum type/host species) across a continuous fire-severity gradient (CBI) and soil horizon using regression and ANOVA, respectively. Initially, we explored models with site as a cofactor and found site not to be a significant factor. Because sites were chosen to represent different fire severities we did not include both site and CBI in the final model. All statistical analyses were performed in JMP 9.0.2 (SAS Institute Inc., 2010) with the exception of the multivariate analysis of fungal communities in PC-ORD 6.0.

## Results

Over 70 % of seedlings survived through the end of the experiment, and percent survivorship was not species-dependent ($T = 1.50$, $P_{(7, 6)} = 0.184$). Compared to controls inoculated with sterilized inoculum, inoculated soils from the ARF reduced log-transformed seedling biomass for both spruce and alder (full model $F_{(236, 2)} = 88.7909$, $P < 0.000$; inoculation $F_{(236, 1)} = 5.642$, $P = 0.018$; species $F_{(236, 1)} = 169.752$, $P < 0.000$) (Fig. 2).

### Treatment effects on seedling biomass

The inoculation treatments given to the seedlings reflected both burn severity and soil horizon of a site.

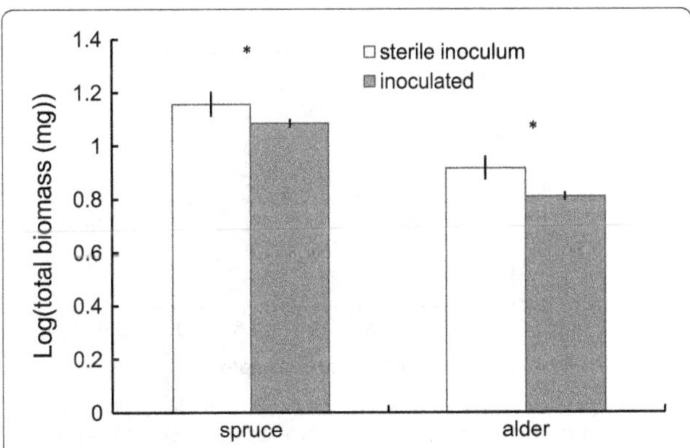


**Fig. 2** Inoculation reduces log-transformed seedling biomass (±1 SE) for both spruce and alder seedlings. *Asterisk* indicates significant differences ($P < 0.01$) between mean log-transformed biomass of seedlings inoculated and seedlings that received the sterile inoculum

In general, fire severity had a stronger effect on seedling biomass than did soil horizon, particularly for alder. For inoculated alder we found a decrease in total seedling biomass with increased fire severity ($F_{(16, 1)} = 4.976$, $P = 0.044$) (Fig. 3a, b). Inoculated spruce log-transformed seedling biomass also decreased with increasing fire severity ($F_{(15, 1)} = 4.175$, $P < 0.0636$) but only for seedlings given mineral-soil inoculum (Fig. 3a, c). In contrast, soil horizon alone had no significant effect on inoculated log-transformed seedling biomass for either alder or spruce (alder $F_{(16, 1)} = 0.728$, $P = 0.409$; spruce $F_{(15, 1)} = 0.704$, $P < 0.418$). There was, however, a significant interaction between soil horizon and fire severity on inoculated spruce log-transformed seedling biomass ($F_{(15, 1)} = 9.340$, $P < 0.010$) (Fig. 3c).

### Fungal composition and effects on seedling biomass

In an exhaustive search for EM root tips, we failed to find any. However, we did observe numerous fungal hyphae and indications of some degree of fungal interaction with plant roots. Microscopic morphological examination (10–40×) of live, turgid root tips that had no root hairs from the root system of each spruce and alder seedling revealed no branched, swollen, or colored root tips with classic EM characteristics. Many of the spruce root tips were smooth with lighter coloration than lateral roots, while alder tips often had white cottony hyphae and dark areas of coloration, some of which appeared necrotic (Additional file 5). We further examined root tips on the compound scope (40–100×) to verify absence of root hairs and a fungal mantle and the presence of dark hyphae or necrosis. These morphological observations were supported by the molecular data. We successfully extracted DNA, amplified, and sequenced fungi from 199 seedlings and found no EM fungi associated with any seedling (Additional file 4). For roots that did not initially amplify, we performed multiple extractions and PCRs to ensure that seedlings indeed had no fungi associated with the root systems. We observed 115 fungal ribotypes across the two host species and obtained sequence identities for 28 ribotypes, including the most abundant ribotypes in our study. Although we were not able to match all ARISA ribotypes with sequence identities, the majority of dominant ARISA ribotypes were identified. Seventy-two percent of the total fluorescence from ARISA electropherograms was identified with matching sequences. The ribotypes that were matched with sequence IDs included a range of functional groups: seven of these are putative endophytes (including DSE), 13 are putative pathogens, two are saprotrophs, and six were identified to a deeper taxonomic level or were associated with multiple functional roles; none belonged to known EM taxa.

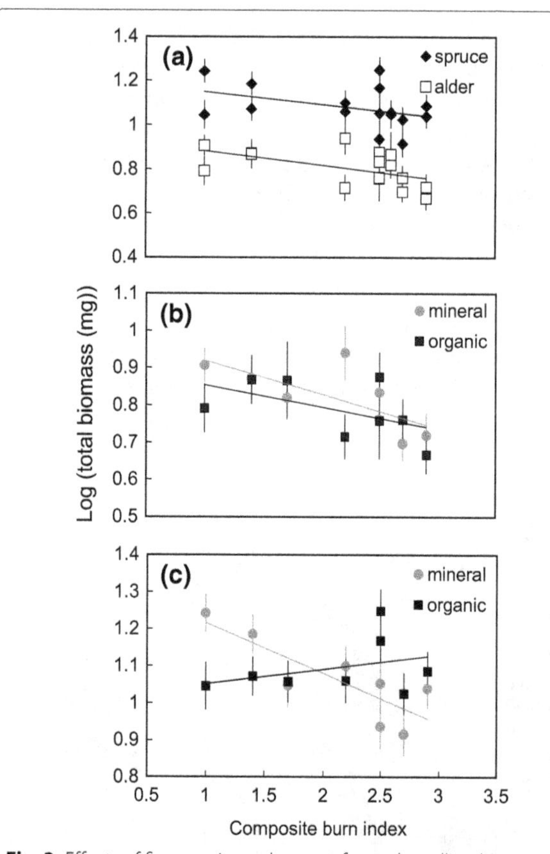


**Fig. 3** Effects of fire severity on log-transformed seedling biomass; **a** treatment means for spruce and alder biomass (±1 SE) decline with increasing fire severity; **b** alder biomass declines with increasing fire severity for seedlings grown with inoculum from mineral and organic soil horizons; **c** spruce biomass declines with increasing fire severity for seedlings grown with inoculum from the mineral soil horizon

NMDS ordinations of all root tips sampled, revealed that fungal composition varied across seedlings and treatment. For inoculated spruce seedlings, fungal communities differed by fire-severity category (MRPP, $A = 0.081$, $P = 0.009$), but not soil horizon (MRPP, $A = -0.012$, $P = 0.859$). In contrast, for inoculated alder seedlings, there was significant variation in fungal composition by soil horizon (MRPP, $A = 0.068$, $P = 0.015$), but not for fire severity category (MRPP, $A = -0.029$, $P = 0.779$).

These results indicate variation in plant–fungal relationships depending on host species, but do not give an indication of the role fungal composition may promote or inhibit growth of seedlings post-fire. To determine the effect of fungal inoculum on log-transformed seedling biomass we used NMDS axes to represent differences in fungal composition across seedlings and treatments (Additional file 6). For future reference, we will refer to these axes as FC (fungal compositional) axes. Stepwise regression indicated that the model containing fungal composition represented by FC axes 1 and 2 generated the best-fit model, regardless of treatment. Log-transformed seedling biomass was significantly related to the fungal composition of inoculum for seedlings regardless of burn severity and soil horizon of a site ($F_{(114, 7)} = 11.051$, $P < 0.000$). We further investigated the effect of fungal composition for each treatment (fire severity and soil-horizon, Additional file 7a, b) on log-transformed seedling biomass for alder and spruce seedlings. For alder FC Axis 1 was the best fit, and was significantly related to biomass for alder seedlings ($F_{(16, 1)} = 6.791$, $P < 0.021$). For spruce the models containing FC axis 1, FC axis 2, or FC axis 3 provided an equal fit to spruce log-transformed seedling biomass, however none were significantly related to spruce log-transformed seedling biomass ($F_{(16, 1)} = 0.506$, $P < 0.489$). Overall, these results illustrate that fungal composition was significantly related to seedling biomass with treatment-level variations depending on seedling species. Alder biomass showed a strong relationship to the variation in fungal composition for individual treatments, while spruce seedling biomass did not.

The proportion of root-associated fungi identified as pathogens was directly related to fire severity across the gradient (Fig. 4, Total $F_{(8\ 1)} = 4.9106$, $P < 0.0686$, $R^2 = 0.45$). Forty-one percent of the total fluorescence from ARISA profiles was attributable to fungal taxa, ribotypes, identified as pathogens. This relationship was stronger for fungi associated with spruce than those associated with alder seedlings (alder $F_{(8, 1)} = 1.2742$, $P < 0.3021$, $R^2 = 0.18$; spruce $F_{(8, 1)} = 6.7790$, $P < 0.0405$, $R^2 = 0.53$). Regardless of treatment, alder log-transformed seedling biomass was negatively correlated with the relative abundance of pathogens, ($F_{(89, 2)} = 5.639$, $P < 0.020$) and showed a marginally significant positive relationship to the relative abundance of the DSE functional group ($F_{(89, 2)} = 3.025$, $P < 0.086$). However, three DSE taxa were positively related to log-transformed seedling biomass: *Phialocephala fortinii* complex (ribotype 25, $F_{(89, 4)} = 5.244$, $P < 0.025$), *Cadophora finlandica* (ribotype 93, $F_{(89, 4)} = 5.029$, $P < 0.024$), and *Phialocephala* sp.(ribotype 90, $F_{(89, 4)} = 15.743$, $P < 0.000$). Relative abundance of these taxa did not vary significantly with fire severity. Together these results show that, as fire severity increases, fungal composition shifted to a greater proportion of known pathogens and log-transformed seedling biomass declined. Although some DSE taxa were positively correlated with seedling growth, pathogens had a strong negative influence on log-transformed seedling biomass.

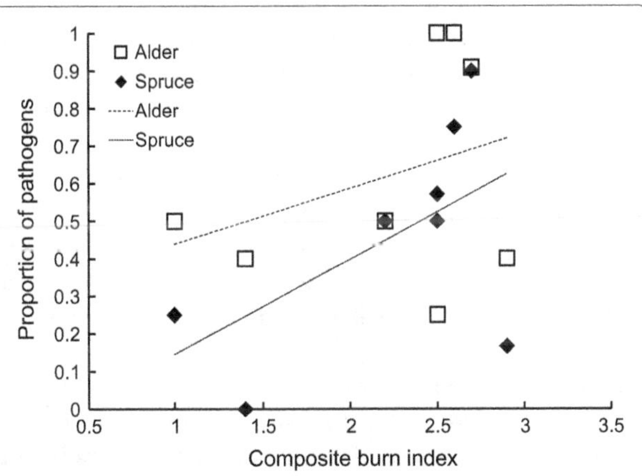

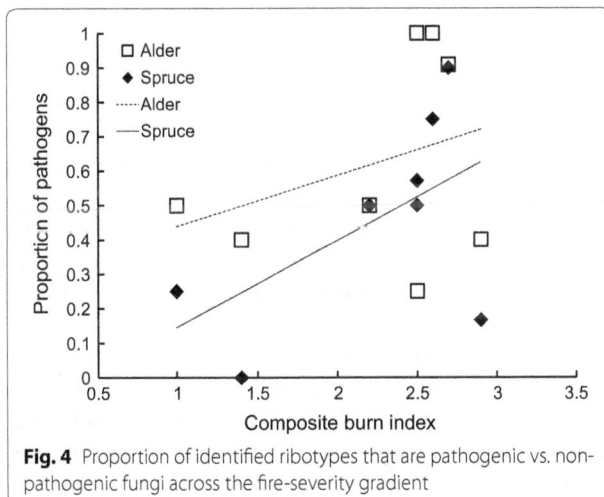

**Fig. 4** Proportion of identified ribotypes that are pathogenic vs. non-pathogenic fungi across the fire-severity gradient

## Discussion

Inoculation with burned soils reduced plant growth, apparently due to pathogenic effects of root-associated fungi. The proportional shift towards more pathogenic fungal symbionts with an increase in fire severity corresponded with reduced growth for alder and a weaker growth response of spruce seedlings inoculated with field soils. The significant colonization of seedlings by pathogens may relate to the ability of pathogens to disperse more widely and more quickly than many EM fungi [50, 51] and potentially DSE. Thirty-eight percent (5/13) of the sequenced fungi that we identified to at least the genus level as pathogens were also observed in soils along a trans-Arctic transect [26]. This suggests that the pathogens found in our study occur across the Arctic, and that pathogen propagules are robust, surviving for long periods in the soil, or distribute rapidly in tundra. The negative effect of pathogens on alder and spruce seedlings are consistent with the hypothesis of Blumenthal et al. [52] that stress-tolerating plant species are similarly susceptible to pathogens inside and outside of their native range, thus suppressing their capacity to invade novel environments.

Although the ARF post-fire soils were not an effective source of mutualist, EM fungi, inoculum for alder or spruce seedlings, they were a source of DSE inoculum. Of the DSE we identified in our bioassay, 100 % (11/11) were also observed in soils across the trans-Arctic transect [26]. DSE are widespread and can be more frequent at high-latitudes than classical mycorrhizal fungi [53]. Their proportional decline across the fire-severity gradient suggests that they are sensitive to fire disturbance, as we previously showed for EM fungi in post-fire treeline and tundra ecosystems [34, 41]. Colonization of roots by DSE is suggested to improve plant nutrition and biomass [13] and has been correlated with reduction of pathogenic root disease intensity [54]. Indeed, we found the relative abundance of three DSE taxa to be positively related to seedling biomass. We also observed an increase in the proportion of pathogenic fungi concurrent with a decline in DSE across the fire-severity gradient.

Because of the rarity of large-scale tundra fire disturbances for at least 11,000 years [22], mutualist soil biota inoculum in Arctic Alaska may lack fire-specialist taxa that are either resilient to fire or are stimulated by fire. The site of the ARF had not burned for the preceding 5000 years [23], in stark contrast to the average fire return interval of 150 years in the boreal forest [55]. Our previous field research suggests that surviving mycorrhizal shrubs are a more likely source of EM inoculum than are spores or sclerotia from the post-fire RPC [31] in soils following tundra fires [34, 41]. Mycelial inoculum sources may be limited in tundra because of low density of EM host plants in many tundra ecosystems. These shrubs can take up to a decade to return to pre-fire densities after wildfire [56]. The effectiveness of the RPC as an inoculum source after fire is often restricted by the availability of spores and sclerotia [57] and by the efficacy of these inoculum sources with certain host plants [58]. Instead of a robust EM RPC community providing beneficial inoculum, we observed DSE colonization in our bioassay, suggesting that other potentially beneficial mycobionts, DSE, may be less sensitive to fire or that they disperse more widely and/or quickly than EM fungi.

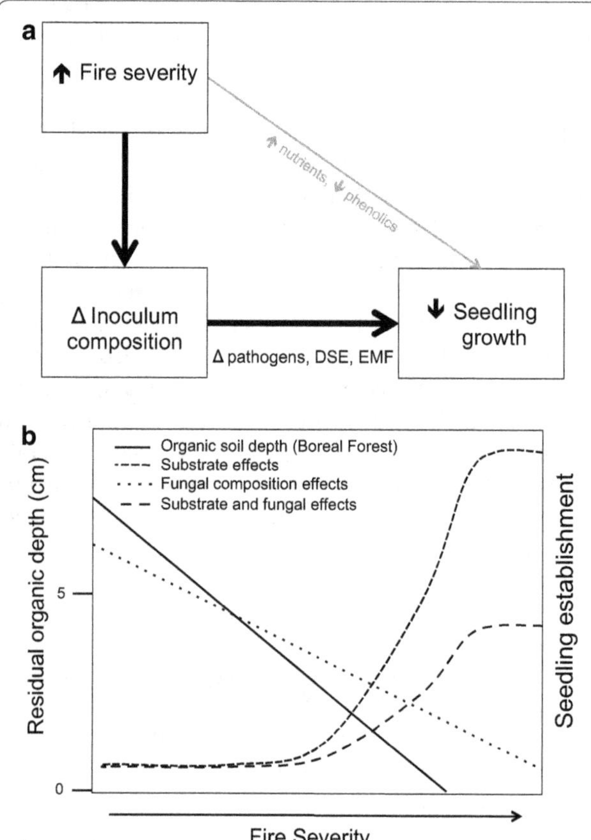

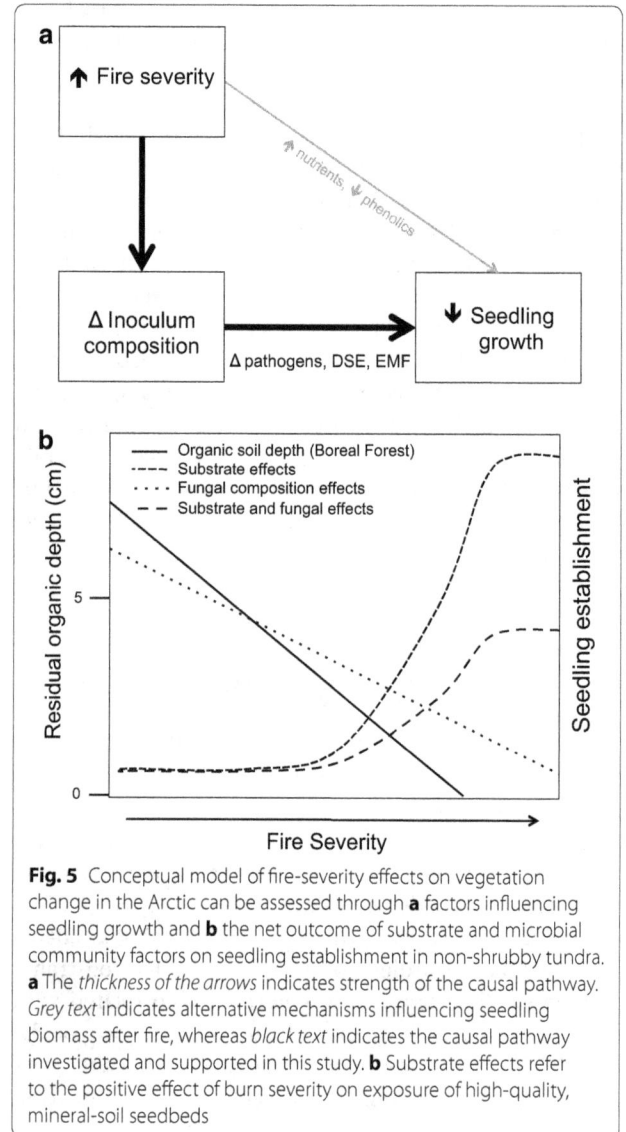

**Fig. 5** Conceptual model of fire-severity effects on vegetation change in the Arctic can be assessed through **a** factors influencing seedling growth and **b** the net outcome of substrate and microbial community factors on seedling establishment in non-shrubby tundra. **a** The *thickness of the arrows* indicates strength of the causal pathway. *Grey text* indicates alternative mechanisms influencing seedling biomass after fire, whereas *black text* indicates the causal pathway investigated and supported in this study. **b** Substrate effects refer to the positive effect of burn severity on exposure of high-quality, mineral-soil seedbeds

Although the lack of mycorrhizal development in our study could also have been an artifact of the growth chamber conditions, other studies have shown that, consistent with our methods, the intermittent application of soluble fertilizer (Castellano et al. 1985) and the storage of soil inoculum (Nunez et al. 2009) or spore slurries (Castellano et al. 1985) for comparable or longer time periods did not inhibit the formation of mycorrhizas on small, first-year seedlings. We conclude that these patterns are ecologically relevant for three main reasons: (1) in this study both biomass and fungal composition were related to fire severity; (2) in companion field studies we observed fire-severity effects on inoculum composition for naturally established tundra shrubs and treeline shrubs and seedlings [34, 41] and treeline seedling biomass was correlated with post-fire fungal composition [41]; and (3) our inoculation and fertilization methods were consistent with other studies where EM formation was observed.

We suggest that the effect of fire severity on seedling biomass in a controlled bioassay was due to fire-severity effects on fungi not variability in available nutrients or inhibitory phenolics in the soil across the fire-severity gradient (Fig. 5a). We dismiss the idea of a nutrient effect because the amount of field soil provided for inoculation was small (12 ml), and all seedlings in the experiment were fertilized. Similarly, phenolic-effects are not likely in this ecosystem because fire consumption of the organic horizon volatilizes inhibitory allelochemicals and phenolics in litter and soil [25], and charcoal sorbs inhibitory compounds [59]. This suggests that a fire effect on fungal community composition is the most likely explanation for the fire-severity effect on seedling biomass. Along these lines, we observed reduced seedling biomass and changes in the proportion of functional groups of root-associated fungi as fire severity increased across the gradient of sites. In addition, seedling biomass was related to fungal composition. Hence, we believe that decreased biomass along the fire-severity gradient is attributable to changes in fungal composition related to fire severity.

The influence of fire severity, however, differed between the two host species. Alder seedling biomass declined as expected with increasing fire severity as the proportion of fungal pathogens increased. There was a strong relationship between fungal composition and mean biomass for each treatment, which appears to drive the effect of fire severity on alder seedling biomass. Spruce seedlings, however, responded to fire severity in a more complex way. Spruce biomass declined with increasing fire severity, though to a lesser degree than alder, with mineral-soil inoculum. In contrast to alder, the mean biomass of spruce seedlings did not significantly correlate to fungal composition, even though spruce seedling biomass declined with increasing fire severity of the inoculum and variability in fungal inoculum differed by fire-severity category. Two potential explanations are that (1) spruce-seedling biomass was reduced by an unidentified landscape factor affecting soil inoculum that was highly correlated with fire severity; or (2) particular fungal taxa instead of overall fungal composition exerted a strong effect on spruce growth. However, analysis of the ribotypes that were matched with sequences did not reveal that the relative abundance of a specific taxon had a strong effect on biomass of spruce seedlings inoculated with organic soils.

Overall, our findings illustrate that in a highly controlled growth chamber setting seedling growth was reduced when grown in soils from sites with increasing tundra fire severity, likely through growth reduction by root-associated fungal pathogens or fungi with pathogenic effects. Although there is uncertainty with translating growth chamber experiments to outcomes in the field, our results suggest that the expected positive effects of fire severity on tree and shrub establishment in tundra after fire [20, 60] may be dampened by changes in soil biota associated with fire severity. Field studies are necessary to determine whether these negative effects of post-fire fungi on seedling growth occur in a complex post-fire field environment, outweigh the potential positive effects due to pathogen release or the benefits of mutualist fungi resilient to tundra fires, and/or constrain seedling establishment beyond the native range in non-shrubby tundra. In a field study in the boreal forest, Johnstone & Chapin [60] observed decreased tree seedling establishment at extremely high-severity sites, despite nearby EM nurse plants that were potential sources of EM inoculum. This observed negative effect of high fire severity on seedling recruitment in the field might reflect a shift in mycobiont composition such as we observed that reduces seedling growth (Fig. 5b).

Understanding the influence of fungi on successful seedling establishment is important to forecasts of spruce migration and alder expansion in Arctic Alaska and associated ecosystem feedbacks. Both spruce migration and alder expansion have large ecosystem impacts due to changes in carbon storage, albedo, ecosystem services, and nutrient cycling. In particular, the expansion of alder has a significant influence on nitrogen and phosphorus cycling because of its role as a nitrogen fixer. On a global scale, Harsch et al. [2] found that 2 of 166 treeline sites receded since 1900 AD and that both of these sites showed evidence of disturbance. These authors infer that disturbance legacies do not likely affect the probability of advance, and instead influence initial recruitment and lag times between warming and treeline advance. Currently 2.3 % of tundra has been converted to forest in Alaska [61]. To extrapolate from our study, we would expect the rate of tree migration and shrub expansion to be constrained by pathogenic effects of soil biota and mutualist limitation after high-severity fires. However, if boreal EM fungi co-migrate or EM tundra shrubs provide surrogate sources of inoculum on the landscape under low and moderate severity fires [34], we would expect that boreal forest mycobionts may then facilitate vegetation change at and beyond current treeline, reducing lag times by facilitating initial recruitment. Thus, mutualistic and pathogenic symbionts may either constrain or facilitate vegetation change depending on fire severity or other contexts.

## Conclusions

This study provides an initial assessment of post-fire plant–fungal interactions in a controlled growth chamber setting for two plant species expected to expand into tundra with future scenarios of warming and wildfire. We found that seedling biomass was related to the composition of root-associated fungi and that fungal composition in our bioassay inoculum shifted across the fire-severity gradient. Decreased seedling biomass across the fire-severity gradient suggests that fire-severity effects on plant–fungal interactions may dampen seedling performance and thus establishment success. However, follow-up field studies are necessary to evaluate the relative importance of the negative effects of plant–fungal interactions on seedling performance within different post-fire contexts.

## Availability of supporting data

The datasets supporting the results of this article are available in the following repositories:

Sequences for each OTU have been archived with GenBank under accession numbers KF660543–KF660580.

Binned ARISA ribotype abundance data have been archived with the Bonanza Creek LTER and the LTER Network Data Portal doi:10.6073/pasta/a131c0d6707b6aa746dfe0265141ad43.

Environmental and fire-severity data are published in Jandt et al. 2012 and available through www.Frames.gov.

## Additional files

**Additional file 1.** Site descriptions for eight sampling sites grouped by fire-severity categories within the Anaktuvuk River Fire burn scar. We classified vegetation communities using the Viereck (1992) vegetation classification: 2C2A = open low scrub mixed shrub-sedge tussock tundra, 2C2H = open low scrub willow-sedge shrub tundra, 3A21 = mesic graminoid herbaceous sedge-birch tundra, 3A2D = mesic graminoid herbaceous tussock tundra, and 3A3 = wet graminoid herbaceous tundra. Fire-severity categories were assigned based on composite burn index and field descriptions.

**Additional file 2.** Detailed methods of molecular techniques used to characterize fungal communities.

**Additional file 3.** Biological key showing ribotype ID, abundance, range of fungal ITS fragment lengths, associated OTU identity, and the best match description as reported in GenBank.

**Additional file 4.** Operational Taxonomic Units at 97 % sequence similarity for fungi associated with seedlings inoculated with soils from the Anaktuvuk River Fire.

**Additional file 5.** Photographs of typical morphologies of seedling root systems: a-c alder; d-f spruce.

**Additional file 6.** Description of final solutions for nonmetric multidimensional scaling ordinations of fungal composition for all seedlings and treatment x species combinations.

**Additional file 7a, b.** Biplots of nonmetric multidimensional scaling ordinations of fungal communities associated with a). alder and b). spruce seedlings inoculated with soils from the Anaktuvuk River Fire and site characteristics. Active layer = depth of the unfrozen soil at time of sampling; dNBR = differenced normalized burn ratio of the sampling site; CBI = composite burn index of the sampling site.

### Abbreviations

ARISA: automated ribosomal intergenic spacer analysis; ARF: Anaktuvuk River fire; CBI: composite burn index; DSE: dark septate endophytes; EM: ectomycorrhizal; ERM: ericoid mycorrhizal; ITS: internal transcribed spacer; MRPP: multiple-response permutation procedures; NMDS: nonmetric multidimensional scaling; RPC: resistant propagule community.

### Authors' contributions

RH, TH, and LT designed the study. RH collected and analyzed data and wrote the first draft of the manuscript. LT advised on molecular techniques and bioinformatics. All authors contributed to interpretation and revisions. All authors read and approved the final manuscript.


### Author details

[1] Institute of Arctic Biology, University of Alaska Fairbanks, Fairbanks, AK 99775, USA. [2] Center for Ecosystem Science and Society, Northern Arizona University, PO Box 5620, Flagstaff, AZ 86011, USA. [3] US Forest Service, Pacific Northwest Research Station, Boreal Ecology Cooperative Research Unit, Fairbanks, AK 99775, USA. [4] Department of Biology, University of New Mexico, Albuquerque, NM 87131, USA.



### Acknowledgements

This research was supported by funding from Alaska EPSCoR (EPS-0701898) and the state of Alaska, a National Science Foundation Graduate Research Fellowship (DGE-0639280 and 1242789), and a Center for Global Change student award to R.E.H. In-kind support was provided by the Bonanza Creek Long-Term Ecological Research site, U.S. Forest Service PNW Research Station, University of Alaska Fairbanks Institute of Arctic Biology greenhouses, and Bureau of Land Management. We thank Randi Jandt, Eric Miller, and David Yokel for field and logistical support, John Alden for our seed source, and Jamie Hollingsworth for creating Fig. 1.


### Competing interests

The authors declare that they have no competing interests.

### References


1. Sturm M, Racine C, Tape K. Increasing shrub abundance in the Arctic. Nature. 2001;411(6837):546–7.
2. Harsch MA, Hulme PE, McGlone MS, Duncan RP. Are treelines advancing? A global meta-analysis of treeline response to climate warming. Ecol Lett. 2009;12(10):1040–9.
3. Rupp TS, Chapin FS, Starfield AM. Response of subarctic vegetation to transient climatic change on the Seward Peninsula in north–west Alaska. Glob Change Biol. 2000;6(5):541–55.
4. Hinzman LD, Bettez ND, Bolton WR, Chapin FS, Dyurgerov MB, Fastie CL, Griffith B, Hollister RD, Hope A, Huntington HP, et al. Evidence and implications of recent climate change in northern Alaska and other arctic regions. Clim Change. 2005;72(3):251–98.
5. Harsch MA, Bader MY. Treeline form—a potential key to understanding treeline dynamics. Global Ecol Biogeogr. 2011;20(4):582–96.
6. Myers-Smith IH, Forbes BC, Wilmking M, Hallinger M, Lantz T, Blok D, Tape KD, Macias-Fauria M, Sass-Klaassen U, Levesque E, et al. Shrub expansion in tundra ecosystems: dynamics, impacts and research priorities. Environ Res Lett. 2011;6(4):045509. doi:10.1088/1748-9326/6/4/045509.
7. Klironomos JN. Feedback with soil biota contributes to plant rarity and invasiveness in communities. Nature. 2002;417(6884):67–70.
8. Horton TR, Bruns TD, Parker VT. Ectomycorrhizal fungi associated with *Arctostaphylos* contribute to *Pseudotsuga menziesii* establishment. Can J Bot. 1999;77(1):93–102.
9. Nunez MA, Horton TR, Simberloff D. Lack of belowground mutualisms hinders Pinaceae invasions. Ecology. 2009;90(9):2352–9.
10. Booth MG. Mycorrhizal networks mediate overstorey-understorey competition in a temperate forest. Ecol Lett. 2004;7(7):538–46.
11. Horton TR, van der Heijden MGA. The role of symbioses in seedling establishment and survival. In: Leck MA, Parker VT, Simpson RL, editors. Seedling ecology and evolution. Cambridge: Cambridge University Press; 2008. p. 189–213.
12. Smith SE, Read DJ. Mycorrhizal Symbiosis. 3rd ed. New York: Academic Press; 2008.
13. Newsham KK. A meta-analysis of plant responses to dark septate root endophytes. New Phytol. 2011;190(3):783–93.
14. Mandyam K, Jumpponen A. Seeking the elusive function of the root-colonising dark septate endophytic fungi. Stud Mycol. 2005;53:173–89.
15. Mandyam KG, Jumpponen A. Mutualism-parasitism paradigm synthesized from results of root-endophyte models. Front Microbiol. 2014;5:776. doi:10.3389/fmicb.2014.00776.
16. Janzen DH. Herbivores and the number of tree species in tropical forests. Am Nat. 1970;104:501–28.
17. Connell JH. On the role of natural enemies in preventing competitive exclusion in some marine animals and in rain forest trees. In: Den Boer PJ, Gradwell GR, editors. Dynamics of population: proceedings of the advanced study institute on dynamics of numbers of populations. Wageningen: Center for Agricultural Publishing and Documentation; 1971. p. 298–312.
18. Mangan SA, Schnitzer SA, Herre EA, Mack KML, Valencia MC, Sanchez EI, Bever JD. Negative plant-soil feedback predicts tree-species relative abundance in a tropical forest. Nature. 2010;466(7307):752–5.
19. Johnstone JF, Chapin FS, Hollingsworth TN, Mack MC, Romanovsky V, Turetsky M. Fire, climate change, and forest resilience in interior Alaska. Can J For Res-Rev Can Rech For. 2010;40(7):1302–12.
20. Lantz TC, Gergel SE, Henry GHR. Response of green alder (*Alnus viridis* subsp *fruticosa*) patch dynamics and plant community composition to fire and regional temperature in north-western Canada. J Biogeogr. 2010;37(8):1597–610.
21. Racine C, Jandt R, Meyers C, Dennis J. Tundra fire and vegetation change along a hillslope on the Seward Peninsula, Alaska, USA. Arct Antarct Alp Res. 2004;36(1):1–10.

22. Higuera PE, Brubaker LB, Anderson PM, Brown TA, Kennedy AT, Hu FS. Frequent fires in ancient shrub tundra: implications of paleorecords for arctic environmental change. PLoS One. 2008;3(3):e0001744.
23. Hu FS, Higuera PE, Walsh JE, Chapman WL, Duffy PA, Brubaker LB, Chapman ML. Tundra burning in Alaska: linkages to climatic change and sea ice retreat. J Geophys Res Biogeosci 2010;115(G04002, doi:10.1029/2009JG001270.):G04002.
24. Landhausser SM, Wein RW. Postfire vegetation recovery and tree establishment at the arctic treeline: climate-change-vegetation-response hypotheses. J Ecol. 1993;81(4):665–72.
25. Neary DG, Klopatek CC, DeBano LF, Ffolliott PF. Fire effects on belowground sustainability: a review and synthesis. For Ecol Manage. 1999;122(1–2):51–71.
26. Timling I, Walker DA, Nusbaum C, Lennon NJ, Taylor DL. Rich and cold: diversity, distribution and drivers of fungal communities in patterned-ground ecosystems of the North American Arctic. Mol Ecol. 2014;23(13):3258–72.
27. Timling I, Dahlberg A, Walker DA, Gardes M, Charcosset JY, Welker JM, Taylor DL. Distribution and drivers of ectomycorrhizal fungal communities across the North American Arctic. Ecosphere 2012;3(11). **(3:art111)**. http://dx.doi.org/110.1890/ES1812-00217.00211.
28. Timling I, Taylor DL. Peeking through a frosty window: molecular insights into the ecology of Arctic soil fungi. Fungal Ecol. 2012;5(4):419–29.
29. Smith JE, McKay D, Brenner G, McIver J, Spatafora JW. Early impacts of forest restoration treatments on the ectomycorrhizal fungal community and fine root biomass in a mixed conifer forest. J Appl Ecol. 2005;42(3):526–35.
30. Treseder K, Mack M, Cross A. Relationships among fires, fungi and soil dynamics in Alaskan boreal forests. Ecol Appl. 2004;14:1826–38.
31. Taylor DL, Bruns TD. Community structure of ectomycorrhizal fungi in a *Pinus muricata* forest: minimal overlap between the mature forest and resistant propagule communities. Mol Ecol. 1999;8:1837–50.
32. Parker TJ, Clancy KM, Mathiasen RL. Interactions among fire, insects and pathogens in coniferous forests of the interior western United States and Canada. Agric For Entomol. 2006;8(3):167–89.
33. Johnstone JF, Hollingsworth TN, Chapin FS, Mack MC. Changes in fire regime break the legacy lock on successional trajectories in Alaskan boreal forest. Glob Chang Biol. 2010;16(4):1281–95.
34. Hewitt RE, Bent E, Hollingsworth TN, Chapin FS, Taylor DL. Resilience of Arctic Mycorrhizal fungal communities after wildfire facilitated by resprouting shrubs. Ecoscience. 2013;20(3):296–310.
35. Lloyd AH, Fastie CL, Eisen H. Fire and substrate interact to control the northern range limit of black spruce (*Picea mariana*) in Alaska. Can J For Res Rev Can Rech For. 2007;37(12):2480–93.
36. Hollingsworth T, Walker M, Chapin F III, Parsons A. Scale-dependent environmental controls over species composition in Alaskan black spruce communities. Can J For Res. 2006;36:1781–96.
37. Molina R, Massicotte H, Trappe JM. Specificity phenomena in mycorrhizal symbioses: community-ecological consequences and practical implications. In: Allen MF, editor. Mycorrhizal functioning: an integrative plant-fungal process. New York: Chapman and Hall; 1992. p. 357–423.
38. Mack MC, Bret-Harte MS, Hollingsworth TN, Jandt RR, Schuur EAG, Shaver GR, Verbyla DL. Carbon loss from an unprecedented Arctic tundra wildfire. Nature. 2011;475(7357):489–92.
39. Jandt RR, Miller AE, Yokel DA, Bret-Harte MS, Kolden CA, Mack MC. Findings of Anaktuvuk River fire recover study. In: BLM technical report. Ft. Wainwright: U.S. Department of the Interior, Bureau of Land Management, Alaska Fire Service; 2012. p. 1–39.
40. Viereck LA, Dyrness CT, Batten AR, Wenzlick KJ. The Alaska vegetation classification. In: General technical report PNW-GTR-286. Portland: U.S. Department of Agriculture, Forest Service, Pacific Northwest Research Station; 1992. p. 278.
41. Hewitt RE. Fire-severity effects on plant–fungal interactions: implications for Alaskan treeline dynamics in a warming climate. Ph.D. thesis. Fairbanks: University of Alaska Fairbanks; 2014.
42. Hewitt RE, Chapin III FS, Hobbie SE, Hollingsworth TN, Taylor DL. Ectomycorrhizal shrub and seedling interactions facilitate tree establishment after wildfire at Alaska arctic treeline. **(in review)**.
43. Bent E, Taylor DL. Direct amplification of DNA from fresh and preserved ectomycorrhizal root tips. J Microbiol Methods. 2010;80(2):206–8.
44. Bent E, Kiekel P, Brenton R, Taylor DL. Root-associated ectomycorrhizal fungi shared by various boreal forest seedlings naturally regenerating after a fire in interior Alaska and correlation of different fungi with host growth responses. Appl Environ Microbiol. 2011;77(10):3351–9.
45. Hewitt R. Fire-severity effects on plant-fungal interactions after a novel tundra wildfire disturbance: implications for arctic shrub and tree migration. Bonanza Creek LTER. Long Term Ecological Research Network. 2016. doi:10.6073/pasta/a131c0d6707b6aa746dfe0265141ad43.
46. Taylor DL, Houston S. A bioinformatics pipeline for sequence-based analyses of fungal biodiversity. In: Xu J-R, Bluhm BH, editors. Fungal genomics, vol. 722. New York: Humana Press; 2011. p. 141–55.
47. Kruskal JB. Nonmetric multidimensional scaling: a numerical method. Psychometrika. 1964;29:115–29.
48. McCune B, Grace JB. Analysis of ecological communities. 2nd ed. Gleneden Beach: MjM Software Design; 2001.
49. Berry KJ, Kvamme KL, Mielke PW. Improvements in the permutation test for the spatial-analysis of the distribution of artifacts into classes. Am Antiq. 1983;48(3):547–53.
50. Aylor DE. Spread of plant disease on a continental scale: role of aerial dispersal of pathogens. Ecology. 2003;84(8):1989–97.
51. Peay KG, Schubert MG, Nguyen NH, Bruns TD. Measuring ectomycorrhizal fungal dispersal: macroecological patterns driven by microscopic propagules. Mol Ecol. 2012;21(16):4122–36.
52. Blumenthal D, Mitchell CE, Pysek P, Jarosik V. Synergy between pathogen release and resource availability in plant invasion. Proc Natl Acad Sci USA. 2009;106(19):7899–904.
53. Newsham KK, Upson R, Read DJ. Mycorrhizas and dark septate root endophytes in polar regions. Fungal Ecol. 2009;2(1):10–20.
54. Tellenbach C, Sieber TN. Do colonization by dark septate endophytes and elevated temperature affect pathogenicity of oomycetes? FEMS Microbiol Ecol. 2012;82(1):157–68.
55. Kasischke ES, French NH, Harrell P, Christensen NL, Ustin SL, Barry D. Monitoring of wildfires in boreal forests using large area AVHRR NDVI composite image data. Remote Sens Environ. 1993;45(1):61–71.
56. Wein RW, Bliss LC. Changes in Arctic *Eriophorum* tussock communities following fire. Ecology. 1973;54:845–52.
57. Castellano MA, Trappe JM, Molina R. Inoculation of container-grown Douglas-fir seedlings with basidiospores of *Rhizopogon vinicolor* and *R. colossus*: effects of fertility and spore application rate. Can J For Res. 1985;15(1):10–3.
58. Ishida TA, Nara K, Tanaka M, Kinoshita A, Hogetsu T. Germination and infectivity of ectomycorrhizal fungal spores in relation to their ecological traits during primary succession. New Phytol. 2008;180(2):491–500.
59. Zackrisson O, Nilsson MC, Wardle DA. Key ecological function of charcoal from wildfire in the boreal forest. Oikos. 1996;77(1):10–9.
60. Johnstone JF, Chapin FS. Effects of soil burn severity on post-fire tree recruitment in boreal forest. Ecosystems. 2006;9:14–31.
61. Chapin FS, Sturm M, Serreze MC, McFadden JP, Key JR, Lloyd AH, McGuire AD, Rupp TS, Lynch AH, Schimel JP, et al. Role of land-surface changes in Arctic summer warming. Science. 2005;310(5748):657–60.

5

# Floral traits influence pollen vectors' choices in higher elevation communities in the Himalaya-Hengduan Mountains

Yan-Hui Zhao[1,2], Zong-Xin Ren[1], Amparo Lázaro[3], Hong Wang[1*], Peter Bernhardt[4], Hai-Dong Li[1,2] and De-Zhu Li[1,5*]

**Abstract**

**Background:** How floral traits and community composition influence plant specialization is poorly understood and the existing evidence is restricted to regions where plant diversity is low. Here, we assessed whether plant specialization varied among four species-rich subalpine/alpine communities on the Yulong Mountain, SW China (elevation from 2725 to 3910 m). We analyzed two factors (floral traits and pollen vector community composition: richness and density) to determine the degree of plant specialization across 101 plant species in all four communities. Floral visitors were collected and pollen load analyses were conducted to identify and define pollen vectors. Plant specialization of each species was described by using both pollen vector diversity (Shannon's diversity index) and plant selectiveness (d' index), which reflected how selective a given species was relative to available pollen vectors.

**Results:** Pollen vector diversity tended to be higher in communities at lower elevations, while plant selectiveness was significantly lower in a community with the highest proportion of unspecialized flowers (open flowers and clusters of flowers in open inflorescences). In particular, we found that plant species with large and unspecialized flowers attracted a greater diversity of pollen vectors and showed higher selectiveness in their use of pollen vectors. Plant species with large floral displays and high flower abundance were more selective in their exploitation of pollen vectors. Moreover, there was a negative relationship between plant selectiveness and pollen vector density.

**Conclusions:** These findings suggest that flower shape and flower size can increase pollen vector diversity but they also increased plant selectiveness. This indicated that those floral traits that were more attractive to insects increased the diversity of pollen vectors to plants while decreasing overlap among co-blooming plant species for the same pollen vectors. Furthermore, floral traits had a more important impact on the diversity of pollen vectors than the composition of anthophilous insect communities. Plant selectiveness of pollen vectors was strongly influenced by both floral traits and insect community composition. These findings provide a basis for a better understanding of how floral traits and community context shape interactions between flowers and their pollen vectors in species-rich communities.

**Keywords:** Diversity of pollen vectors, Floral display, Flower abundance, Flower shape, Flower size, Flowering duration, Plant-pollen vector interactions, Pollen vector density, Plant selectiveness, Specialization

## Background

Studies show that specialization of plant species within a community ranges from extremely generalized to highly specialized [1, 2]. Community level investigations on how floral traits influence plant specialization provide us with a broader understanding of floral trait evolution and adaptation [3–10]. Plant species with larger flowers/inflorescences tend to attract a greater diversity of pollinators [3, 11]. Larger flowers may be preferred by pollinators because they are easier to see and minimize the time foragers need to locate them [12]. In addition, flower size

*Correspondence: wanghong@mail.kib.ac.cn; dzl@mail.kib.ac.cn
[1] Key Laboratory for Plant Diversity and Biogeography of East Asia, Kunming Institute of Botany, Chinese Academy of Sciences, Kunming 650201, People's Republic of China
Full list of author information is available at the end of the article

often correlates positively with reward amounts and production, including pollen and/or nectar [13, 14].

Differences in reward accessibility of flowers may also play an important role in the specialization of some pollination systems or syndromes. Open flowers (dish shaped with shallow floral tubes) and compacted inflorescences (e.g. head or spicate) are easily accessible to most pollinators and may attract a greater diversity of floral foragers [11, 15]. In contrast, individual flowers with bilateral symmetry (e.g. gullet or flag-shaped) [16, 17], flowers that are inverted on their pedicels [18] and flowers that produce elongated tubes or spurs with narrow sinuses [19], are more likely to restrict access to the majority of resident pollinators. These flowers are more likely to be pollinated by animals with specialized mouthparts and/or specialized modes of foraging [4, 20]. Additional studies have shown that plant species with large floral displays and remain in bloom for long periods can increase visitation frequencies and/or the number of visiting pollinator species [3, 21–23].

Floral visitation to a plant species is also related to the community composition of both plants and flower visitors [11, 24, 25]. Plant species with abundant flowers commonly interact more frequently with more pollinator species than plant species showing depauperate flowering [4, 10]. Floral visitation also increases with increasing pollinator abundance [11]. These findings were consistent with the neutrality hypothesis that states that, the occurrence of interactions results from random encounters among individual plants and pollinators [26]. Besides, the abundance and composition of co-flowering species could also influence the patterns and rates of floral visitation of a plant species [25]. Sympatric insect-pollinated plants can either compete for pollinators or mutually attract and share the same pollinators (pollination facilitation), depending on the relative abundance, accessibility and diversity of their floral rewards [24, 25, 27, 28]. In addition, the potential for an indirect influence via shared pollinators was also related to the phylogenetic distance among co-flowering species [25]. Plant and pollinator community composition often show systematic variation along elevational gradients [29, 30] and this may result in corresponding variation of pollen vector choices to co-flowering plant species.

Plant species could benefit by attracting a greater diversity of pollinators to increase reproductive success [31]. However, pollinator sharing among plant species may lead to declines in fitness due to competition for pollinators and the increased incidence of interspecific pollen transfer [32]. Although the evolution of divergence in pollen placement on pollinators' bodies has helped to minimize interspecific pollen transfer, this mechanism does not reduce all reproductive interference [33]. Koski et al. [10] found that plants decreased overlap in their use of flower visitors by increasing flower sizes across a meta-community of five serpentine seeps in California. Therefore, to achieve optimal reproductive output, a successful strategy for pollinator-dependent plant species could be to produce enough rewards and occur at such a relatively high abundance to attract a high diversity of pollinators while decreasing the need to share pollinators.

Empirical evidence showing how floral traits and community context influence plant-pollinator interactions at the community level remains uncommon. Such studies tend to be restricted to regions where plant diversity is low [3, 5, 9, 11]. We studied the interactive effects of floral traits and pollen vector community composition in highly diverse and temperate communities within a Himalayan floristic province. Specifically, we measured floral traits and plant specialization in 101 herbaceous species found in four communities (elevation from 2725 to 3910 m) in the Yulong Mountain in Lijiang, SW China. We used two species-specific indices, pollen vector diversity (Shannon's diversity index) and selectiveness (d' index), to describe the specialization of each plant species. We addressed the following questions: (1) Does plant specialization differ among the four study communities? (2) How do floral traits and pollen vector community composition (i.e., pollen vectors diversity and density) influence specialization in plant species?

## Methods

### Study systems

The study was conducted on the Yulong Mountain in the Himalaya-Hengduan Mountains, SW China. The study communities were located at the Lijiang Forest Ecosystem Research Station operated by the Kunming Institute of Botany, Chinese Academy of Sciences. We selected four 1.5 ha subalpine/alpine meadows on the eastern slope of the mountain. All meadows were at high elevation but with a difference of 1185 m between the lowest and highest community: 1) Yushuizhai (YSZ), 2725 m above sea level (a.s.l.), 27°00′10″N, 100°12′05″E; (2) Haligu (HLG), 3235 m a.s.l., 27°00′09″N, 100°10′57″E; (3) Yakou (YK), 3670 m a.s.l., 27°00′56″N, 100°10′17″E; and (4) Diyifeng (DYF, above tree line), 3910 m a.s.l., 27°01′41″N, 100°11′03″E. The linear distance between neighboring communities was ca. 2.0 km. Additionally, there was variation in landscape characters based on local land use and construction. The YSZ site was adjacent to a tourist center while HLG was adjacent to a major water reservoir below the field station. Sites YK and DYF were at higher elevations with less anthropogenic impact. All sites received some grazing by cattle, yaks or horses. The vegetation cover within a 2 km radius also differed among sites from forests dominated by *Pinus* species to

*Abies* and *Rhododendron* species. Flowering duration and pollinator activity periods tended to decrease with increasing elevation. Mean temperatures (from 11 May to 29 September, 2012) among the four communities varied from low to high elevation; they were 16.6, 12.9, 9.6 and 8.9 °C respectively (recorded with Temperature/Relative Humidity Data Loggers, HOBO U23-001, Onset Computer Corporation, Bourne, MA, USA).

### Field surveys of flower visitors and measurements of floral traits

We collected flower visitors from 101 insect-pollinated, herbaceous species at the four communities from early May to early October in 2012. These collections nearly covered the entire flowering periods of all four communities. Two creeping shrubs (*Cotoneaster adpressus* and *Rhododendron fastigiatum*) were found infrequently in our quadrats but we excluded them from data analyses, because both species grew as overlapping clumps and it was not possible to segregate and evaluate individual floral displays (see below). We conducted nine surveys at 2-week intervals for each community. Floral visitors were collected by walking along arranged transects approximately 150 m in length and 2 m in width from 9:00–17:00 h on either sunny days or during sunshine gaps on cloudy or foggy days. Only insects that contacted plant reproductive organs or were foraging for nectar and/or pollen were classified as legitimate visitors and collected. We calculated flower visitor density by using the average number of insect visits recorded per observation period for each survey. Insect specimens were netted and euthanized in small jars with fumes of ethyl acetate prior to pinning and identification. Voucher specimens were deposited in the Kunming Institute of Botany, Chinese Academy of Sciences.

Flower abundance, floral display and flowering duration were estimated by using 30, $1 \times 1$ $m^2$ quadrats at each community. These quadrats were spaced within the 1.5 ha plot. The minimum distance between two neighboring quadrats was 10 m. The number of flowering individuals per species and the number of open flowering units produced by each individual (floral display, hereafter) inside the quadrats were recorded in each survey. Flowering units were defined as either individual flowers or whole inflorescences depending on species. For species with densely compact inflorescences (e.g. Asteraceae and Apiaceae) each inflorescence was counted as a single flowering unit [22]. We used the mean number of flowering units of each survey to describe flower abundance of a plant species. Floral display of each species was defined as the mean number of flowering units per individual in each survey. The flowering duration of a species was defined as the number of weeks that the plant was recorded in bloom in the quadrats.

Flower and inflorescence shape for each species (referred to here as flower shape, hereafter) was subdivided into unspecialized and specialized flowers based on corolla traits or inflorescence architecture [11]. The unspecialized flowers or inflorescences were held erect, had easily accessible floral rewards in bowl-shaped perianths or in short tubes produced by single flowers (e.g. radially symmetrical members of the Rosaceae) or inflorescences (e.g. Asteraceae). Specialized flowers produced corollas that hid their rewards in elongated tubes or gullets (e.g. Lamiaceae) or spurs, restricting foragers with short mouthparts.

We used different formulae to calculate the mean flower size (unit area) of 10–20 randomly selected flowering units for each species according to the shape of flowers/inflorescences. In radially symmetrical (actinomorphic), shallow or flat flowers, and the head inflorescences of the Asteraceae, the flowering area was calculated as a circle (formula: $\pi r^2$). In zygomorphic/stereomorphic flowers (bilateral symmetry; e.g. *Pedicularis* spp.) the flowering area was calculated as a rectangle based on flower length and width (formula: $L \times W$) [3]. When inflorescences produced architectures that were nearly cylindrical (e.g. *Polygonum coriaceum*) or spherical (e.g. *Trifolum* spp.), we calculated their areas as $2\pi rd + \pi r^2$ and $4\pi r^2$, respectively [34].

### Plant specialization indices

Over the flowering season, a total of 5855 flower visitor individuals were collected from the 101 plant species in the four communities. From the collected insect specimens, a total of 2992 specimens representing 355 insect taxa were examined for pollen loads to determine if they carried the host plant pollen grains. One to five insect specimens were chosen from each plant-flower visitor pair at each survey and community for pollen analysis. Each pollen sample was viewed under a Hitachi S-4800 scanning electron microscope. Pollen grains were identified by comparing them to a reference library of pollen based on grains removed from field-collected flowers. If one of the specimens of a plant-flower visitor pair carried the host plant pollen we presumed that all the remaining specimens in that insect's morphotype were also effective pollen vectors. If all the specimens of a plant-flower visitor pair failed to carry the host plant pollen, we presumed that this insect morphotype was ineffective as a pollen vector on that particular plant species. However, as we did not test the pollination effectiveness of a flower visitor by experiments such as analyzing pollen deposition on stigmas per visit [35], pollen vectors recorded in this study must be regarded as putative or prospective pollinators.

We constructed a weighted plant-pollen vector network for each survey in each of the four communities by excluding visitation interactions made by inefficient flower visitors. This resulted in a reduction of 6.3–30.4 % of total interactions for each survey of the four communities (Zhao et al. unpublished data). We calculated two plant specialization indices for each plant species in the 36 plant-pollen vector networks (4 communities × 9 surveys): pollen vector diversity (Shannon's diversity index) and selectiveness (d' index). Shannon's pollen vector diversity for each plant species was calculated as $H = \sum_{k=0}^{n} p_i \ln p_i$, where $p_i$ is the proportion of visits by pollen vector i to the focal plant species [36]. In this study, we used the total number of pollen vector individuals on a plant species to calculate $p_i$. The values of a plant Shannon's diversity index increased with the number of species and evenness of pollen vectors. The d' index (selectiveness) expressed the relative deviation in the actual interaction frequencies of a focal plant species from a null model which assumed that all pollen vectors were used in proportion to their availability [37]. Its value ranged from 0 (minimum selectiveness) to 1 (maximum selectiveness). According to this specialization index, a high selective plant species is characterized by little overlap in its pollen vector exploitation with its co-occurring and co-blooming species. Both specialization indices were calculated in the bipartite package [38] in R [39].

### Statistical analysis

We tested for the effects of elevation, floral traits, and pollen vector community composition on plant specialization. Specialization indices and floral traits of plant species with related phylogenies may be similar due to common ancestry and hence are not statistically independent. To account for phylogenetic non-independence, we applied phylogenetic generalized linear mixed models (PGLMM) with Markov chain Monte Carlo techniques (MCMCglmm) [40]. This approach allows control for phylogenetic co-variation among species by implementing the phylogenetic tree as a random factor into the model [40]. A maximum likelihood phylogeny for all the plant species of the four communities was reconstructed from DNA sequences of internal transcribed spacer (ITS), ribulose-bisphosphate carboxylase (*rbc*L) and Maturase K (*mat*K) (Zhao et al. unpublished data).

We tested for differences in plant specialization indices (pollen vector diversity and selectiveness) among the four study communities, with individual plant species as the sampling unit. We treated elevation as a categorical fixed factor, and included survey, species identity and phylogeny as random factors. Then we tested the effects of floral traits (flower shape, flower size, floral display, flowering duration and flower abundance), pollen vector community composition (pollen vector richness and pollen vector density), as well as their two-way interactions on plant specialization. We used community identity, survey, species identity and phylogeny as random factors. We removed non-significant interaction terms by backward elimination. Models were compared based on deviance information criterion (DIC), with ΔDIC values >2 taken to indicate a significantly improved model fit [40].

For all PGLMMs, we used an inverse-Wishart prior (V = 1, nu = 0.002) for random effects according to the package guidelines (MCMCglmm Course Notes; https://cran.r-project.org/web/packages/MCMCglmm/vignettes/CourseNotes.pdf). The PGLMM models were run for 5,000,000 iterations with a burn-in of 10,000 iterations and a thinning interval of 500 iterations. Prior to all analyses, all continuous response and predictor variables were scaled to a mean of zero and a standard deviation of 1 to allow the use of regression estimates as effect sizes [41]. Estimates of the posterior mean with 95 % credible intervals (lower and upper CI) and *P* values ($P_{MCMC}$) were reported. Associations between two variables were considered significant when the 95 % CI excluded zero, and $P_{MCMC} \leq 0.05$.

## Results

There were 101, insect-pollinated, herbaceous species belonging to 63 genera, representing 26 families in the four communities. Plant assemblages included 40, 30, 33 and 27 plant species from YSZ to DYZ, respectively.

Of the 355 insect taxa collected from the four communities, 328 carried the host plant pollen and were assumed to be the effective pollen vectors. These pollen vector taxa belonged to 51 families in five insect Orders. The number of pollen vector taxa in the study communities was 163, 121, 85 and 56 from YSZ to DYF, respectively. Ten functional groups (according to presumed similarities in the selection pressures on floral traits pollen vectors exert) were detected for each community. Pollen vectors were mainly comprised of long-tongued bees and beetles at YSZ. Pollen vectors at HLG were dominated by hover flies (Syrphidae) and long-tongued bees. At YK and DYF the majority of pollen vectors were short-tongued, muscid flies (Muscidae) and long-tongued bees (Table 1).

### Variation of plant specialization indices among communities

We found significant differences in plant specialization indices (pollen vector diversity and selectiveness) among the four communities. Specifically, the pollen vector diversity tended to be higher at the low elevation

**Table 1 Pollen vector assemblages at four communities on the Yulong Mountain, SW China**

| Functional groups | YSZ | HLG | YK | DYF |
|---|---|---|---|---|
| Long-tongued bees | 20.1 | 19.8 | 27.1 | 31.2 |
| Short-tongued bees | 9.8 | 9.5 | 13.2 | 7.8 |
| Other hymenoptera | 5.2 | 7.6 | 3.2 | 6.9 |
| Muscoid flies | 12.1 | 16.1 | 30.8 | 37.4 |
| Hover flies | 14.2 | 34.1 | 12.4 | 5.7 |
| Beeflies | 4.8 | 1.1 | 0.1 | 0.0 |
| Butterflies | 13.8 | 3.0 | 2.6 | 0.5 |
| Moths | 0.5 | 0.2 | 1.2 | 3.6 |
| Beetles | 19.2 | 8.0 | 0.4 | 1.4 |
| Hemiptera | 0.4 | 0.5 | 8.9 | 5.6 |

Numbers represent the percentage of visits conducted by each pollen vector functional group in each community

*YSZ* 2725 m above sea level, *HLG* 3235 m above sea level, *YK* 3670 m above sea level, *DYF* 3910 m above sea level

communities (Fig. 1a). Plant species were less selective in their exploitation of pollen vectors in the YK community which also had the highest proportion of unspecialized flowers (Fig. 1b).

### Effects of floral traits and community composition on pollen vector diversity

PGLMM analysis showed that plant species with unspecialized flowers had a greater pollen vector diversity than those with specialized flowers across the four communities (Table 2). Our analysis further revealed a highly significant and positive correlation between pollen vector diversity and flower size (Table 2). However, pollen vector diversity was not significantly influenced by floral display, flowering duration, flower abundance, or pollen vector richness and pollen vector density in the communities (Table 2). None of the two-way interactions between pollen vector composition and floral traits were significant indicating a similar pollen vector diversity response to floral traits among communities showing a different pollen vector richness and density (results not shown).

### Effects of floral traits and community composition on selectiveness

The selectiveness of plant species in their use of pollen vectors was also related to flower shape and flower size. Plant species with unspecialized and larger flowers/inflorescences showed a higher selectiveness (Table 2). We found that plant species with larger floral displays and a greater flower abundance showed a higher degree of selectiveness (Table 2). Moreover, there was a significant negative relationship between plant selectiveness and pollen vector density (Table 2). In this case, plant species in communities with a higher pollen vector density showed lower selectiveness. By contrast, flowering duration and pollen vector richness had no significant effects on plant selectiveness (Table 2). In addition, all two-way interactions between pollen vector composition in the communities and floral traits were not significant (results not shown).

## Discussion

In this study we showed that there were significant differences in plant specialization indices among the four communities. Our results also indicated that the differences

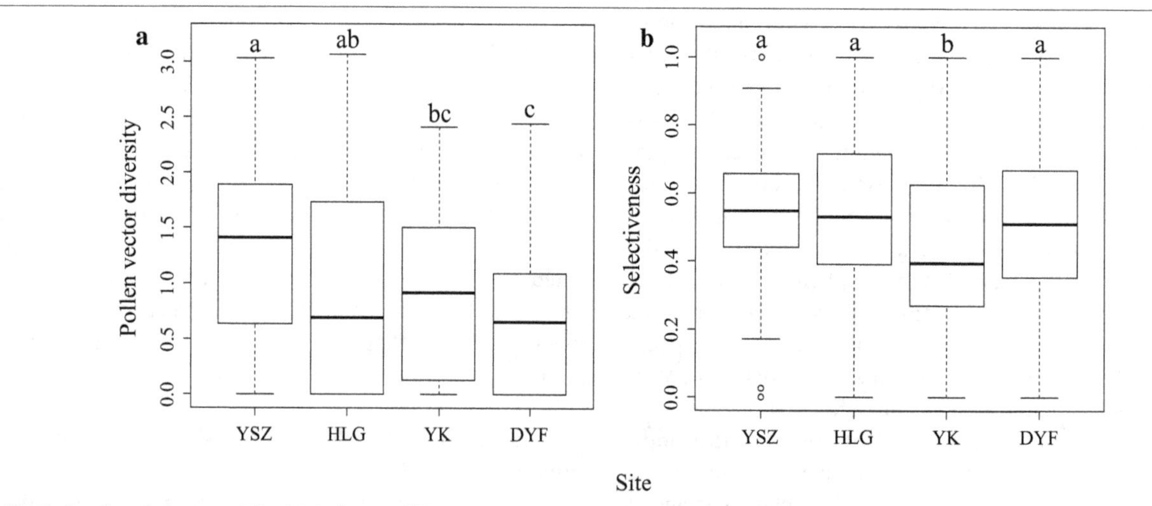


**Fig. 1** *Boxplots* of plant specialization indices at different communities on the Yulong Mountain, SW China. *YSZ* 2725 m above sea level; *HLG* 3235 m above sea level; *YK* 3670 m above sea level; *DYF* 3910 m above sea level. The *bottom* and *top* limits of each *box* are the lower and upper quartiles, respectively. The *horizontal black lines* across *boxes* are medians. *Error bars* represent the 95 % confidence interval of the median. **a** Pollen vector diversity (Shannon's diversity index); **b** selectiveness (d' index)

**Table 2 Results of the phylogenetic generalized linear mixed model (PGLMM) for evaluating pollen vector diversity or selectiveness (d') of 101 plant species in relation to floral traits and pollen vector community composition variables at four communities on the Yulong Mountain, SW China**

| Specialization index | Variable | Posterior mean | Lower CI | Upper CI | $P_{MCMC}$ |
|---|---|---|---|---|---|
| Pollen vector diversity | Intercept | 0.165 | −0.351 | 0.733 | 0.515 |
| | Flower shape | −0.728 | −1.082 | −0.362 | <0.001 |
| | Flower size | 0.146 | 0.012 | 0.280 | 0.034 |
| | Floral display | −0.003 | −0.113 | 0.114 | 0.962 |
| | Flowering duration | 0.014 | −0.149 | 0.176 | 0.868 |
| | Flower abundance | 0.003 | −0.201 | 0.197 | 0.978 |
| | Pollen vector richness | 0.205 | −0.011 | 0.420 | 0.088 |
| | Pollen vector density | 0.062 | −0.177 | 0.312 | 0.614 |
| Selectiveness | Intercept | −0.031 | −0.584 | 0.561 | 0.917 |
| | Flower shape | −0.464 | −0.841 | −0.109 | 0.009 |
| | Flower size | 0.208 | 0.079 | 0.337 | 0.003 |
| | Floral display | 0.122 | 0.008 | 0.233 | 0.032 |
| | Flowering duration | −0.031 | −0.194 | 0.125 | 0.705 |
| | Flower abundance | 0.222 | 0.025 | 0.407 | 0.027 |
| | Pollen vector richness | 0.195 | −0.018 | 0.417 | 0.084 |
| | Pollen vector density | −0.416 | −0.641 | −0.181 | 0.001 |

among plant species in pollen vector diversity were exclusively explained by flower shape and flower size. However, the differences in plant selectiveness were related not only to flower shape and flower size, but also to floral display, flower abundance and pollen vector density in the communities.

### Variation of plant specialization indices among communities

Olesen and Jordano [42] showed that the mean number of interacting pollinators per plant species decreased with increasing elevation. In this study we reported a similar finding. Plant species at lower elevations were visited by a greater diversity of pollen vectors than plant species growing at higher elevations. The decrease in pollen vector diversity for plant species with increasing elevation may be the result of a decrease in the number and abundance of insect taxa, in general, as elevation increased [43, this study]. However, it is difficult to tell whether elevation affects pollen vector diversity as our sampling was restricted to only one site at each elevation. Additional studies with a greater number of replicate sites at each elevation will be needed to confirm this current pattern of pollen vector diversity vs. elevation. This could also be tested by selecting a greater number of transects through a continuous elevation gradient.

The d' index is supposed by Blüthgen et al. [37] to be more appropriate to compare specialization or selectiveness of species within or across networks because it has the advantage of not being affected by network size and sampling intensity. In our study, plant selectiveness showed no systematic variation with elevation. The plant species showed lower selectiveness within a mid-elevation community (YK), indicating that the overlap in pollen vector use among plant species was higher at this community. One possible reason for this pattern is that visitation by pollen vectors to plant species were high due to the high proportion of unspecialized flowers in this community (YSZ: 45.5 %, HLG: 41.1 %, YK: 78.9 %, DYF 69.2 %).

### Effects of floral traits on pollen vector diversity

Flower shape and flower size influenced pollen vector diversity in our study. Compared to species with specialized flowers, species with unspecialized flowers were visited by a greater diversity of pollen vectors. This relationship between floral shape and pollen vector diversity has also been found in Norway [3, 5, 11]. This is to be anticipated as unspecialized flowers are accessible to the vast majority of flower visitors regardless of their physical size, foraging behavior or proboscis length. In contrast, specialized flowers are far more likely to restrict access to their edible rewards (see above) and are pollinated primarily by pollinator guilds with canalized morphologies and behaviors [15, 17, 19, 44].

For flower size, a series of studies in Norway showed that flower size correlated positively with an increase in pollinator diversity [3, 11]. In our study, we also found a positive relationship between flower size and pollen vector diversity. This effect might be attributed primarily to the greater attractiveness of larger flowers and/or floral displays [3]. In addition to these floral traits, we cannot rule out that other floral cues (e.g. scent and pigmentation patterns) [5–8], not included in this study, also affected pollen vector diversity in some of our plant species.

### Effects of floral traits on selectiveness

Plant species could benefit from enhanced attractiveness to increase visitation rates by a greater diversity of pollinators. However, enhanced attractiveness could also have negative consequences if foraging bouts include visits to multiple plant species and these foragers transfer the pollen of one species to the stigmas of others [45]. Koski et al. [10] found that the positive relationship between flower size and plant selectiveness suggested that plants with larger flower sizes showed less overlap in flower visitor use compared with other species with smaller flowers. In addition to flower size, we also found that plant selectiveness related positively with floral display and flower abundance. This indicated that these floral traits could ultimately decrease overlap in pollen vector exploitation among co-flowering plant species.

Our finding that species with unspecialized flowers were more selective than species with specialized flowers contradicted the common assumption that specialized flower shapes, especially zygomorphic flowers, must always receive fewer pollen vector species [15]. One possible explanation is that some species with bilateral or asymmetric flowers (e.g. *Pedicularis, Lotus, Prunella, Clinopodium, Roscoea*) are pollinated almost exclusively by a few native bumblebees (*Bombus*) species. Such sharing of the same pollen vector by several plant species should ultimately decrease the selectiveness of specialized flowers. In these systems, however, interspecific pollen transfer is reduced by depositing the pollen of each co-blooming species on very isolated parts (e.g. head vs. dorsum of thorax vs. dorsum of abdomen etc.) on the same insect's body [46].

### Effects of community composition on plant specialization

Plant selectiveness, but not pollen vector diversity, was related to pollen vector community composition in this study. Plant selectiveness decreased consistently with the increase in pollen vector density. Plant species in communities with low pollen vector density should be more selective than plant species in communities with high pollen vector density because interspecific competition among plant species for limited pollinator resources should reduce pollen vector overlap [47].

## Conclusions

Our combined analyses of the effects of floral traits and community composition in a species-rich region showed that floral traits play important roles in pollen vector diversity and selectiveness. Specifically, flower shape and flower size increased both pollen vector diversity and selectiveness. These findings indicate that some floral traits, that make the plants more attractive to insects, also increase the diversity of pollen vectors while decreasing overlap in pollen vector exploitation when plant species have overlapping flowering periods. Additionally, plant selectiveness was also associated with local pollen vector composition. This study can help us further to understand the effect of floral traits and community composition on specialization of plant species.

## Additional file

**Additional file 1.** Raw data on plant specialization indices (pollen vector diversity and selectiveness), floral traits (flower shape, flower size, floral display, flowering duration and flower abundance) and pollen vector community composition (pollen vector richness and pollen vector density) at four communities on the Yulong Mountain, SW China.

#### Abbreviations

CI: credible interval; DIC: deviance information criterion; ITS: internal transcribed spacer; *mat*K: maturase K gene; MCMCglmm: phylogenetic generalized linear mixed models with Markov chain Monte Carlo techniques; PGLMM: phylogenetic generalized linear mixed models; *rbc*L: ribulose-bisphosphate carboxylase gene.

#### Authors' contributions

HW, DZL, and YHZ conceived and designed the experiments; YHZ, HDL and ZXR performed experiments; YHZ and HW analyzed the data and wrote draft. All authors read and approved the final manuscript.

#### Author details

[1] Key Laboratory for Plant Diversity and Biogeography of East Asia, Kunming Institute of Botany, Chinese Academy of Sciences, Kunming 650201, People's Republic of China. [2] Kunming College of Life Sciences, University of Chinese Academy of Sciences, Kunming 650201, People's Republic of China. [3] Mediterranean Institute for Advanced Studies, c/Miquel Marquès 21, 07190 Esporles, Spain. [4] Department of Biology, Saint Louis University, Saint Louis 63103, MO, USA. [5] Germplasm Bank of Wild Species, Kunming Institute of Botany, Chinese Academy of Sciences, Kunming 650201, People's Republic of China.

#### Acknowledgements

We are grateful to ZB Tao, MY Zhang, H Tang for their help with fieldwork, to JB Yang with lab work, and to Dr. ZK Wu for identifying plant specimens. Prof. KK Huo and Dr. G-Y Yang are thanked for identifying partial insect specimens. We also thank all staff of the Lijiang Forest Ecosystem Research Station for giving full support.

#### Competing interests

The authors declare that they have no competing interests.

#### Funding

This study was supported by Grants from the National Key Basic Research Program of China (2014CB954100), the Joint Fund of the National Natural Science Foundation of China -Yunnan Province (U1502261), the Major

International Joint Research Project of National Natural Science Foundation of China (31320103919), the National Natural Science Foundation of China (31470323), and the Key Research Program of the Chinese Academy of Sciences (KJZD-EW-L07).

## References


1. Vázquez DP, Aizen MA. Null model analyses of specialization in plant-pollinator interactions. Ecology. 2003;84:2493–501.
2. Waser NM, Ollerton J. Plant-pollinator interactions: from specialization to generalization. Chicago: University of Chicago Press; 2006.
3. Hegland SJ, Totland Ø. Relationships between species' floral traits and pollinator visitation in a temperate grassland. Oecologia. 2005;145:586–94.
4. Stang M, Klinkhamer PGL, van der Meijden E. Size constraints and flower abundance determine the number of interactions in a plant-flower visitor web. Oikos. 2006;112:111–21.
5. Lázaro A, Hegland SJ, Totland Ø. The relationships between floral traits and specificity of pollination systems in three Scandinavian plant communities. Oecologia. 2008;157:249–57.
6. Junker RR, Höcherl N, Blüthgen N. Responses to olfactory signals reflect network structure of flower-visitor interactions. J Anim Ecol. 2010;79:818–23.
7. Gong YB, Huang SQ. Temporal stability of pollinator preference in an alpine plant community and its implications for the evolution of floral traits. Oecologia. 2011;166:671–80.
8. Junker RR, Blüthgen N, Brehm T, Binkenstein J, Paulus J, Schaefer HM, et al. Specialization on traits as basis for the niche-breadth of flower visitors and as structuring mechanism of ecological networks. Funct Ecol. 2013;27:329–41.
9. Lázaro A, Totland Ø. The influence of floral symmetry, dependence on pollinators and pollination generalization on flower size variation. Ann Bot. 2014;114:157–65.
10. Koski MH, Meindl GA, Arceo-Gómez G, Wolowski M, LeCroy KA, Ashman T-L. Plant-flower visitor networks in a serpentine metacommunity: assessing traits associated with keystone plant species. Arthropod Plant Interact. 2015;9:9–21.
11. Lázaro A, Jakobsson A, Totland Ø. How do pollinator visitation rate and seed set relate to species' floral traits and community context? Oecologia. 2013;173:881–93.
12. Spaethe J, Tautz J, Chittka L. Visual constraints in foraging bumblebees: flower size and color affect search time and flight behavior. Proc Natl Acad Sci USA. 2001;98:3898–903.
13. Cresswell JE, Galen C. Frequency-dependent selection and adaptive surfaces for floral character combinations: the pollination of *Polemonium viscosum*. Am Nat. 1991;138:1342–53.
14. Gómez JM, Bosch J, Perfectti F, Fernández JD, Abdelaziz M, Camacho JPM. Spatial variation in selection on corolla shape in a generalist plant is promoted by the preference patterns of its local pollinators. Proc R Soc B Biol Sci. 2008;275:2241–9.
15. Fenster CB, Armbruster WS, Wilson P, Dudash MR, Thomson JD. Pollination syndromes and floral specialization. Ann Rev Ecol Evol Syst. 2004;35:375–403.
16. Macior LW. Plant community and pollinator dynamics in the evolution of pollination mechanisms in *Pedicularis* (Scrophulariaceae). In: Armstrong JA, Powell JM, Richards AJ, editors. Pollination and evolution. Sydney: Royal Botanic Gardens; 1983. p. 29–45.
17. Westerkamp C, Claßen-Bockhoff R. Bilabiate flowers: the ultimate response to bees? Ann Bot. 2007;100:361–74.
18. Bernhardt P, Montalvo EA. The pollination ecology of *Echeandia macrocarpa* (Liliaceae). Brittonia. 1979;31:64–71.
19. Goldblatt P, Manning JC, Bernhardt P. Pollination biology of *Lapeirousia* subgenus *Lapeirousia* (Iridaceae) in southern Africa; floral divergence and adaptation for long-tongued fly pollination. Ann Mo Bot Gard. 1995;82:517–34.
20. Campbell AJ, Biesmeijer JC, Varma V, Wäckers FL. Realising multiple ecosystem services based on the response of three beneficial insect groups to floral traits and trait diversity. Basic Appl Ecol. 2012;13:363–70.
21. Olesen JM, Bascompte J, Elberling H, Jordano P. Temporal dynamics in a pollination network. Ecology. 2008;89:1573–82.
22. Lázaro A, Nielsen A, Totland Ø. Factors related to the inter-annual variation in plants' pollination generalization levels within a community. Oikos. 2010;119:825–34.
23. Mitchell RJ, Karron JD, Holmquist KG, Bell JM. The influence of *Mimulus ringens* floral display size on pollinator visitation patterns. Funct Ecol. 2004;18:116–24.
24. Dietzsch AC, Stanley DA, Stout JC. Relative abundance of an invasive alien plant affects native pollination processes. Oecologia. 2011;167:469–79.
25. Carvalheiro LG, Biesmeijer JC, Benadi G, Fründ J, Stang M, Bartomeus I, et al. The potential for indirect effects between co-flowering plants via shared pollinators depends on resource abundance, accessibility and relatedness. Ecol Lett. 2014;17:1389–99.
26. Krishna A, Guimarães PR Jr, Jordano P, Bascompte J. A neutral-niche theory of nestedness in mutualistic networks. Oikos. 2008;117:1609–18.
27. Moeller DA. Facilitative interactions among plants via shared pollinators. Ecology. 2004;85:3289–301.
28. Ghazoul J. Floral diversity and the facilitation of pollination. J Ecol. 2006;94:295–304.
29. Hoiss B, Krauss J, Potts SG, Roberts S, Steffan-Dewenter I. Altitude acts as an environmental filter on phylogenetic composition, traits and diversity in bee communities. Proc R Soc B Biol Sci. 2012;279:4447–56.
30. Qian H, Hao Z, Zhang J. Phylogenetic structure and phylogenetic diversity of angiosperm assemblages in forests along an elevational gradient in Changbaishan. China. J Plant Ecol. 2014;7:154–65.
31. Greenleaf SS, Kremen C. Wild bees enhance honey bees' pollination of hybrid sunflower. Proc Natl Acad Sci USA. 2006;103:13890–5.
32. Ashman TL, Arceo-Gómez G. Toward a predictive understanding of the fitness costs of heterospecific pollen receipt and its importance in co-flowering communities. Am J Bot. 2013;100:1061–70.
33. Fang Q, Huang SQ. A directed network analysis of heterospecific pollen transfer in a biodiverse community. Ecology. 2013;94:1176–85.
34. Gong YB, Huang SQ. Floral symmetry: pollinator-mediated stabilizing selection on flower size in bilateral species. Proc R Soc B Biol Sci. 2009;276:4013–20.
35. Ballantyne G, Baldock KCR, Willmer PG. Constructing more informative plant–pollinator networks: visitation and pollen deposition networks in a heathland plant community. Proc R Soc B. 2015;282:20151130.
36. Shannon CE. A mathematical theory of communication. Bell Syst Tech J. 1948;27:623–56.
37. Blüthgen N, Menzel F, Blüthgen N. Measuring specialization in species interaction networks. BMC Ecol. 2006;6:9.
38. Dormann CF, Gruber B, Fründ J. Introducing the bipartite package: analysing ecological networks. R news. 2008;8:8–11.
39. R Core Team. R: a language and environment for statistical computing. Vienna: R foundation for statistical computing. 2014. http://www.R-project.org/. Accessed 22 Jan 2015.
40. Hadfield JD. MCMC methods for multi-response generalised linear mixed models: the MCMCglmm R package. J Stat Softw. 2010;33:1–22.
41. Schielzeth H. Simple means to improve the interpretability of regression coefficients. Methods Ecol Evol. 2010;1:103–13.
42. Olesen JM, Jordano P. Geographic patterns in plant-pollinator mutualistic networks. Ecology. 2002;83:2416–24.
43. Bingham RA, Ort AR. Efficient pollination of alpine plants. Nature. 1998;391:238–9.
44. Stebbins GL. Variation and evolution in plants. New York: Columbia University Press; 1950.
45. Mitchell RJ, Flanagan RJ, Brown BJ, Waser NM, Karron JD. New frontiers in competition for pollination. Ann Bot. 2009;103:1403–13.
46. Huang SQ, Shi XQ. Floral isolation in *Pedicularis*: how do congeners with shared pollinators minimize reproductive interference? New Phytol. 2013;199:858–65.
47. Pauw A. Can pollination niches facilitate plant coexistence? Trends Ecol Evol. 2013;28:30–7.

6

# Correspondence between the habitat of the threatened pudú (Cervidae) and the national protected-area system of Chile

Melissa Pavez-Fox[1,2*] and Sergio A. Estay[1,3]

**Abstract**

**Background:** Currently, many species are facing serious conservation problems due to habitat loss. The impact of the potential loss of biodiversity associated with habitat loss is difficult to measure. This is particularly the case with inconspicuous species such as the threatened pudú (*Pudu puda*), an endemic Cervidae of temperate forests of Chile and Argentina. To evaluate the effectiveness of the Chilean protected-area system in protecting the habitat of the pudú, we measured the congruence between this specie's potential distribution and the geographical area occupied by the protected areas in central and southern Chile. The measurements of congruency were made using the Maxent modeling method.

**Results:** The potential habitat of the pudú was found to be poorly represented in the system (3–8 %) and even the most suitable areas for the species are not currenly protected. According to these results, the protected area network cannot be considered as a key component of the conservation strategy for this species.

**Conclusions:** The results presented here also serve as a guide for the reevaluation of current pudú conservation strategies, for the design of new field studies to detect the presence of this species in human-disturbed areas or remaining patches of native forest, and for the implementation of corridors to maximize the success of conservation efforts.

**Keywords:** Conservation, Niche modeling, Protected-area networks, Temperate rain forest, Threatened species

## Background

Several emblematic species are currently facing serious conservation problems due to the loss and degradation of their habitat caused by the expansion of the human population [1]. The impact of habitat loss on biodiversity cannot always be estimated. For example, the effect of habitat loss on inconspicuous species, which are difficult to detect or inhabit inaccessible places, cannot be easily measured. Furthermore, the lack of basic information about their life histories of inconspicuous species makes planning for conservation yet more difficult [2].

The pudú, an inconspicuous species endemic to the temperate forests of Chile and Argentina (*Pudu puda*), is one of the smallest deers in the world and one of the least studied mammals of Chilean forest fauna [3]. According to the IUCN [4], the conservation status of the pudú is Vulnerable with an estimated 10,000 individuals distributed from 36–49°S in Chile [5] and from 39–43°S in Argentina [6]. Given their evasive behavior, this species remains unstudied in its natural habitat. Furthermore, it has been suggested that the pudú is being affected by landscape fragmentation and forest loss, predation by domestic dogs, competition with exotic species, and poaching activities [4, 7].

Generally spanning 16–23 ha [8], the home range of this species is quite restricted. The pudú feeds on several species of native shrubs and trees, eating the most nutritious parts including young leaves, buds, fruits, and

*Correspondence: melissa.pavez.fox@gmail.com
[2] Magíster en Ciencias Biológicas mención Neurociencia, Facultad de Ciencias, Universidad de Valparaíso, Valparaíso 2360102, Chile
Full list of author information is available at the end of the article

flowers [7, 9]. As the only specialist deer of the temperate rainforest, this species likely plays a key role as a seed disperser [10, 11]. The restricted home range and its role as a seed disperser have led some authors to suggest that it is possible that viable pudú populations could be maintained within natural reserves [12, 13]; thus, some believe that the pudú is a viable target for conservation efforts.

Nationally managed protected areas have been an invaluable tool for in situ conservation [14]. These systems of protected areas have proven their effectiveness at protecting ecosystems and species with respect to the significant pressures of land-use change and land clearing [15]. The Chilean National System of Protected Areas (SNASPE in Spanish) has been a key instrument in determining wildlife conservation strategies in the country. Currently, the SNASPE is comprised of 100 management units distributed among 36 national parks, 49 national reserves, and 15 natural monuments. These units cover an area of 14.5 million hectares corresponding to 19.2 % of mainland Chile; thus, Chile's current system of protected areas is above the 10 % protected ecosystems per country target threshold set by the Convention on Biological Diversity of the United Nations [16].

However, this system appears to be insufficient for the conservation of the fauna associated with temperate forests [17]. Since over 90 % of protected areas in Chile are concentrated at high latitudes (>43°), the areas with the highest species richness (35.6–41.3° S) remain largely outside the system. This, in turn, increases the risk of extinction not only due to the lack of protected areas in temperate forests but also because these areas are isolated within a mosaic of tree plantations, agricultural landscapes, and urban areas [17]. Although the pudú is commonly cited as the most common herbivore in national parks [18], there are no systematic records of its presence in these areas [7].

The conservation of a rare species is a difficult task considering the lack of information about current and potential habitats or habitat requirements [19]. Subsequently, predictive models, such as niche-based modeling, can be useful for obtaining reliable distribution maps to assess the suitability of proposed sites for conservation [20]. In this regard, the emergence of new mathematical methods for estimating potential distributions has complemented the lack of data [2, 21–23]. These methods often use the environmental characteristics of areas where a species is known to inhabit in order to estimate the environmental suitability of regions that currently lack records [24].

Currently, mathematical modeling of species distributions has several applications in conservation science including the prediction of geographic ranges of threatened or rare species [25–27], the identification of priority areas for conservation efforts [28, 29], the evaluation of extinction risk and/or suitable sites for reintroduction programs, and the implementation of wildlife corridors [20]. Considering the ecological importance of the pudú and the lack of up to date information concerning the distribution and conservation status of this species, in this study we estimate the distribution of the pudú within the National System of Protected Areas of Chile. Additionally, we evaluate the effectiveness of the system for the conservation of this species, define the priorities for new areas to protect, and evaluate the reduction of potential pudú habitat due to landscape fragmentation.

## Methods

### Study area

Although originally encompassing an area of 300,000 km$^2$ [30, 31], the geographical range of the temperate rainforest of Chile and Argentina has been considerably reduced at geological time scales as a result of the advance and retreat of mountain glaciers during the Pleistocene [32], and in recent history due to large-scale human impacts on the landscape [33]. Now remnant vegetation covers just 30 % of the original area [31]. The current distribution of the temperate rainforest ranges from 38–49°S on the Chilean side of the Andes and adjacent areas of the provinces of Neuquén, Río Negro, and Chubut in Argentina [34]. The temperate rainforest has a Mediterranean pluviseasonal bioclimate and a temperate and sub-Mediterranean hyper-oceanic bioclimate [35].

The temperate rainforest of Chile encompasses the Valdivian Rainforest Ecoregion, which has been listed among the most endangered ecoregions of the world and has a critical conservation status [36, 37]. The high number of endemic birds (2 %), mammals (16 %), reptiles (62 %), amphibians (54 %) and plant species (49 %) makes this region biologically valuable. Hence, the Valdivian Rainforest Ecosystem is considered a hotspot of biodiversity and, therefore, a region of high conservation priority [31, 38, 39].

### Occurrence data

Records of occurrence of the pudú were collected from several sources found in the literature [6, 8, 40, 41], as well as from databases of museum collections like MaNIS [42]. Points were considered only within the distribution range normally cited for Chile and Argentina [4, 5], since records of the southern area are controversial and their validity is unclear [41, 43]. In total, we considered 135 points of occurrence, of which 73 are located in Chile and 62 are located in Argentina (Fig. 1). In the event that the occurrences were not georeferenced (only location names), a standard procedure was followed whereby coordinates were assigned using the Gazetter GeoNames (http://www.geonames.org). Details of the presence

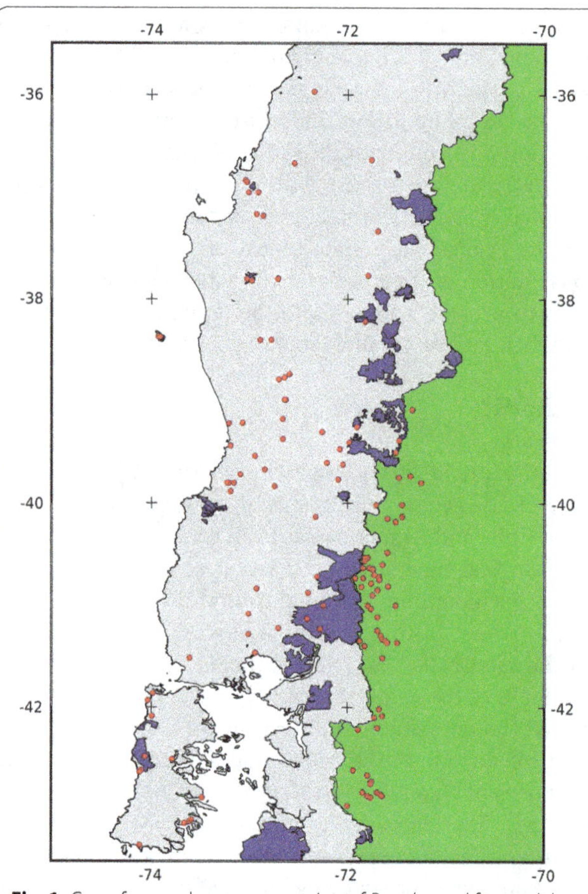


**Fig. 1** Georeferenced occurrence points of *P. puda* used for model fitting (*red*) in Chile (*light grey*) and Argentina (*green*). *Dark blue* polygons represent protected areas of Chile within the study area

points including source, year, and coordinates can be found in the Additional file 1.

#### Environmental and geographic data

Global layers of the current weather conditions were obtained from the WorldClim database; the data have a spatial resolution of 2.5 arc—minutes (~5 km; [44]). These layers contain grouped variables collected monthly from 1950 to 2000. The variables used were selected according to previous modeling studies of other species of deer (e.g. *Odocoileus hemionus*; [45]). The following variables were used: average annual temperature (Ann T°), mean diurnal temperature range (MDR), temperature seasonality (T° S), maximum temperature of the warmest month (T° Max), minimum temperature of the coldest month (T° Min), annual precipitation (Ann Pp), seasonal precipitation (Pp S), precipitation over the wettest quarter (Pp Wet), precipitation of the driest quarter (Pp Dri), precipitation of the warmest quarter (Pp War), and precipitation of the coldest quarter (Pp Col). In addition to these variables, altitude (Alt) [44] was also incorporated in the analysis. The layers of the pudú distribution and the Chilean system of protected are were obtained from the IUCN [4] and the Chilean Forest Service (CONAF), respectively. Considering records of individuals moving up to 20 km in Argentina [46], and the fact that the pudú has an evolutionary history in the region, the study area used to train the model was defined by bounding the observed presences with a buffer of 100 km as a reasonable proxy of the area that has been accessible and probably explored [47, 48] by this small cervid. Processing of the environmental layers was performed in QGIS 2.10 and GRASS7.

#### Statistical methods

A practical way to estimate the geographic distribution of a species is by characterizing the environmental conditions that are currently suitable for its persistence [49] and then identifying those areas where such conditions may be found [50]. A group of quantitative modeling approaches, known collectively as species distribution models (SDM), have been widely used to predict the potential geographic distribution of several animal species [24, 25, 49, 50]. Species distribution models are numerical tools that combine observations of species (either presence or presence and absence data) in a set of locations with environmental variables to obtain ecological and evolutionary insights and to predict distributions across landscapes [11, 15].

Considering that only presence data could be gathered to estimate the pudú geographical distribution, a maximum entropy approach was implemented in the Maxent 3.3.3 k software [51–53] as our ecological niche modeling approach. The Maxent model is a probability distribution selected by maximizing the entropy subject, which is constrained in that the expected value of each environmental variable under this uniform distribution should match the empirical value [51, 52]. The logistic model output represents the degrees of "habitat suitability", ranging from 0 (not suitable) to 1 (suitable) [49]. The Maxent model was fitted using the default settings, and then it was evaluated using the AUC of the ROC curve and the "regularized training gain". The ROC curve corresponds to the graph between 1—specificity (false positive rate) versus sensitivity (rate of true positives, [52]). The AUC measures the ability (probability) of the Maxent model to discriminate between presence sites and background sites [51, 54, 55]. The relative importance of each variable was estimated using the jack-knife method. First, the decrease in gain is calculated by adjusting the model using all variables except the focal variable and comparing this value with the gain of the full model (including all variables). Then, the model is fit using only the focal variable and comparing the gain with respect to the full model. The Maxent model results correspond to the average value of 20 replicas using a cross-validation framework [54, 56]. We used 20 replicates instead of the standard 10 just to increase our evaluation of the model's predictive ability. The cross-validation scheme divides the dataset into 20 subsets. In each step, the model is fitted using 19 subsets and using the last dataset (independent) to test (validate) the fit. This procedure is repeated 20 times, and the AUC and jackknife values reported correspond to the average value of the 20 testing procedures.

### Post-processing

Our results are focused and restricted to Chilean habitats and protected areas. The fitted model, trained in the study area, was later projected to Chile, to estimate the distribution of the species. The original map was converted to a binary map (0 = no-suitable, 1 = suitable), applying a threshold that maximizes sensitivity and specificity [57] to obtain a balance between commission and omission errors. Then, the percentage of the area contained in the public protected areas was calculated for the distribution map (Fig. 1). In addition, the percentage of area containing the pudú distribution proposed by the IUCN was calculated.

## Results

The most suitable areas for the pudú were restricted to the central valley (between the Andes and the Cordillera de la Costa) at low altitudes in the Andes from 38.67 to 39.81°S and in some isolated patches from 42.02°S southwards (Fig. 2a). The AUC for the model was 0.818, indicating a high predictive capacity [55]. The most important variables, according to their effect on the training gain (decrease), were seasonal temperature and mean diurnal temperature range (Table 1). Furthermore, the analysis of the models including only one variable showed that the models fitted with only seasonal temperature, precipitation of the driest quarter, and mean diurnal temperature range, had the highest values of gain (Table 1). The threshold that maximizes sensitivity and specificity was 0.3591, which was used to obtain the binary maps.

The pudú distribution predicted by the Maxent model had an estimated area of 79,047 km$^2$ (33,934 km$^2$ in Chile, 43 %), 37,722 km$^2$ (29.4 %) of which matched the IUCN estimated distribution of 128,278 km$^2$ (Fig. 2b). The area of the model-predicted distribution contained in Chilean protected areas is almost entirely located at high altitudes of the Andes. There was only 4644 km$^2$ of overlap between the predicted distribution of the pudú and area currently being protected in Chile; this overlap represents only 5.87 % of the complete distribution of the pudú (Fig. 2c).

## Discussion

The estimated distribution of the pudú calculated here is almost entirely contained within the area described by the IUCN for this species [4], however the predicted distribution is less extensive in size. The predicted distribution lies within the Valdivian Rainforest Ecoregion, which is consistent with the habitat preferences previously described for this temperate rainforest associated species [5, 6, 39]. Studies have noted that this species prefers rainforest habitats where it can find shelter and highly nutritional food [7, 9]. The most suitable habitats for the pudú were located in the central valley (central depression) of Chile. This area is currently a highly fragmented landscape with isolated populations that are vulnerable to strong anthropogenic pressures [58]. Furthermore, in this area the system of public protected areas (SNASPE) has low coverage making the viability of these remnant populations more difficult (Fig. 2c). Landscape fragmentation in the central valley has resulted in the degradation of the most favorable pudú habitats. As has been found for other species, this fragmentation has forced relict populations to persist in the periphery of their historical geographic range [59].

The model supports the presence of the pudú in the 14 protected areas cited in the literature: the Vicente Perez Rosales, Puyehue, Villarrica, Tolhuaca, Conguillo, and Chiloé National Parks; and the Nonguen, Ñuble, Altos de Pemehue, Isla Mocha, Huerquehue, Mocho Choshuenco, Alerce Costero, and Futaleufu National Reserves (Figs. 1, 2c). The percentage of the predicted pudú distribution in protected areas was low, and in most cases these were not considered to be the most suitable areas for the species according to our models. Due to the fact that the majority of the overlay between pudú favorable habitat and protected area being located in the Andes, together with the low representation of protected areas in the central valley and the coast, imply a major threat to the species as a result of isolation between these populations [60–62]. Suitable areas outside the protected parks represent an opportunity for the conservation of this cervid. Approximately 14,000 km$^2$ of the area around the parks houses broadleaved forests in different stages of conservation [63] that are currently being used for pudú corridors and refuges. However, the high fragmentation of these lands makes coordinating conservation efforts difficult.

The most important variables affecting the estimated distribution of the pudú were the mean diurnal temperature range and temperature seasonality (these variables were selected using two different criteria). According to the models, mean diurnal temperate range was negatively related with habitat suitability. This suggests that the pudú is intolerant to sudden changes in temperature throughout the day and is probably better adapted to tolerate

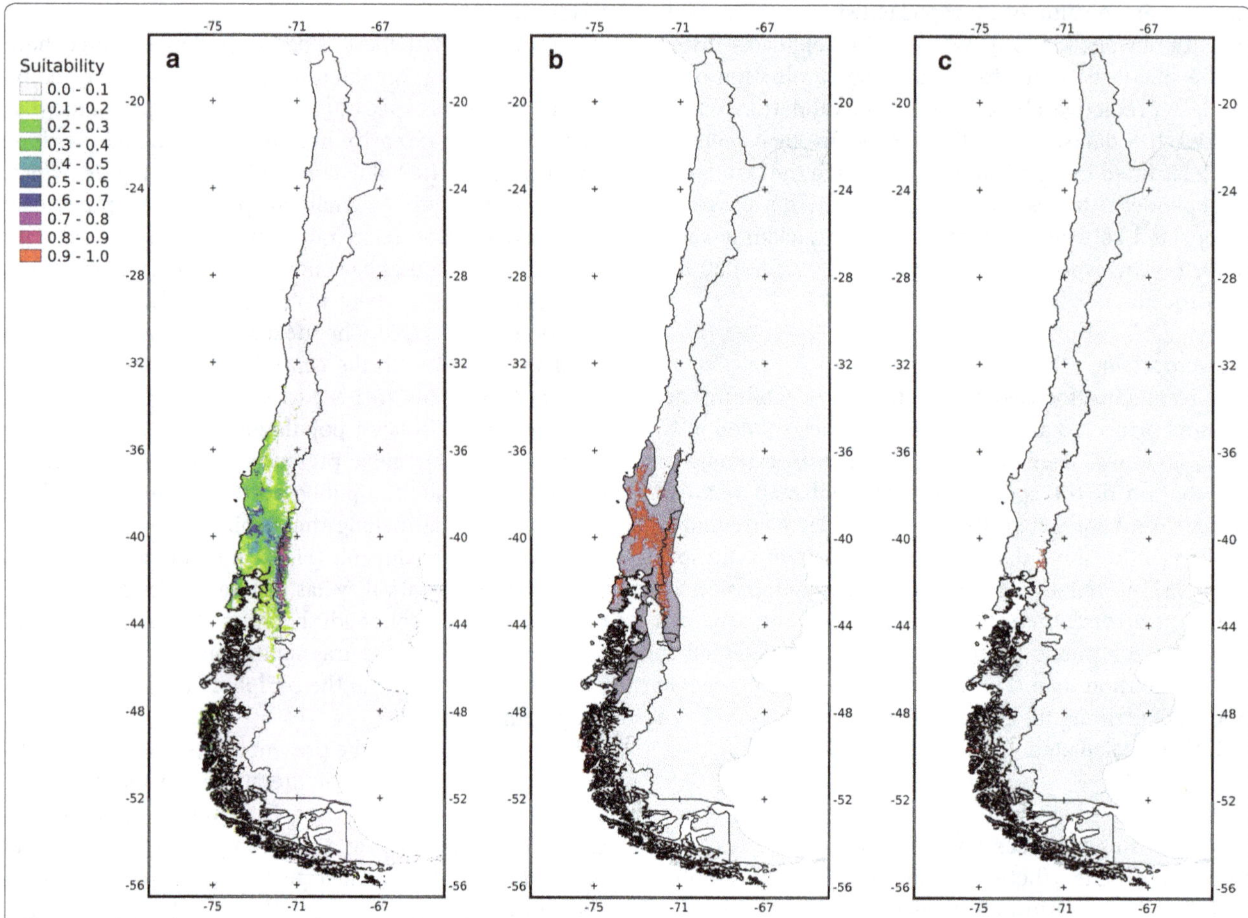


**Fig. 2** Model results: **a** Projection of the model fitted for the Chilean territory. *Colors* represent the suitability of each pixel for *P. puda* habitat. **b** Binary map of the projection of the model fitted for the Chilean territory (*red*) with respect to the distribution determined by the IUCN (*dark grey*) for *P. puda*. **c** Overlap areas between *P. puda* suitable areas and protected areas according to the model (*red*)

low rather than high temperatures. This is in agreement with previous studies, which indicate that the pudú is less active when the sun's intensity is highest and instead is more active during sunrise and sunset [8]. The same negative effect of sunlight or temperature has been described for the kudu (*Tragelaphus strepsiceros*), elk (*Cervus canadensis*), and other deer species (i.e. *Odocoileus* spp.) [64]. Moreover, some authors suggest that ungulate species inhabiting temperate climates would have a lower tolerance to high temperatures than ungulates inhabiting non-temperate climates [64, 65]. On the other hand, the relationship between temperature seasonality and habitat suitability found in this study was irregular, but in general it has been suggested that the pudú can tolerate moderate seasonal variation in temperature, which could be due to the relationship between temperature and the availability of vegetation that it feeds on [66, 67].

The current condition of the pudú within the Chilean system of protected areas remains unknown; however, it has been suggested that even within national parks and reserves, there are still anthropogenic threats such as the presence of feral dogs and domestic livestock [68]. The categorization of the species as Vulnerable according to the IUCN seems to be appropriate given the low representation of suitable habitat within protected areas and the limited amount of information on the status of wild populations. It is therefore crucial to implement new protected areas within the central valley, which could serve as corridors to reduce the rate of species extinctions and increase the likelihood of the re-colonization of parks [60, 67]. The importance of the pudú extends beyond the species' ecological role. As the pudú is a small charismatic species, this deer is a good candidate flagship species that could help attract public attention and sympathy to the conservation of these important habitats, and this, in turn, would likely be of benefit to other species [68].

**Table 1 Jackknife statistics of model performance and relative importance of each variable**

| Environmental variables | Training gain |
|---|---|
| Alt | 0.942†–0.120 |
| Ann T° | 0.938†–0.101 |
| Ann Pp | 0.955–0.295 |
| Pp S | 0.946–0.240 |
| Pp Wet | 0.952–0.254 |
| Pp Dri | 0.955–0.325‡ |
| Pp War | 0.956–0.326‡ |
| Pp Col | 0.949–0.257 |
| MDR | 0.903†–0.328‡ |
| T° S | 0.879†–0.326‡ |
| T° Max | 0.948–0.118 |
| T° Min | 0.951–0.068 |

For each variable, the first value corresponds to the gain of a model fitted using all variables except the focal one. The most important variables according to this criterion are marked with †

The second value corresponds to the gain of a model fitted using only the focal variable. The most important variables according to this criterion are marked with ‡ (see "Methods" for details)

## Conclusions

Overall, the results of this study indicate that the habitat of this endangered cervid is poorly represented in the Chilean system of nationally protected areas. Habitat within this system only represents marginal (less suitable) sites of the pudú's original distribution. Currently, the more suitable areas, in the central valley of Chile, are highly fragmented and used for agricultural, forestry, or other human activities. This highly fragmented land may be a major obstacle for conservation efforts. Initiatives such as new protected areas (public and private), feral dog control, habitat conservation, among others, will contributed to the viability of the small remnant populations of pudú in the central valley.

### Abbreviations

IUCN: International Union for Conservation of Nature; SNASPE: Chilean National System of Protected Areas; MaNIS: The Mammal Networked Information System; AUC: area under the curve; ROC: receiver operating characteristic.

### Authors' contributions

Both authors contributed to the conception and design of the study, interpretation of analyses, and critical revision of the manuscript. MPF collected data, carried out the analyses, and drafted the paper. Both authors read and approved the final manuscript.


### Author details

[1] Instituto de Ciencias Ambientales y Evolutivas, Facultad de Ciencias, Universidad Austral de Chile, Valdivia, Chile. [2] Magíster en Ciencias Biológicas mención Neurociencia, Facultad de Ciencias, Universidad de Valparaíso, Valparaíso 2360102, Chile. [3] Center of Applied Ecology and Sustainability (CAPES), Facultad de Ciencias Biológicas, Pontificia Universidad Católica de Chile, Santiago 6513677, Chile.



### Acknowledgements

We thank Stella Januario and Daniela López for their help during the preparation of the manuscript. Funded by CAPES-CONICYT grant FB-0002, line 4.


### Competing interests

The authors declare that they have no competing interests.

### References


1. Ehrlich PR. The loss of diversity: Causes and consequences. In: Wilson O, editor. Biodiversity. Washington: National Academic Press; 1988. p. 21–7.
2. Papes M, Gaudet P. Modelling ecological niches from low numbers of ocurrences: assessment of the conservation status of poorly known viverrids (Mammalia, Carnivora) across two continents. Divers Distrib. 2007;13:890–902.
3. Weber M, Gonzalez S. Latin American deer diversity and conservation: a review of status and distribution. Ecoscience. 2003;10:443–54.
4. Jiménez J, Ramilo E. *Pudu puda*. In: IUCN. IUCN Red List of Threatened Species. 2013. www.iucnredlist.org. Accessed 25 Sept 2014.
5. Miller S, Rottmann J, Taber R. Dwindling and endangered ungulates of Chile: *Vicugna*, *Lama*, *Hippocamelus*, and *Pudu*. TN Am Wildl Nat Res. 1973;38:55–68.
6. Meier D, Merino ML. Distribution and habitat features of southern pudu (*Pudu puda* Molina, 1782) in Argentina. Mamm Biol. 2007;72:204–12.
7. Jiménez JE. Southern pudu *Pudu puda* (Molina 1782). In: Barbanti JM, González S, editors. Neotropical Cervidology: biology and medicine of latin american deer. Jaboticabal: Funep and IUCN; 2013. p. 140–50.
8. Eldridge WD, MacNamara MM, Pacheco NV. Activity patterns and habitat utilization of pudus (*Pudu puda*) in South-Central Chile. In: Wemmer CM, editor. Biology and management of the Cervidae. Washington: Smithsonian Institution Press; 1987. p. 352–70.
9. Pavez-Fox M, Pino M, Corti P. Muzzle morphology and food consumption by pudu (*Pudu puda* Molina 1782) in south-central Chile. Stud. Neotrop. Fauna Environ. 2015; 1–6.
10. Novillo A, Ojeda RA. The exotic mammals of Argentina. Biol Invasions. 2008;10:1333–44.
11. Stoner KE, Riba-Hernández P, Vulinec K, Lambert JE. The role of mammals in creating and modifying seedshadows in Tropical forests and some possible consequences of their elimination. Biotropica. 2007;39:316–27.
12. Shaffer M. Minimum viable populations: doping with uncertainty. In: Soulé M, editor. Viable populations for conservation. Cambridge: Cambridge University Press; 1987. p. 69–86.
13. Simonetti JA, Mella JE. Park size and the conservation of Chilean mammals. Rev Chil Hist Nat. 1997;70:213–20.
14. Chape S, Harrison J, Spalding M, Lysenko I. Measuring the extent and effectiveness of protected areas as an indicator for meeting global biodiversity targets. Proc R Soc B. 2005;360:443–55.
15. Bruner AG, Gullison RE, Rice RE, da Fonseca GAB. Effectiveness of Parks in protecting tropical biodiversity. Science. 2001;291:125–8.
16. UNEP-WCMC. United Nations Environment Programme—World Conservation Monitoring Centre: Green Report 2009. Cambridge: UNEP-WCMC; 2010.
17. Armesto JJ, Rozzi R, Smith-Ramírez C, Arroyo MTK. Conservation targets in South American temperate forests. Science. 1998;282:1271–2.
18. CONAMA. Biodiversidad de Chile: Patrimonio y desafíos. Santiago: MMA; 2008.
19. Silva-Rodríguez EA, Verdugo C, Aleuy OA, Vianna JA, Vidal F, Jiménez JE. Priorities for the conservation of the pudu (*Pudu puda*) in southern South America. Anim Prod Sci. 2011;51:375–7.
20. Chefaoui RM, Hortal J, Lobo JM. Potential distribution modelling, niche characterization and conservation status assessment using GIS tools: a case study of Iberian *Copris* species. Biol Cons. 2005;122:327–38.

21. Gaubert P, Papes M, Peterson AT. Natural history collections and the conservation of poorly known taxa: ecological niche modeling in central African rainforest genets (*Genetta spp.*). Biol Cons. 2006;130:106–17.
22. Peterson AT, Soberón J, Sánchez-Cordero V. Conservatism of ecological niches in evolutionary time. Science. 1999;285:1265–7.
23. Phillips SJ, Dudik M, Schapire RE. A maximum entropy approach to species distribution modeling. In: Proceedings of the twenty-first international conference on Machine learning. New York: ACM Press; 2004. p. 655–662.
24. Anderson RP, Peterson AT, Gómez-Laverde M. Using niche-based GIS modeling to test geographic predictions of competitive exclusion and competitive release in South American pocket mice. Oikos. 2002;98:3–16.
25. Anderson RP, Martínez-Meyer E. Modeling species' geographic distributions for preliminary conservation assessments: an implementation with the spiny pocket mice (*Heteromys*) of Ecuador. Biol Cons. 2002;116:167–79.
26. Loiselle BA, Howell CA, Graham CH, Goerck JM, Brooks T, Smith KG, et al. Avoiding pitfalls of using species-distribution models in conservation planning. Conserv Biol. 2003;17:1–10.
27. Raxworthy CJ, Martínez-Meyer E, Horning N, Nussbaum RA, Schneider GE, Ortega-Huerta MA, et al. Predicting distributions of known and unknown reptile species in Madagascar. Nature. 2004;426:837–41.
28. Peterson AT, Egbert SL, Sánchez-Cordero V, Price KP. Geographic analysis of conservation priority: endemic birds and mammals in Veracurz. Mexico Biol Cons. 2000;93:85–94.
29. Rondini C, Stuart S, Boitani L. Habitat suitability models and the shortfall in conservation planing for African vertebrates. Cons Biol. 2005;19:1488–97.
30. Alaback PB. Comparative ecology of temperate rainforests of the Americas along analogous climatic gradients. Rev Chil Hist Nat. 1991;64:399–412.
31. Myers N, Mittermeier RA, Mittermeier CG, da Fonseca GAB, Kent J. Biodiversity hotspots for conservation priorities. Nature. 2000;403:853–8.
32. Armesto J, Rozzi R, Caspersen J. Temperate forests of North and South America. In: Chapin F, Sala O, Huber-Sannwald E, editors. Global biodiversity in a changing environment. Scenarios for the 21st Century. New York: Springer; 2011. p. 223–49.
33. Armesto J, Villagrán C, Donoso C. Desde la era glacial a la industria: la historia del bosque templado chileno. Ambiente y Desarrollo. 1994;10:66–72.
34. Lücking R, Wirth V, Ferraro LI, Cáceres MES. Foliicolous lichens from Valdivian temperate rain forest of Chile and Argentina: evidence of an austral element, with the description of seven new taxa. Global Ecol Biogeogr. 2003;12:21–36.
35. Luebert F, Pliscoff P. Sinopsis bioclimática y vegetacional de Chile. Santiago: Editorial Universitaria; 2004.
36. Dinerstein EDM, Olson DM, Graham DJ, Webster AL, Primm SA, Bookbinder MP, Ledec G. A conservation assessment of the terrestrial ecoregions of Latin America and the Caribbean. Washington: The World Bank; 1995.
37. Olson DM, Dinerstein E. Issues in International Conservation: the Global 200: a representation approach to conserving the Earth's most biologically valuable ecoregions. Cons Biol. 1998;12:502–15.
38. Smith-Ramírez C. The Chilean coastal range: a vanishing center of biodiversity and endemism in South American temperate rainforests. Biodivers Conserv. 2004;13:373–93.
39. Ormazábal CS. The conservation of biodiversity in Chile. Rev Chil Hist Nat. 1993;66:383–402.
40. Fuentes-Hurtado M, Marín JC, González-Acuña D, Verdugo C, Vidal F, Vianna JA. Molecular divergence between insular and continental Pudu deer (*Pudu puda*) populations in the Chilean Patagonia. Stud Neotrop Fauna E. 2011;46:23–33.
41. Hershkovitz P. Neotropical deer (Cervidae). Part I. Pudus, genus *Pudu* Gray. Fieldana Zool New series. 1982;11:1–86.
42. Stein BA, Wieczorek J. Mammals of the world: MaNIS as an example of data integration in a disturbed network environment. Biodiversity Informatics. 2004;1:14–22.
43. Muñoz-Pedreros A, Yañez, J. *Mamíferos de Chile*. Temuco Chile: Centro de Estudios Agrarios & Ambientales (CEA) and Museo Nacional de Historia Natural de Chile; 2000.
44. Hijmans RJ, Cameron SE, Parra JL, Jones PG, Jarvis A. Very high resolution interpolated climate surfaces for global land areas. Int J Climatol. 2005;25:1965–78.
45. Pease KM, Freedman AH, Pollinger JP, McCormack JE, Buermann W, Rodzen J, et al. Landscape genetics of California mule deer (*Odocoileus hemionus*): the roles of ecological and historical factors in generating differentiation. Mol Ecol. 2009;18:1848–62.
46. Jiménez JE. Southern pudu *Pudu puda* (Molina 1782). In: Barbanti Duarte, JM, Gonzalez S, editors. Neotropical Cervidology: Biology and Medicine of Latin American Deer. Funep & IUCN, Jaboticabal & Gland. 2010. p. 140–150.
47. Peterson AT, Soberón J, Pearson RG, Anderson RP, Martínez-Meyer E, Nakamura M, Araújo MB. Ecological Niches and Geographic Distributions (MPB-49). Princeton University Press. 2011.
48. Peterson AT. Mapping disease transmission risk: enriching models using biogeography and ecology. Baltimore: JHU Press; 2014.
49. Pearson RG, Raxworthy CJ, Nakamura M, Peterson AT. Predicting species distributions from small numbers of ocurrence records: a test case using cryptic geckos in Madagascar. J Biogeography. 2007;34:102–17.
50. Franklin J. Mapping species distribution: spatial inference and prediction. Cambridge: Cambridge University Press; 2009.
51. Phillips SJ, Anderson RP, Schapire RE. Maximum entropy modeling of species geographic distributions. Ecol Model. 2006;190:231–59.
52. Phillips SJ, Dudik M. Modeling of species distributions with Maxent: new extensions and a comprehensive evaluation. Ecography. 2008;31:161–75.
53. Elith J, Phillips SJ, Hastie T, Dudík M, En Chee Y, Yates CJ. A statistical explanation of MaxEnt for ecologists. Divers Distrib. 2011;17:43–57.
54. Fielding AH, Bell JF. A review of methods for the assessment of prediction errors in conservation presence/absence models. Environ Conserv. 1997;24:38–49.
55. Hosmer DW, Lemeshow S. Applied Logistic Regression. New York: Wiley and Sons; 1989.
56. Hijmans RJ. Cross-validation of species distribution models: removing spatial sorting bias and calibration with a null model. Ecology. 2012;93:679–88.
57. Liu C, White M, Newell G. Selecting thresholds for the prediction of species occurrence with presence-only data. J Biogeogr. 2013;40:778–89.
58. Bello MA. Ecología del pudú (*Pudu pudu*, Molina 1782) y su valor como especie focal, en la Provincia de Valdivia, Ecorregión Valdiviana. Temuco: Thesis Universidad Católica de Temuco; 2003.
59. Naves J, Wiegand T, Revilla E, Delibes M. Endangered species constrained by natural and human factors: the case of brown bears in Northern Spain. Cons Biol. 2003;17:1276–89.
60. Harris LD. The fragmented forest: island biogeography theory and the preservation of biotic diversity. Chicago: The University of Chicago Press; 1984.
61. Bennett AF. Linkages in the landscape: the role of corridors and connectivity in wildlife conservation. 2nd ed. Cambridge UK: IUCN; 2003.
62. Carey C, Nigel D, Stolton S. Squandering paradise? The importance and vulnerability of the world's protected areas. Gland Switzerland: World Wildlife Found; 2000.
63. USGS. Global land cover characterization. South America land cover characteristics data base version 2.0. 2008. http://edc2.usgs.gov/glcc/sadoc2_0.php#usgs. Accessed 20 Jan.
64. Belovsky GE, Slade JB. Time budgets of grassland herbivores: body size similarities. Oecologia. 1986;70:56–62.
65. Owen-Smith N. How high ambient temperature affects the daily activity and foraging time of a subtropical ungulate, the greater kudu (*Tragelaphus strepsiceros*). J Zool. 1998;246:183–92.
66. Ogutu JO, Piepho HP, Dublin HT, Bhola N, Reid RS. Rainfall influences on ungulate population abundance in the Mara-Serengeti ecosystem. J Anim Ecol. 2008;77:814–29.
67. Ogutu JO, Owen-Smith N. ENSO, rainfall and temperature influences on extreme population declines among African savanna ungulates. Ecol Lett. 2003;6:412–9.
68. Silva-Rodríguez EA, Verdugo C, Aleuy OA, Sanderson JG, Ortega-Solís GR, Osorio-Zúñiga F, et al. Evaluating mortality sources for the vulnerable pudu *Pudu puda* in Chile: implications for the conservation of a threatened deer. Oryx. 2010;44:97–103.

# 7

# Heterogeneous distributional responses to climate warming: evidence from rodents along a subtropical elevational gradient

Zhixin Wen[1†], Yi Wu[2†], Deyan Ge[1], Jilong Cheng[1,3], Yongbin Chang[1,3], Zhisong Yang[4], Lin Xia[1] and Qisen Yang[1*]

**Abstract**

**Background:** Understanding whether species' elevational range is shifting in response to directional changes in climate and whether there is a predictable pattern in that response is one of the major challenges in ecology. However, so far very little is known about the distributional responses of subtropical species to climate change, especially for small mammals. In this study, we examined the elevational range shifts at three range points (upper and lower range limits and abundance-weighted range centre) of rodents over a 30-year period (1986 to 2014–2015), in a subtropical forest of Southwest China. We also examined the influences of four ecological traits (body mass, habitat breadth, diet and daily activity pattern) on the upslope shifts in species' abundance-weighted range centres.

**Results:** Despite the warming trend between 1986 and 2015, the 11 rodent species in analysis displayed heterogeneous dynamics at each of the three range points. Species which have larger body sizes and narrower habitat breadths, show both diurnal and nocturnal activities and more specialized dietary requirements, are more likely to exhibit upslope shifts in abundance-weighted range centres.

**Conclusions:** Species' distributional responses can be heterogeneous even though there are directional changes in climate. Our study indicates that climate-induced alleviation of competition and lag in response may potentially drive species' range shift, which may not conform to the expectation from climate change. Difference in traits can lead to different range dynamics. Our study also illustrates the merit of multi-faceted assessment in studying elevational range shifts.

**Keywords:** Climate change, Heterogeneity, Range shift, Rodent, Species traits, Subtropical

## Background

The past 30 years have seen an accelerating increase in the global average surface temperature, and the warming trend is still continuing [1]. One of the most striking biological impacts of ongoing climate warming is the upslope range shifts of organisms, especially when the vegetation and food resources they rely on occur successively at higher elevations [2–4]. There is a high risk of extinction for the species which are unable to keep pace with the climate change or cross the range-shift gaps [5], thus invoking a keen interest of ecologists and conservationists in elevational range shifts over recent decades.

Mountains are perhaps the best systems to investigate the interplay between climate change and species' ranges because researchers can benefit from studying shifts in both range limits (i.e. upper and lower) of a species over relatively short spatial distances. For this reason, a substantial number of empirical studies (usually carried out at one or several mountain ranges) and meta-analyses [6, 7] have been conducted to explore species' elevational range shifts, with the focal species including almost all the biotic groups on earth. Obviously, these studies have shown great variability in the observed responses. Notwithstanding a significant increase in temperature,

*Correspondence: yangqs@ioz.ac.cn
†Zhixin Wen and Yi Wu contributed equally to this work
[1] Key Laboratory of Zoological Systematics and Evolution, Institute of Zoology, Chinese Academy of Sciences, Beichen West Road, Beijing 100101, China
Full list of author information is available at the end of the article

species may show static distributions due to lag effect [8, 9], low mobility [10], acclimate to unfavorable climates [11] and behavioral thermoregulation instead of range shift [12]; and even counterintuitive downslope range shifts which are frequently explained by alleviation of species competition [13], water availability change [14], habitat modification [15, 16] and decreased precipitation [17]. Moreover, it is indicated that species' range dynamics are trait-dependent. Taking small mammals as an example, some life-history and ecological traits such as longevity [18], habitat preference [19] and diet [4] have been shown to affect the responses. Considering the complexity of potential mechanisms underlying elevational range shifts, it remains a great challenge to predict the shifting direction of species. The ability to make an accurate prediction will be valuable to the evaluation of future species assemblage structures at different elevations of a mountain.

Small mammals are sensitive to environmental change [20]. To our knowledge, assessments of small mammal elevational range shifts by systematically resurveying historical sites were only seen in four elevational gradients of North America. Researchers found that small mammals showed much greater heterogeneity in range dynamics than that expected from warming [19, 21, 22]. In contrast to many other taxa, there has been little attention paid to small mammals in tropical and subtropical mountains, despite the fact that biotas here are more threatened by climate due to their generally narrower thermal niches [7, 23, 24]. Certainly, evidence from tropical and subtropical mountains is indispensable to gain an insightful understanding of the elevational range shifts of small mammals worldwide and compare the shifting directions and rates among different regions.

In this study, by revisiting historical sampling sites, we examined spatial shifts in rodent elevational ranges between historical (1986) and modern (2014–2015) times in a subtropical forest of Southwest China. Following Lenoir and Svenning [2], species' range shift was simultaneously assessed at the upper range limit, lower range limit and range centre as their responses to climate change may differ [25, 26]. In addition, we related four species traits (body mass, habitat breadth, diet and daily activity pattern) to the upslope shifts in species' abundance-weighted range centres. Our aims were to test (1) if species' range shifts follow the same pattern as predicted from local climate change; and (2) if the selected traits could explain the difference in distributional responses among species.

## Methods

### Study area and climate data

The study area was an extensive elevational gradient (1550–3500 m) in the Wolong Nature Reserve (102°52′–103°24′E, 30°45′–31°25′N), Sichuan Province. Five vegetation types dominate at different elevations: evergreen broad-leaf forest (<1600 m); evergreen and deciduous mixed broad-leaf forest (1600–2100 m); coniferous and broad-leaf mixed forest (2100–2600 m); coniferous forest (2600–3100 m) and subalpine shrub and meadow (3100–3500 m). Most areas of this gradient have been fully protected since the early 1980s. Because of the extremely limited meteorological records (Dengsheng ecological station at 2800 m a.s.l., climate record available only from 1999 to 2008) in this reserve, we used climate data from the Dujiangyan meteorological station (698 m, approximately 68 km east of our study sites) to estimate the mean annual temperature (MAT) and total annual precipitation (TAP) trends between 1986 and 2015. During this period, MAT increased from 14.2 to 16.4 °C while TAP fluctuated greatly (Additional file 1: Fig. S1). Although the climate data came from a station outside our study area and were measured at a lower elevation than the sampling sites, they were the best available data in this region by far. It is also noted that MAT of Dengsheng ecological station increased by 0.6 °C between 1999 and 2008, with TAP showing little difference over time.

### Historical and modern surveys

The historical survey was conducted at eight sites (Fig. 1) by Wu et al. [27] from March to October (each site was surveyed once every month), 1986, and the survey covered all the vegetation types along the gradient. The elevational range over which our sampling sites were distributed was 1550–3500 m. In total, 725 rodent individuals representing 18 species were captured during 11,430 trap-nights (snap traps). The skulls and voucher specimens are deposited in the Zoological Museum, China West Normal University. By examining these materials, we validated the species identification according to the taxonomic system of Wilson and Reeder [28].

In 2014–2015, we resurveyed the original sites in the same seasons (2014: July to October; 2015: March to June) and by applying the same sampling protocol and technique as in 1986. Yi Wu who led the 1986 survey contributed to the 2014–2015 resurvey by locating the sampling sites and designing the field sampling protocols. At each of the eight study sites, the difference in trapping effort (number of trap-nights) between the historical and modern surveys was less than 200 trap-nights (average difference across eight sites was 116.3 ± 19.2 trap-nights, mean ± SE). Altogether, our resurvey produced 10,500 trap-nights which resulted in the capture of 710 individuals representing 17 species. The skulls and specimen are now preserved in the Institution of Zoology, Chinese Academy of Sciences (IOZCAS). The detailed sampling information of 1986 and 2014–2015 surveys are given in Table 1.

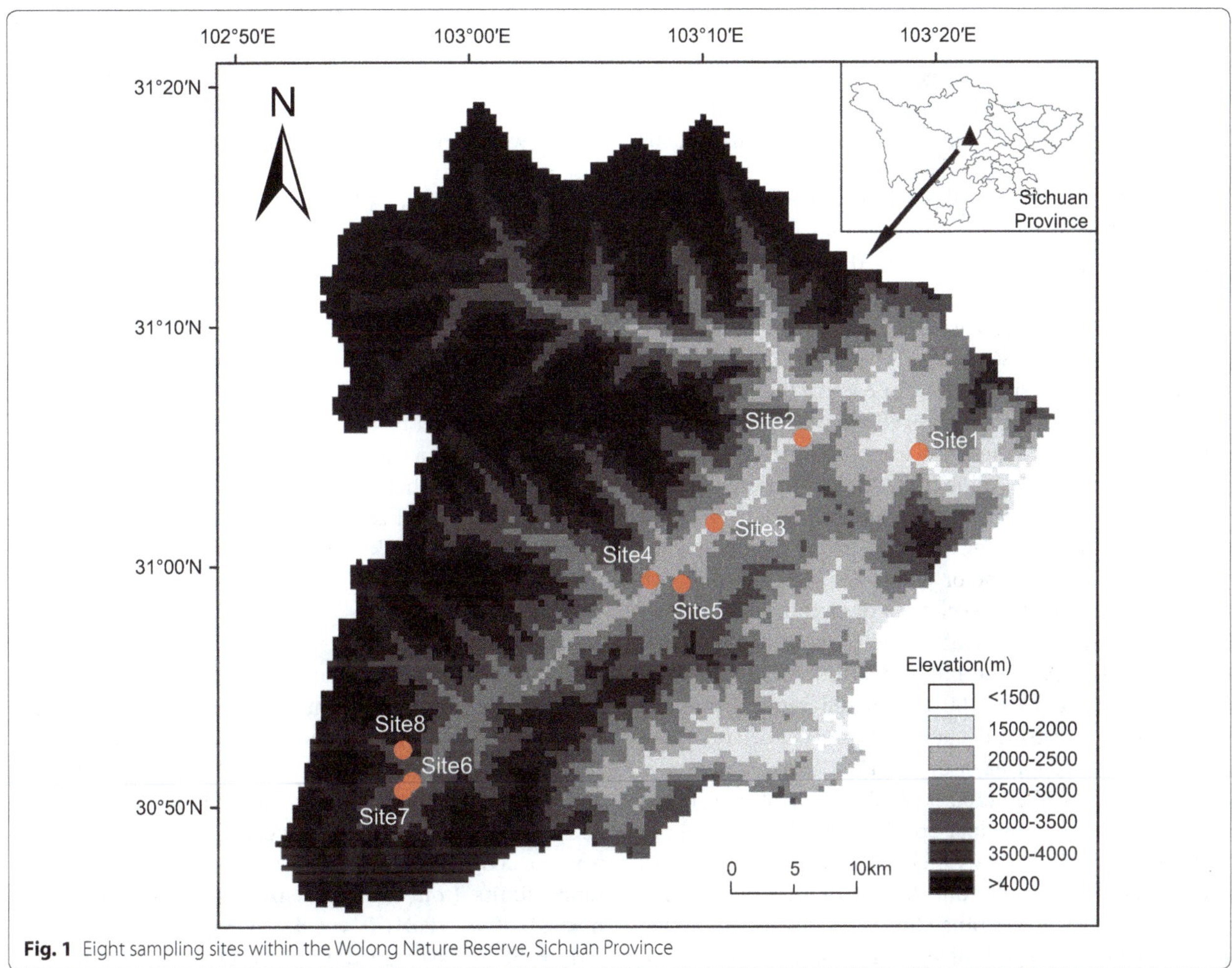


**Fig. 1** Eight sampling sites within the Wolong Nature Reserve, Sichuan Province

### Evaluating range shifts

We examined the shifts in upper range limit, lower range limit and abundance-weighted range centre of the 11 most common species (species with at least five captures in both periods, which formed the historical and modern datasets for comparison) between 1986 and 2014–2015. The abundance-weighted range centre of each period was calculated as:

**Table 1 Detail information of sampling sites and sampling summary in 1986 and 2014–2015**

| Sampling sites | Elevation (m) | Vegetation type | Trap-nights | | All individuals (species) | | Eleven species individuals (species) | |
|---|---|---|---|---|---|---|---|---|
| | | | 1986 | 2014–2015 | 1986 | 2014–2015 | 1986 | 2014–2015 |
| 1 | 1550 | EB | 2430 | 2300 | 161 (10) | 193 (12) | 147 (8) | 186 (9) |
| 2 | 1800 | EDMB | 1200 | 1200 | 81 (7) | 179 (9) | 73 (5) | 176 (7) |
| 3 | 1930 | EDMB | 1200 | 1100 | 134 (6) | 73 (7) | 133 (5) | 72 (6) |
| 4 | 2200 | CBM | 1220 | 1100 | 70 (7) | 43 (5) | 67 (4) | 43 (5) |
| 5 | 2500 | CBM | 1200 | 1100 | 65 (5) | 39 (5) | 65 (5) | 39 (5) |
| 6 | 2800 | CF | 1250 | 1100 | 107 (7) | 100 (6) | 105 (6) | 99 (5) |
| 7 | 3050 | CF | 1250 | 1100 | 43 (5) | 37 (5) | 40 (4) | 36 (4) |
| 8 | 3500 | SSM | 1680 | 1500 | 64 (2) | 46 (2) | 64 (2) | 46 (2) |

Vegetation type abbreviation: *EB* evergreen broad-leaf forest, *EDMB* evergreen and deciduous mixed broad-leaf forest, *CBM* coniferous and broad-leaf mixed forest, *CF* coniferous forest, *SSM* subalpine shrub and meadow

$$\Sigma_{m,n} E_i \times P_{Ai}$$

where $m$ and $n$ were the range limits of species $A$, $E_i$ was the elevation (m) of site $i$ and $P_{Ai}$ was the proportion of species $A$ individuals in site $i$ in its total individuals caught along the whole gradient [29].

To standardize the sampling effort, we randomly resampled (100 times without replacement) the historical and modern datasets to generate the identical number of individuals between periods at each site (e.g. 147 from the 186 individuals of 2014–2015 at site one, 72 from the 133 individuals of 1986 at site three; Table 1) [25], in the R environment (version 3.2.2). For each of the three range points, the average values derived from the 100 random resamplings of two periods were compared (modern value minus historical value) to evaluate the range shift.

To test whether an observed range shift at a given range point (range limits or centre) for a given rodent species was due to chance alone or it was a significant shift, we performed the species-level tests of significance regarding the magnitude of the observed range shift. Based on the 100 replicates of the initial datasets of two surveys, we first calculated the 100 paired differences (modern survey minus historical survey) for each of the three range points and for each of the 11 species, which produced a total of 3300 (100 × 3 × 11) elevational range shift values. For individual species, we then used three boxplots to illustrate its range shift results at different range points separately (one figure for the lower range limit, one for the lower range limit and one for the abundance-weighted range centre). Finally, each of the three boxplot was displayed against the zero reference line (i.e. no change over time) and a Student's $t$ test was conducted to determine whether the altitude of the focal point varied significantly between 1986 and 2014–2015.

### Species traits in explaining upslope shifts in range centres

We used linear regression models to examine the effect of four species traits on the upslope shifts (downslope shifts were denoted by negative values) in species' abundance-weighted range centres, which were body mass (average mass of adults captured in 1986 and 2014–2015), habitat breadth (number of habitat types for a species; obtained from IUCN [30]), diet (categorical variable: zero for herbivores or carnivores and one for omnivores; obtained from the MammalDIET dataset by Kissling et al. [31]) and daily activity pattern [zero for obligately diurnal or nocturnal (be active only in the daytime or only at night) rodents and one for facultatively diurnal (be active mostly at night but occasionally in the daytime) rodents]. The data of species traits are provided in Additional file 1: Table S1. These traits are associated with the climate-induced range shifts for a wide range of animals, including butterflies [32], birds [33, 34] and mammals [4, 18]. Some additional traits such as litters per year, longevity and adult mobility were not tested here because data were unavailable to include these traits which were investigated in other studies [22]. Including these four traits as independent variables resulted in 15 possible models (Additional file 1: Table S2), and the best subset of models were selected by comparing their Akaike's information criterion corrected for small sample size ($AIC_C$) [35]. Because top-ranking models received nearly equivalent support (i.e., little difference in $AIC_C$ values), we performed model averaging of coefficients on the models with $\Delta AIC_C \leq 2$ from the best model. This approach enabled us to assess the relative importance of each variable in predicting the upslope shift in range centre, according to their model-averaged standardized coefficients [36]. Model selection and model averaging were performed using the R package "MuMIn" [37].

## Results

### Species' range shifts

Despite the general warming trend between 1986 and 2015, species' movements at the upper range limits were heterogeneous (four upslope, six stasis, one downslope; Wilcoxon signed-rank test: $n = 11$, $Z = -1.36$, $P = 0.17$), and the average upslope and downslope changes were 256 ± 86 and 250 m, respectively. Similarly, there was no constant trend in the movements of lower range limits (four upslope, six stasis, one downslope; $n = 11$, $Z = -0.81$, $P = 0.42$; average upslope change: 101 ± 58 m, average downslope change: 250 m). Different species also showed different dynamics at the abundance-weighted range centres between periods (six upslope, one stasis, four downslope; $n = 11$, $Z = -0.66$, $P = 0.51$; average upslope change: 204 ± 37 m, average downslope change: 259 ± 146 m) (Fig. 2).

Patterns of elevational range shift varied among species. The Student's $t$ tests of range shift for individual species showed that for the upper range limit, three (*Niviventer andersoni*, *Eothenomys melanogaster* and *Rattus norvegicus*) of six upslope shifts were found to be significant, and the only one downslope shift (*Micromys minutus*) was significant. For the lower range limit, two (*Caryomys eva* and *Microtus oeconomus*) of four upslope shifts were significant, and so was the only one downslope shift of *Apodemus latronum*. For the abundance-weighted range centre, all of the six upslope shifts were significant, while only three of four downslope shifts were significant between periods (Figs. 3, 4).

### Species traits in explaining upslope shifts in range centres

The model with the lowest $AIC_C$ contained only the trait body mass. There were three models with $\Delta AIC_C$

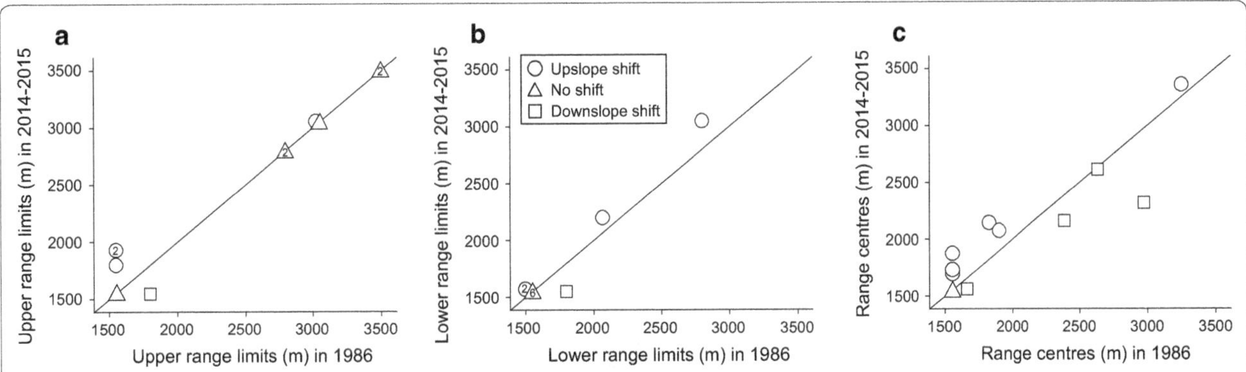


**Fig. 2** Elevational shifts of **a** upper range limits, **b** lower range limits and **c** abundance-weighted range centres of 11 rodent species between 1986 and 2014–2015. *Values* inside graphs represent the number of species showing the same range dynamics over time

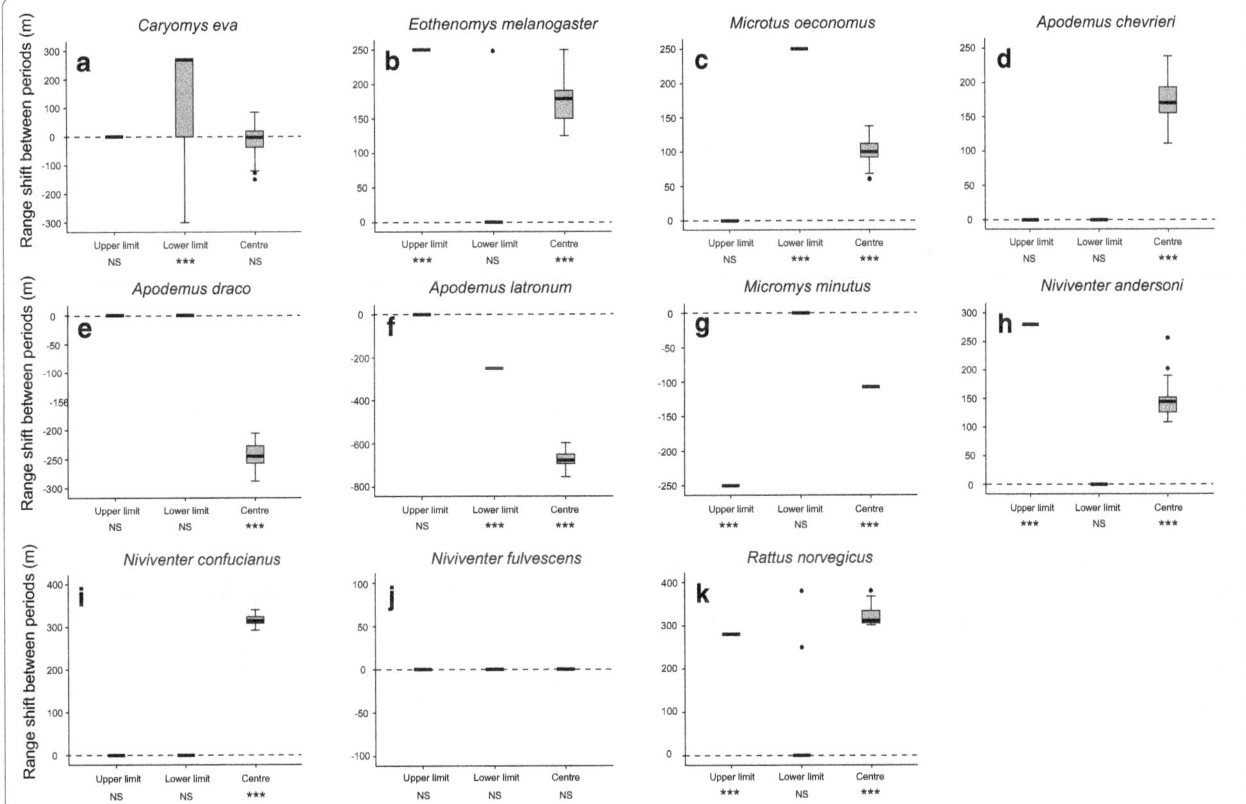


**Fig. 3** Boxplots (median and first and third quartile values are shown, outliers are denoted by *filled circles*) illustrating the elevational shifts of upper range limit, lower range limit and abundance-weighted range centre for each of 11 rodent species (**a** *Caryomys eva*, **b** *Eothenomys melanogaster*, **c** *Microtus oeconomus*, **d** *Apodemus chevrieri*, **e** *Apodemus draco*, **f** *Apodemus latronum*, **g** *Micromys minutus*, **h** *Niviventer andersoni*, **i** *Niviventer confucianus*, **j** *Niviventer fulvescens*, **k** *Rattus norvegicus*) between 1986 and 2014–2015. Elevational range shift values (n = 100) were calculated for each range point of each species as the 100 paired differences in elevation between the modern and historical surveys, based on the 100 replicates of the initial datasets of two periods. Each boxplot was displayed against the zero reference line (i.e. no shift between periods, *dotted line*) and the significance of shift was examined using a Student's *t* test (***$P < 0.001$; *NS* not significant)

≤2 from the best model: the model contained habitat breadth alone, followed by the model contained daily activity pattern alone and the one containing diet alone (Table 2). Similarly, model averaging indicated that body mass had the highest relative importance in predicting the upslope shifts in species' abundance-weighted range

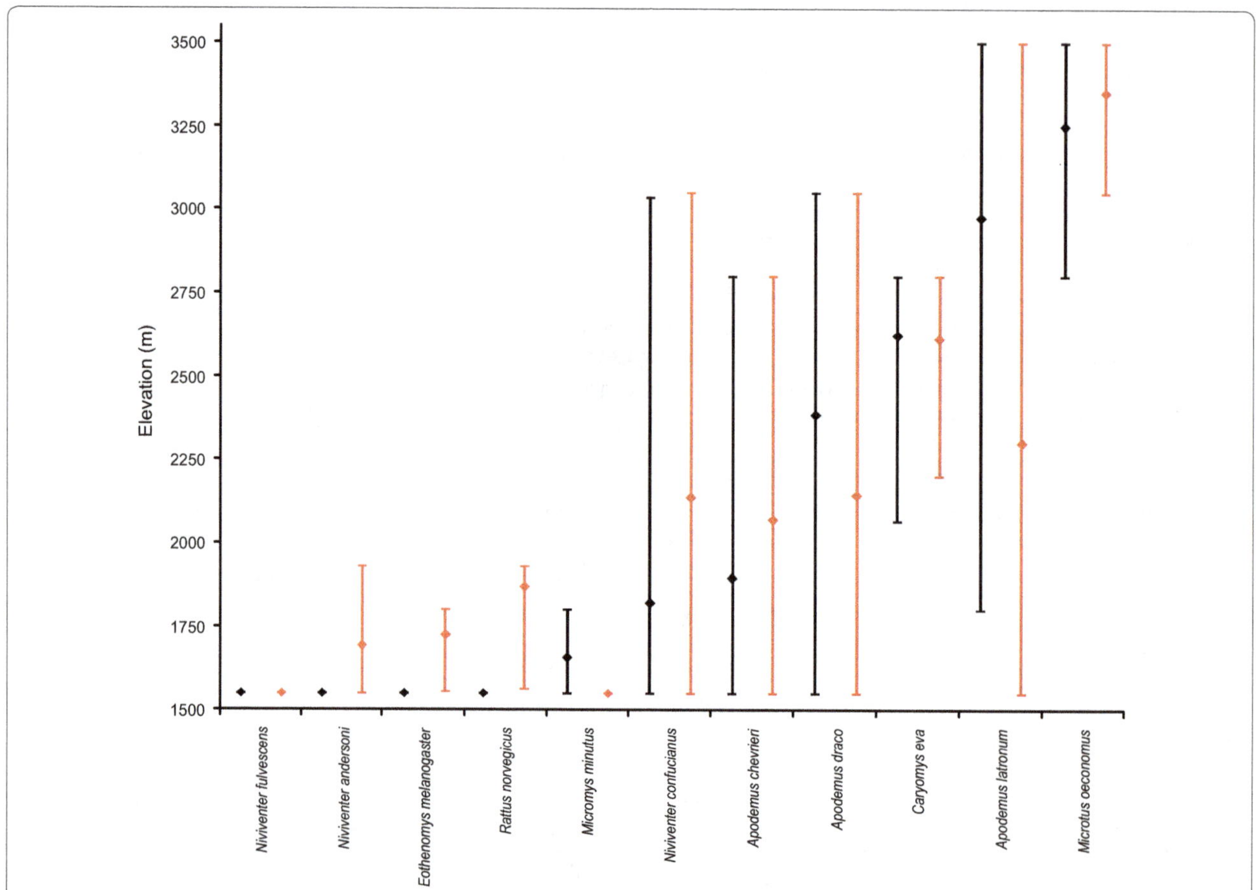


**Fig. 4** Elevational range shifts of 11 rodent species between 1986 (*black*) and 2014–2015 (*red*). The *short horizontal lines* represent the range limits and *diamonds* represent the abundance-weighted range centres. Species are arranged in ascending order of historical abundance-weighted range centre

**Table 2 Model selection and model averaging results of models relating the upslope shifts (m) of 11 rodent species' abundance-weighted range centres to four species traits (body mass, habitat breadth, diet and daily activity pattern), in the Wolong Nature Reserve between 1986 and 2014–2015**

| Model selection results | | | | | Model-averaged standardized coefficients (95% CI) | | | |
|---|---|---|---|---|---|---|---|---|
| Parameter in model | $AIC_C$ | $\Delta AIC_C$ | $AIC_C$ weight | $R^2$ | Body | Habitat | Diet | Activity |
| Body | 162.33 | 0 | 0.274 | 0.138 | 0.372 | −0.314 | −0.159 | 0.290 |
| Habitat | 162.83 | 0.5 | 0.213 | 0.099 | −0.328 to 1.072 | −1.03 to 0.402 | −0.432 to 1.012 | −0.903 to 0.586 |
| Activity | 163.0 | 0.67 | 0.195 | 0.084 | | | | |
| Diet | 163.69 | 1.36 | 0.139 | 0.025 | | | | |

The relationships between range shifts and different sets of trait variables were examined with generalized linear regression models, with models sorted by increasing Akaike's information criterion ($AIC_C$). Only models with $\Delta AIC_C \leq 2$ from the best model are shown in the table. Model-averaged standardized coefficients indicate the relative importance of four traits in predicting the upslope shifts in species' range centres. The 95% confidence intervals are given below the standardized coefficients

centres, followed by habitat breadth, daily activity pattern and diet (Table 2). Therefore, species which have larger body sizes and narrower habitat breadths, show both diurnal and nocturnal activities and more specialized dietary requirements, were more likely to shift their range centres towards higher elevations, although none of the traits exhibited a significant relationship with the magnitude of upslope shift in range centre.

## Discussion

### Heterogeneous range shifts and potential causes

There has been a growing concern about the capacity of montane biotas to track the displacement of their climatic optima, but the evidence from subtropical zones is rare [2]. Here, by comparing rodent elevational distributions across a 30-year interval, we demonstrate that despite a warming trend in a subtropical forest of China, there is heterogeneity in species' distributional responses. Diverse patterns of range shift were revealed, including stasis, upslope and downslope range shifts. Many previous studies which reported species moving towards higher elevations in response to climate warming have also detected a portion of the investigated taxa showed no changes or downslope movements. In a global meta-analysis, Parmesan and Yohe [38] found that approximately 20% of the plant and animal species displayed downslope and southward range shifts. For plants, Lenoir et al. [39] demonstrated that in Northeast France, 53 of 171 (31%) species shifted their optimum elevations downslope between 1905–1985 and 1986–2005; and Wolf et al. [40] recently reported that merely 15% of the Californian plants' mean elevations were higher than a century ago, and those of the rest species showed little or downslope movements. For birds, Tingley et al. [17] found that only half of the species in Sierra Nevada showed upslope shifts in upper or lower range boundary after a century of warming, possibly due to a relatively narrow elevational range over which bird eggs will hatch. As for small mammals which is our focal taxa, idiosyncratic patterns of elevational range shift have been borne out by several studies in North America [19, 21]. The disagreement between the observed range dynamics of rodents in Wolong and climatic expectations can be driven by a number of factors that shall be discussed below.

The upslope expansions of *N. andersoni*, *E. melanogaster* and *R. norvegicus*, and upslope contractions of *C. eva* and *M. oeconomus*, was probably a direct reaction to the increased temperature. As a non-native species, *R. norvegicus* may benefit from climate change in colonizing new habitats. When the temperature increases, non-native species usually tend to shift their ranges upslope to occupy the expanded potential niche space at higher elevations [40]. Temporal changes in competitive species interaction may also result in the upslope shifts. In 1986, the upper limits of *N. andersoni*, *E. melanogaster* and *R. norvegicus* were situated at 1550 m where the rodents had the highest species richness (eight species) along the gradient, implying an intense interspecific competition at this elevation with the competitors most likely to consist of ecologically similar species (e.g. three *Niviventer* species [41]; also see Additional file 1: Table S1). The competition can be provisionally alleviated by climate warming, enabling species to fill their potential distribution areas by conducting an upslope (or downslope) range shift [13]. By examining the dynamic of abundance-weighted range centre, it is possible to gain a more subtle insight into species' distributional response. Although the change in range centre is closely linked with the change in range limits, upslope displacement of centre may manifest even though both boundaries remain unchanged, as found in *Apodemus chevrieri*. We would expect an imminent upslope range shift of the species provided the warming trend continues.

In our study, stasis was the most common dynamic (six of 11) at both range limits. By comparison, Moritz et al. [22] found that 36% of the small mammals in Yosemite National Park of California shifted their lower limits upslope. The different results between this and their research may be due to the different sampling intervals, which was a century in Moritz et al. [22]. In Wolong, the increased temperature between 1986 and 2015 may be within the tolerance ranges of some thermotolerant species. Alternatively, less vagile species may need more than 30 years to display an evident upslope shift in lower limit (i.e. lag effect [9, 25]). The stasis of the upper limits of two species, *A. latronum* and *M. oeconomus*, deserve particular attention because they were historically located at the mountaintop. The impossibility to move "higher" may have strongly affected their range dynamics and made them more vulnerable to local extinction [42]. Another interpretation of the static distributions is that species can use behavioral thermoregulation to prevent overheating, such as altering daily activity (e.g. reduce diurnal activity [21]) and hiding under vegetation shade [12]. Besides, if there are desirable habitats around at the same elevation, rodents may prefer to migrate to these nearby refugia rather than conduct an arduous shift [43], as the North American elk (*Cervus elaphus*) did in an Idaho desert [44]. Indeed, there are many cool environments in Wolong such as caves and shady valleys. These sites could facilitate the local persistence of species so that they can avoid long-distance vertical migration.

In line with many previous studies, we also observed downslope range shifts. The downslope expansion of *A. latronum* could be explained by the climate-induced alleviation of species competition [13], as long as predicable food and suitable habitats are available downwards. For *Micromys minute*, alleviated competition may be the same factor causing its downslope contraction, while changes in non-temperature factors like water availability could also underlie the movement [14]. Our result was consistent with that reported in the Great Basin, where the downslope shifts of rodents were attributed to climate-induced floristic change and land use [19].

### Range shifts related to species traits

The importance of species' ecological traits in estimating and explaining climate-induced range shifts has been long recognized [18]. We observed a positive, albeit not statistically significant, relationship between the body mass of rodents and upslope shift in species' abundance-weighted range centre. This finding supports the idea that compared to smaller species, larger mammal species are generally more mobile [4, 45] and characterized by better abilities to colonize a new region on the basis of higher fecundities and larger home-range sizes [46]. Intriguingly, differing from some earlier studies [18, 46, 47], habitat and dietary specialists were found showing greater upslope displacements at the range centres than generalists. We attribute our result to the strong dependence of specialists on specific prey and habitat. For example, the upslope shift in range centre of *M. oeconomus* (herbivore) could be driven by the upslope shift of its favored habitat (e.g. forest-meadow ecotone [48]) and plants. The phenomenon that habitat and dietary specialists shift upslope/northward more than generalists has also been observed in North American [33] and Central European [34] birds.

## Conclusions

Our study represents one of the first attempts to explore the climate-induced elevational range shifts of subtropical small mammals. Using a multi-faceted assessment of range shifts (upper and lower range limits, abundance-weighted range centre) for 11 rodent species, we demonstrate that the distributional responses are heterogeneous despite a general warming trend, with stasis, upslope and downslope movements all being detected. The heterogeneity is possibly due to the difference in species traits such as body mass and habitat breadth. Climate-induced alleviation of competition and lag in response may potentially drive species' range shift, which may not conform to the expectation from climate change.

## Additional file

**Additional file 1: Table S1.** Body mass, habitat types, diet and daily activity pattern of the 11 rodent species in range shift analysis. **Table S2.** Model selection results of all 15 models relating the upslope shifts of 11 rodent species' abundance-weighted range centres to four species traits. **Figure S1.** Changes in mean annual temperature and total annual precipitation between 1986 and 2015.

#### Abbreviations

$AIC_C$: Akaike's information criterion corrected for small sample size; IOZCAS: Institution of Zoology, Chinese Academy of Sciences; MAT: mean annual temperature; TAP: total annual precipitation.

#### Authors' contributions

ZXW, YW and QSY conceived the idea and designed the research. DYG, JLC, YBC and ZSY conducted the field sampling. ZXW and LX analyzed the data, and ZXW and YW wrote the manuscript. All authors read and approved the final manuscript.

#### Author details

[1] Key Laboratory of Zoological Systematics and Evolution, Institute of Zoology, Chinese Academy of Sciences, Beichen West Road, Beijing 100101, China. [2] College of Life Sciences, Guangzhou University, Guangzhou 510006, China. [3] Graduate University of Chinese Academy of Sciences, Yuquan Road, Beijing 100049, China. [4] Institute of Rare Animals and Plants, China West Normal University, Nanchong 637009, China.

#### Acknowledgements

We are grateful to Wolong Nature Reserve Authority for permission to conduct the study. We thank Xiaogang Shi, Huanhuan Zhang, Yu Yang, Li Zhang, Lei Song and Yunfei Mu for their kind assistance in the field work.

#### Competing interests

The authors declare that they have no competing interests.

#### Funding

Our study was supported by the National Science Fund for Fostering Talents in Basic Research (Special Subjects in Animal Taxonomy, NSFC-J1210002) and the Fund for Exploring the Biodiversity and Natural Resources of Wolong Nature Reserve (Y590911135).

#### References

1. IPCC. Climate Change 2013: the physical science basis. In: Working group I contribution to the fifth assessment report of the Intergovernmental Panel on Climate Change. New York: Cambridge University Press; 2013.
2. Lenoir J, Svenning J-C. Climate-related range shifts—a global multidimensional synthesis and new research directions. Ecography. 2015;38:15–28.
3. Parmesan C. Ecological and evolutionary responses to recent climate change. Annu Rev Ecol Evol Syst. 2006;37:637–69.
4. Santos MJ, Thorne JH, Moritz C. Synchronicity in elevation range shifts among small mammals and vegetation over the last century is stronger for omnivores. Ecography. 2015;38:556–68.
5. Colwell RK, Brehm G, Cardelús CL, Gilman AC, Longino JT. Global warming, elevational range shifts, and lowland biotic attrition in the wet tropics. Science. 2008;322:258–61.
6. Chen I-C, Hill JK, Ohlemüller R, Roy DB, Thomas CD. Rapid range shifts of species associated with high levels of climate warming. Science. 2011;333:1024–6.
7. Gibson-Reinemer DK, Sheldon KS, Rahel FJ. Climate change creates rapid species turnover in montane communities. Ecol Evol. 2015;5:2340–7.
8. Popy S, Bordignon L, Prodon R. A weak upward elevational shift in the distributions of breeding birds in the Italian Alps. J Biogeogr. 2009;37:57–67.
9. Renwick KM, Rocca ME. Temporal context affects the observed rate of climate-driven range shifts in tree species. Glob Ecol Biogeogr. 2015;24:44–51.

10. Davis AJ, Jenkinson LS, Lawton JH, Shorrocks B, Wood S. Making mistakes when predicting shifts in species range in response to global warming. Nature. 1998;391:783–6.
11. Thuiller W, Lavorel S, Araújo MB. Niche properties and geographical extent as predictors of species sensitivity to climate change. Glob Ecol Biogeogr. 2005;14:347–57.
12. Kearney M, Shine R, Porter WP. The potential for behavioral thermoregulation to buffer "cold-blooded" animals against climate warming. Proc Natl Acad Sci USA. 2009;106:3835–40.
13. Lenoir J, Gégout J-C, Guisan A, Vittoz P, Wohlgemuth T, Zimmermnn NE, Dullinger S, Pauli H, Willner W, Svenning J-C. Going against the flow: potential mechanisms for unexpected downslope range shifts in a warming climate. Ecography. 2010;33:295–303.
14. Crimmins SM, Dobrowski SZ, Greenberg JA, Abatzoglou JT, Mynsberge AR. Changes in climatic water balance drive downhill shifts in plant species' optimum elevations. Science. 2011;331:324–7.
15. Archaux F. Breeding upwards when climate is becoming warmer: no bird response in the French Alps. Ibis. 2004;146:138–44.
16. Cannone N, Sgorbati S, Guglielmin M. Unexpected impacts of climate change on alpine vegetation. Front Ecol Environ. 2007;5:360–4.
17. Tingley MW, Koo MS, Moritz C, Rush AC, Beissinger SR. The push and pull of climate change causes heterogeneous shifts in avian elevational ranges. Glob Change Biol. 2012;18:3279–90.
18. Angert AL, Crozier LG, Rissler LJ, Gilman SE, Tewksbury JJ, Chunco AJ. Do species' traits predict recent shifts at expanding range edges? Ecol Lett. 2011;14:677–89.
19. Rowe RJ, Finarelli JA, Rickart EA. Range dynamics of small mammals along an elevational gradient over an 80-year interval. Glob Change Biol. 2010;16:2930–43.
20. Pűettker T, Pardini R, Meyer-Lucht Y, Sommer S. Responses of five small mammal species to micro-scale variations in vegetation structure in secondary Atlantic Forest remnants, Brazil. BMC Ecol. 2008;8:9.
21. Rowe KC, Rowe KMC, Tingly MW, Koo MS, Patten JL, Conroy CJ, Perrine JD, Beissinger SR, Moritz C. Spatially heterogeneous impact of climate change on small mammals of montane California. Proc R Soc B. 2015;282:20141857.
22. Moritz C, Patton JL, Conroy CJ, Parra JL, White GC, Beissinger SR. Impact of a century of climate change on small-mammal communities in Yosemite National Park, USA. Science. 2008;322:261–4.
23. Freeman BG, Freeman AMC. Rapid upslope shifts in New Guinean birds illustrate strong distributional responses of tropical montane species to global warming. Proc Natl Acad Sci USA. 2014;111:4490–4.
24. Janzen DH. Why mountain passes are higher in the tropics. Am Nat. 1967;101:233–49.
25. Chen I-C, Hill JK, Shiu H-J, Holloway JD, Benedick S, Chey VK, Barlow HS, Thomas CD. Asymmetric boundary shifts of tropical montane Lepidoptera over four decades of climate warming. Glob Ecol Biogeogr. 2011;20:34–45.
26. Maggini R, Lehmann A, Kéry M, Schmid H, Beniston M, Jenni L, Zbinden N. Are Swiss birds tracking climate change? Detecting elevational shifts using response curve shapes. Ecol Model. 2011;222:21–32.
27. Wu Y, Hu J-C, Li H-C, Qu M-C. Research on community structure of small rodents in Wolong Nature Reserve. J China West Norm Univ. 1988;9:95–102.
28. Wilson DE, Reeder DM. Mammal species of the world: a taxonomic and geographic reference. vol 2. 3rd ed. Baltimore: Johns Hopkins University Press; 2005.
29. Menéndez R, Gonzélez-Megías A, Jay-Robert P, Marquéz-Ferrando R. Climate change and elevational range shifts: evidence from dung beetles in two European mountain ranges. Glob Ecol Biogeogr. 2014;23:646–57.
30. IUCN. http://www.iucnredlist.org/. Accessed Feb 2016.
31. Kissling WD, Dalby L, Fløjgaard C, Lenoir J, Sandel B, Sandom C, Trøjelsgaard K, Svenning J-C. Establishing macroecological trait datasets: digitalization, extrapolation, and validation of diet preferences in terrestrial mammals worldwide. Ecol Evol. 2014;4:2913–30.
32. Pöyry J, Luoto M, Heikkinen RK, Kuussaari M, Saarinen K. Species traits explain recent range shifts of Finnish butterflies. Glob Change Biol. 2009;15:732–43.
33. Auer SK, King DI. Ecological and life-history traits explain recent boundary shifts in elevation and latitude of western North American songbirds. Glob Ecol Biogeogr. 2014;23:867–75.
34. Reif J, Flousek J. The role of species' ecological traits in climatically driven altitudinal range shifts of central European birds. Oikos. 2012;121:1053–60.
35. Burnham KP, Anderson DR. Model selection and multimodel inference: a practical information-theoretic approach. 2nd ed. New York: Springer; 2002.
36. Johnson JB, Omland KS. Model selection in ecology and evolution. Trends Ecol Evol. 2004;19:101–8.
37. Bartoń K. MuMIn: multi-model inference. R package version 1.15.6.2016. http://CRAN.R-project.org/package=MuMIn. Accessed May 2016.
38. Parmesan C, Yohe G. A globally coherent fingerprint of climate change impacts across natural systems. Nature. 2003;421:37–42.
39. Lenoir J, Gégout JC, Marquet PA, de Ruffray P, Brisse H. A significant upward shift in plant species optimum elevation during the 20th century. Science. 2008;320:1768–71.
40. Wolf A, Zimmerman NB, Anderegg WRL, Busby PE, Christensen J. Altitudinal shifts of the native and introduced flora of California in the context of 20th-century warming. Glob Ecol Biogeogr. 2016;25:418–29.
41. Wang Y-Z, Hu J-C. The imitatively-colored pictorial handbook of the mammals of Sichuan. 1st ed. Beijing: China Forestry Publishing House; 1999.
42. Elsen PR, Tingley MW. Global mountain topography and the fate of montane species under climate change. Nat Clim Change. 2015;5:772–6.
43. Scherrer D, Körner C. Topographically controlled thermal-habitat differentiation buffers alpine plant diversity against climate warming. J Biogeogr. 2011;38:406–16.
44. Long RA, Bowyer RT, Porter WP, Mathewson P, Monteith KL, Kie JG. Behavior and nutritional condition buffer a large-bodied endotherm against direct and indirect effects of climate. Ecol Monogr. 2014;84:513–32.
45. Schloss CA, Nuñez TA, Lawler JJ. Dispersal will limit ability of mammals to track climate change in the Western Hemisphere. Proc Natl Acad Sci USA. 2012;109:8606–11.
46. Sunday JM, Pecl GT, Frusher S, Hobday AJ, Hill N, Holbrook NJ, Edgar GJ, Stuart-Smith R, Barrett N, Wernberg T, Watson RA, Smale DA, Fulton EA, Slawinski D, Feng M, Radford BT, Thompson PA, Bates AE. Species traits and climate velocity explain geographic range shifts in an ocean-warming hotspot. Ecol Lett. 2015;18:944–53.
47. Betzholtz P-E, Pettersson LB, Ryrholm N, Franzén M. With that diet, you will go far: trait-based analysis reveals a link between rapid range expansion and a nitrogen-favoured diet. Proc R Soc Lond B Biol Sci. 2013;280:20122305.
48. Liang E-Y, Wang Y-F, Piao S-L, Lu X-M, Camarero JJ, Zhu H-F, Zhu L-P, Ellison AM, Ciais P, Peñuelas J. Species interactions slow warming-induced upward shifts of treelines on the Tibetan Plateau. Proc Natl Acad Sci USA. 2016;113:4380–5.
49. Wen Z-X, Wu Y, Ge D-Y, Cheng J-L, Chang Y-B, Yang Z-S, Xia L, Yang Q-S. Data from: heterogeneous distributional responses to climate warming: evidence from rodents along a subtropical elevational gradient. Dryad Digit Repos. 2017. doi:10.5061/dryad.1q413.

# 8

# A test of priority effect persistence in semi-natural grasslands through the removal of plant functional groups during community assembly

Kenny Helsen[1,2*], Martin Hermy[3] and Olivier Honnay[1]

**Abstract**

**Background:** It is known that during plant community assembly, the early colonizing species can affect the establishment, growth or reproductive success of later arriving species, often resulting in unpredictable assembly outcomes. These so called 'priority effects' have recently been hypothesized to work through niche-based processes, with early colonizing species either inhibiting the colonization of other species of the same niche through niche preemption, or affecting the colonization success of species of different niches through niche modification. With most work on priority effects performed in controlled, short-term mesocosm experiments, we have little insight in how niche preemption and niche modification processes interact to shape the community composition of natural vegetations. In this study, we used a functional trait approach to identify potential niche-based priority effects in restored semi-natural grasslands. More specifically, we imposed two treatments that strongly altered the community's functional trait composition; removal of all graminoid species and removal of all legume species, and we compared progressing assembly with unaltered control plots.

**Results:** Our results showed that niche preemption effects can be, to a limited extent, relieved by species removal. This relief was observed for competitive grasses and herbs, but not for smaller grassland species. Although competition effects acting within functional groups (niche preemption) occurred for graminoids, there were no such effects for legumes. The removal of legumes mainly affected functionally unrelated competitive species, likely through niche modification effects of nitrogen fixation. On the other hand, and contrary to our expectations, species removal was after 4 years almost completely compensated by recolonization of the same species set, suggesting that priority effects persist after species removal, possibly through soil legacy effects.

**Conclusions:** Our results show that both niche modification and niche preemption priority effects can act together in shaping community composition in a natural grassland system. Although small changes in species composition occurred, the removal of specific functional groups was almost completely compensated by recolonization of the same species. This suggests that once certain species get established, it might prove difficult to neutralize their effect on assembly outcome, since their imposed priority effects might act long after their removal.

**Keywords:** Emergent groups, Functional traits, Graminoids, Historical contingency, Legumes, Niche modification, Niche preemption, Plant-soil feedback, Size-asymmetric competition, Soil legacies

*Correspondence: kenny.helsen@ntnu.no
[1] Plant Conservation and Population Biology, Department of Biology, University of Leuven, Arenbergpark 31, 3001 Heverlee, Belgium
Full list of author information is available at the end of the article

## Background

Evidence continues to build that plant community assembly is rarely predictable at the species level, strongly challenging the traditional deterministic view of Clements [1] on succession [2, 3]. Indeed, many studies have shown that assembly outcome is not solely determined by abiotic conditions, but is partly unpredictable, often resulting in multiple alternative end states of the assembly process [4–6]. These observations support the view of stochastic community assembly, first discussed by Gleason [7] and Diamond [8], where assembly is expected to be, up to a certain extent, contingent upon historical processes [2, 9]. Historical contingency is hypothesized to affect assembly through multiple pathways, such as land-use legacies, interannual variation in (a)biotic conditions, historical landscape connectivity and priority effects [3, 10, 11]. In recent years, the importance of priority effects on plant assembly outcome has been given much attention [12–14]. Priority effects occur when early colonizing species inhibit or facilitate the establishment, growth or reproductive success of later arriving species, and are, for plants, hypothesized to be mainly caused either directly by size-asymmetric competition effects (inhibitory) [15, 16], or indirectly, by soil legacies (inhibitory or facilitative) [12, 17]. In the latter case the priority effects are effectuated by changes in nutrient availability, soil microbial communities or the buildup of allelochemicals [12, 16]. Soil legacies may persist for a long time after the causal species had disappeared from the community through the effects of plant-soil feedbacks [13, 17].

Even though priority effects are sometimes considered to be independent of species identity (neutral theory) [10, 18, 19], an increasing number of studies on both asymmetric competition effects and plant-soil feedbacks strongly suggests the opposite, with certain species seemingly strongly affected by priority effects exerted by specific species, while others remain unaffected by the same initial species [20–24]. It has, for this reason, been argued that the occurrence and severity of both inhibitory and facilitative priority effects are often strongly dependent upon a species niche. Fukami [14] more specifically hypothesizes that priority effects can be governed by two alternate niche-based processes, namely niche preemption (inhibitory) and niche modification (either inhibitory or facilitative). According to this framework, priority effects governed by niche preemption processes, such as size-asymmetric competition, will only affect species within niches, while niche modification based priority effects, such as soil legacies, will primarily act across niches [14]. Support for this niche preemption hypothesis has been found during experimental assembly of bacterial communities, where strong priority effects, and hence multiple community states, only emerged when species pools contained species with great niche overlap [25], and for nectar-inhabiting microorganisms, where priority effect size was significantly related to the extent of niche overlap [26]. Although direct evidence for this hypothesis is currently lacking for plant assembly, it has been experimentally shown that—at least certain—plant communities are most inhibitive to invasion of new species with niche requirements that are similar to those of species already present in those communities [27, 28]. Other studies, however, suggest that this limiting similarity process might not be universally applicable for plant communities [29].

Nonetheless, since functional plant traits are considered to be directly linked to a species niche, the functional group identity of species will likely greatly improve our predictive ability of priority effect presence and strength. Indeed, priority effects through size-asymmetric competition are expected to only occur within functional groups [26], while soil legacies likely affect species both within and among functional groups [14]. This potential predictive power of functional traits has already been illustrated by the observation of deterministic assembly at the functional trait (niche) level, as opposed to contingent assembly at the species identity level [30–32]. In these studies, the presence of multiple species with similar functional traits within a species pool are assumed to explain the occurrence of strong inhibitory priority effects at the species level through niche preemption, within each of the present niche spaces [33]. Nevertheless their potential, functional traits have only rarely been included as predictive variables in priority effect research during plant community assembly (but see [23, 34]).

The current knowledge on priority effects has been mainly gained through largely controlled mesocosm experiments looking either only at size-asymmetric competition (e.g., [10, 19, 24]) or at soil legacies (e.g., [22, 34, 35]). However, in natural systems, both inhibitory niche preemption and inhibitory or facilitative niche modification based priority effects may be simultaneously shaping community structures, making generalizations from these mesocosm experiments difficult [16]. Similarly, the priority effects observed in these mesocosm studies are usually surprisingly strong, likely because of optimal growing conditions and relatively short studied time scales (1–2 years after initial colonization) [10, 36]. Although little information is available on long-term priority effects, Hawkes et al. [22] have shown that experimental plant soil feedbacks can become increasingly negative for many species after 4 years. Two studies have furthermore observed indications of persistent priority effects at somewhat larger timescales (4–5 years) in grassland systems [32, 37]. A study of vernal pool plant communities in a more natural system, on the other hand, found the disappearance of priority effects after 7 years [20]. In conclusion, we can say that there is need for more in situ research to adequately quantify the importance of priority effects on long-term assembly progress and outcome in natural systems [19].

In this study, we want to fill part of this knowledge gap by evaluating potential priority effects at the functional trait level, using natural dry grasslands as a model system. More specifically, we evaluated small-scale plant community composition during the early stages of grassland development, following restoration practices. We imposed two treatments that severely altered the functional trait composition of the community; removal of all graminoid species and removal of all nitrogen fixating species (legumes). Additionally we also included a control treatment. The experiment was performed in four different restored semi-natural grasslands on the French-Belgian border, with four 5 × 5 m replicates (plots) of each treatment in each grassland, and was followed up during four consecutive years. In this experiment, we expected graminoids to mainly impose local inhibitory niche preemption (competition) effects on the community, since graminoids are often highly competitive. Legumes on the other hand, are known to fixate nitrogen, thus altering soil nutrient content. Taking into account the relatively weak competitive abilities of the legume species in our study system, we hypothesized that legumes mainly impose local facilitative niche modification (soil legacy) effects on the community. By comparing the changes in the species composition of both the treatment functional groups (graminoids and legumes) and previously defined functional trait groups (emergent groups) among the three treatments, we tried to verify the following hypotheses:

1. Removal of graminoids will relief local inhibitory within-niche competition effects, resulting in the local colonization of species with the same functional trait set as the removed species (inhibitory niche preemption).
2. Removal of legumes will likely result in the local colonization of species adapted to high nutrient availability, thus resulting in the colonization of species with a different functional trait set as the removed legume species (facilitative niche modification).
3. Small-scale plot level changes in the community composition caused by the graminoids and legumes will still be visible after 4 years, through newly enforced priority effects of secondary colonized species.

## Methods

### Study area

The study was performed in four recently restored semi-natural grassland patches, on the French-Belgian border (c. 50°N, 4.5°E). These patches are part of four larger, isolated grassland fragments, which are embedded in a matrix consisting of a mixture of arable land and forests, surrounded by several other grassland fragments. The four studied grassland patches were all restored from forest or shrub encroachment in 2007, and are adjacent to mature grassland within the grassland fragment. Initial restoration practices consisted of the complete removal of all aboveground vegetation and litter, after which spontaneous colonization of the bare soil was allowed. Soil characteristics were not directly altered, nor were plant species or seeds deliberately introduced to the restored sites. The follow up management of these grasslands consists of annual grazing by a migratory sheep flock. The grazing management prevents domination by woody species and also allows the dispersal of plant species through zoochory. Note that this setup strongly reduces any effect of dispersal limitation at the plot level, since all plots are imbedded within one of four larger grassland patches.

### Experimental design

To test for possible priority effects on grassland community assembly we used an experimental design consisting of three (functional identity) conditions. More specifically, these conditions consisted of the removal of all nitrogen fixation species or Fabaceae (*condition L; legumes*), removal of all Poaceaea, Cyperaceae and Juncaceae species (*condition G; graminoids*) and a standard condition with no manipulation of assembly (*condition C; control*). Manipulation for all conditions was performed within separate 5 × 5 m plots by carefully applying very small and targeted amounts of glyphosphate to the target species in the summer of 2010 (July), with a follow up in September of 2010 to remove the dead aboveground biomass and to make sure all treatment species were successfully killed. This set-up was replicated over four grassland patches. Each treatment was spatially randomly replicated for four times within each grassland patch, adding up to a total of 48, 5 × 5 m plots. After initial manipulation in 2010, these communities were allowed to follow spontaneous community assembly. Species composition and abundance (% cover) was collected for all plants (Tracheophytes) in each plot during the summer (July) of four consecutive years (2010–2013), with the 2010 data collected before initial manipulation. This field study was performed on public land. As our manipulations did not directly involve, nor did indirectly affect any endangered species we did not require special permission. The full plots × species dataset can be found at [38].

### Functional traits and emergent groups

For the evaluation of functional trait patterns, we used the emergent groups (EGs) that were defined by Helsen et al. [39] for the species pool of similar dry semi-natural grasslands. More precisely, Helsen et al. [39] delineated seven EGs based on twenty-eight functional plant traits using a minimum variance clustering method based on Gower's similarity (Table 1). These traits were selected based on their relevance for community assembly, through their effects on species dispersal, establishment and persistence (cf. [40]) and were obtained from different databases (Additional file 1). Additionally, we calculated the community weighted means (CWM, as defined by [41]) for the three binary functional traits (thus the weighted proportions): nitrogen fixation, graminoid morphology and clonality. The values for these three functional traits were collected from the Ecoflora and Biolflor databases [42, 43].

### Diversity metrics

Species richness (S) and Pielou's evenness index (E) were calculated for each plot, including all species, generalist species only, and specialist species only. Specialist species were defined as species confined to dry semi-natural grasslands in Belgium [44, 45] (Additional file 2). Species richness and total plant cover (%) were calculated twice for each EG separately, once including all species within each EG, and once excluding all treatment species (all Fabaceae, Poaceae, Cyperaceae and Juncaceae species).

### Species identity turnover

To evaluate changes in the species composition within treatment functional groups (graminoids and legumes), we quantified the species replacement of both species groups between the first (pre-treatment, year 0) and last year (year 3) of the experiment. Comparing species replacement of both species groups among the different treatments allows better insight in how priority effects are shaping the community composition. Species replacement was calculated as the 'relativized species replacement' $R_{rel}$ based on presence-absence data [46] and as the 'relativized abundance replacement' $^{a}R_{rel}$ based on abundance data [47] for both treatment functional groups separately, resulting in four species replacement measures.

### Statistical analysis

Differences in overall species diversity and species diversity of individual emergent groups between the three treatments were assessed using repeated measures linear mixed models (RMLMMs). More specifically, we constructed a separate RMLMM for each diversity metric as a dependent variable, including treatment, time (year) and the interaction between time and treatment as fixed factors, grassland identity as a random factor ('variance components' covariance type) and time (year) as a random repeated measure ('unstructured' covariance type). The model included both fixed and random (ID) intercepts and was based on restricted maximum likelihood (REML). Analogous RMLMMs were constructed for the CWM of nitrogen fixation, graminoid morphology and clonality. Prior to statistical analyses, several response variables were transformed to obtain normal distributions of the model residuals. In particular, all three measures of evenness (for all species, specialists and generalists) and the clonality CWM were squared, the cover of EG 1 was log transformed and we took the square root for the nitrogen fixation CWM, graminoid morphology CWM and the cover of all individual EGs (except EG 1) and total cover.

Among-treatment differences in species replacement within treatment functional groups were tested using linear mixed models (LMMs). More precisely, we

**Table 1 Overview of the emergent groups (EGs) as defined in Helsen et al. [39]**

| Emergent group | Group name | Characteristics |
|---|---|---|
| 1 | Megaphanerophytes | Long lived, shade-tolerant species, early flowering, wind pollinators, large seeds, transient seed bank, allogamous, anemo- and dysochores. Species of nutrient rich soils |
| 2 | Forest/shrub species | Long lived, shade-tolerant herbaceous and woody (understory) species, insect pollinated, transient seed bank, mixed mating system, few and heavy seeds, dysochores, large leaves. Species of nutrient rich soils, shade tolerant |
| 3 | Orchids | Many, small seeds, mycorrhiza-dependent |
| 4 | Small grassland herbs | Allogamous, shade intolerant, small herbs, autochores and zoochores, nitrogen fixators, semi-rosette species, specialists |
| 5 | Large herbs and grasses | Semi-rosette species, late flowering, large seeds, large species, large leaves, hemero- and zoochores, competitives. Species of nutrient rich soils |
| 6 | Sedges and shallow soil specialists | Mixed mating system, long seedbank longevity, small and light seeds, auto- and anemochores, mycorrhiza-independent |
| 7 | Annuals | Early flowering, autogamous, short-lived, small seeds and plants, zoochores, ruderals |

For every group the name and typical functional trait values (characteristics) are given. Note that EG 3 (orchids) was not used in this study since too few species of this group were observed constructed a separate LMM for each of the four calculated species replacement measures ($R_{rel}$ and $^{a}R_{rel}$) as a dependent variable, including treatment as a fixed factor and grassland identity as a random factor ('variance components' covariance type). The model included a fixed intercept and was based on restricted maximum likelihood (REML). Semi-partial $R^2_\beta$ coefficients were calculated for each covariate in all performed RMLMMs and LMMs, using the method of Edwards et al. [48].

## Results

Total species richness increased through time for all plots, independent of the treatment. This increase through time was found to be mainly driven by a strong increase in specialist species, with a decrease of generalist species (Table 2; Fig. 1a). The total evenness per plot, on the other hand, decreased through time, a pattern also observed for specialist species, and, although less pronounced, for generalist species (Fig. 1b). Evenness of specialist species was also affected by treatment, with a higher evenness in the G treatment (β G, Table 2). Total vegetation cover per plot strongly increased through time, but was reduced for the G treatment compared to the C treatment, in the first and second year following vegetation manipulation. At the third year following vegetation manipulation, no significant difference in total cover remained between the three treatments (significant interaction term, Table 2; Fig. 1c).

The CWM for nitrogen fixation did, overall, not change through time. However, when contrasting the three treatments, the CWM for nitrogen fixation decreased in the first year following vegetation manipulation in the L treatment, resulting in a significantly lower value for the L treatment compared to the C and G treatments in the first 2 years following vegetation manipulation. However, in the last year, no significant difference remained in the CWM for nitrogen fixation among the three treatments (Table 2; Fig. 1d). A comparable pattern was observed for the CWM of graminoid morphology, with an initial decrease for the G treatment compared to the C and L treatments in the first year following vegetation manipulation, and a gradual recovery of the CWM of graminoid morphology through time. Interestingly, unlike the CWM for nitrogen fixation, the CWM of graminoid morphology showed a gradual overall increase through time, independent of treatment (Table 2; Fig. 1e). The CWM for clonality showed the same patterns as the CWM for graminoid morphology, although the differences between the G treatment on the one hand and the C and L treatments on the other hand were less pronounced (Table 2; Fig. 1f).

An overview of the seven EGs defined by Helsen et al. [39] is presented in Table 1. Group names are based on the groups' trait composition: megaphanerophytes (group 1), forest/shrub species (group 2), orchids (group 3), small grassland herbs (group 4), large herbs and grasses (group 5), sedges and shallow soil specialists (group 6) and annuals (group 7). Since too few species of EG 3 (orchids) were present in this study, it was removed from further analyses. Comparing the treatment functional groups (graminoids and legumes) with the EGs revealed that most legumes belong to EG 4 (small grassland

**Table 2 Parameter estimates of the performed repeated measures linear mixed models on diversity measures and CWMs**

| | Time | | | | | | Treatment | | | | | Interaction | |
|---|---|---|---|---|---|---|---|---|---|---|---|---|---|
| | F | $R^2_\beta$ | β T0 | β T1 | β T2 | β T3 | F | $R^2_\beta$ | β C | β L | β G | F | $R^2_\beta$ |
| S | 12.70*** | 0.22 | 34.63 | 33.13 | 35.82 | 36.13 | 0.15 | <0.01 | 36.51 | 35.88 | 36.13 | 0.39 | 0.01 |
| S spec. | 18.74*** | 0.29 | 10.56 | 10.87 | 11.37 | 12.25 | 0.16 | <0.01 | 12.81 | 12.56 | 12.25 | 1.38 | 0.03 |
| S gen. | 4.65** | 0.09 | 24.07 | 22.25 | 24.44 | 23.88 | 0.01 | <0.01 | 23.69 | 23.32 | 23.88 | 0.56 | 0.01 |
| E[a] | 28.16*** | 0.38 | 0.73 | 0.75 | 0.64 | 0.69 | 1.02 | 0.03 | 0.68 | 0.68 | 0.69 | 0.53 | 0.01 |
| E spec.[a] | 9.25*** | 0.17 | 0.73 | 0.79 | 0.68 | 0.62 | 13.67*** | 0.25 | 0.53 | 0.56 | 0.62 | 1.14 | 0.02 |
| E gen.[a] | 34.00*** | 0.43 | 0.69 | 0.70 | 0.57 | 0.65 | 1.70 | 0.05 | 0.69 | 0.66 | 0.65 | 0.54 | 0.01 |
| Cover[b] | 9.62*** | 0.18 | 13.62 | 13.33 | 14.15 | 15.44 | 1.43 | 0.04 | 15.25 | 14.97 | 15.44 | 3.42** | 0.07 |
| CWM N fix[b] | 0.41 | 0.01 | 0.22 | 0.27 | 0.35 | 0.24 | 2.88° | 0.08 | 0.24 | 0.22 | 0.24 | 8.26*** | 0.20 |
| CWM graminoids[b] | 35.80*** | 0.44 | 0.43 | 0.33 | 0.39 | 0.51 | 9.65*** | 0.21 | 0.58 | 0.57 | 0.51 | 12.86*** | 0.22 |
| CWM clonality[a] | 36.46*** | 0.45 | 0.35 | 0.34 | 0.45 | 0.51 | 1.87 | 0.04 | 0.53 | 0.52 | 0.51 | 2.47* | 0.05 |

Beta-coefficient, test statistic and semi-partial $R^2_\beta$ given for time, treatment and the interaction term

*S* species richness, *E* Pielou's evenness, *CWM* community weighted mean, *T* time since treatment (year), *C* control treatment, *L* legumes treatment, *G* graminoids treatment

° $0.10 \geq P > 0.05$ * $0.05 \geq P > 0.01$; ** $0.01 \geq P > 0.001$; *** $0.001 \geq P$

[a] Squared transformation

[b] Square root transformation

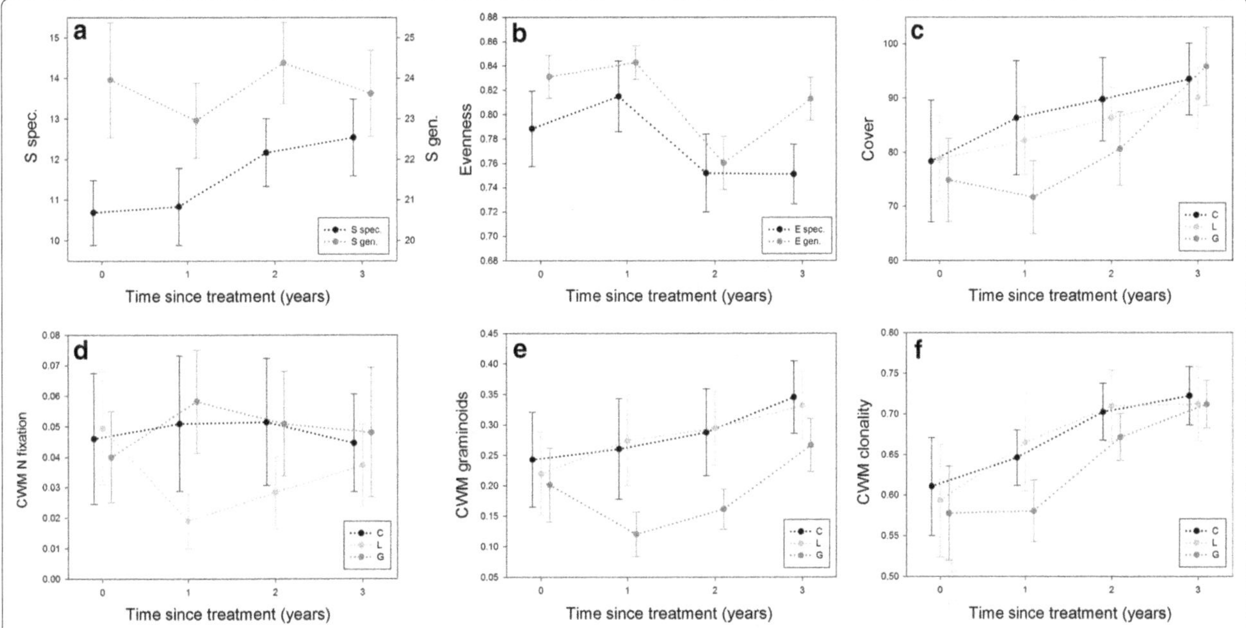

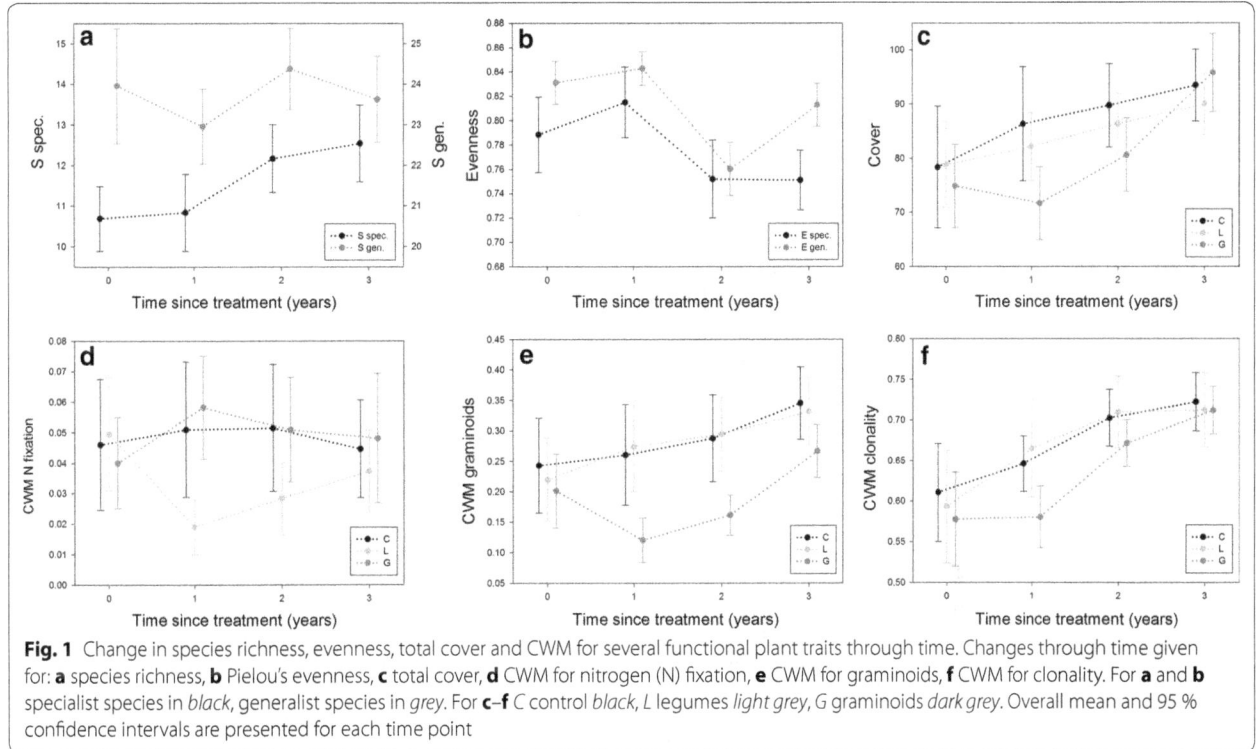

**Fig. 1** Change in species richness, evenness, total cover and CWM for several functional plant traits through time. Changes through time given for: **a** species richness, **b** Pielou's evenness, **c** total cover, **d** CWM for nitrogen (N) fixation, **e** CWM for graminoids, **f** CWM for clonality. For **a** and **b** specialist species in *black*, generalist species in *grey*. For **c–f** *C* control *black*, *L* legumes *light grey*, *G* graminoids *dark grey*. Overall mean and 95 % confidence intervals are presented for each time point

herbs), and to lesser extent to EG 7 (annuals). Graminoids occur mainly in EGs 4 (small grassland herbs), 5 (large herbs and grasses) and 6 (sedges and shallow soil specialists), with also three species in EG 7 (annuals) (Additional file 2). Consequently, the RMLMMs for EG species richness and cover were only performed twice for these EGs (i.e., 4, 5, 6 and 7), with and without the treatment species (graminoids and legumes). For EGs 1 and 2, the RMLMMs were performed once for both species richness and cover (Table 3).

Five of the six studied EGs were observed to change through time, both in terms of species richness and total cover, with a decrease of megaphanerophytes and annuals, and an increase in species richness of small grassland herbs and sedges and shallow soil specialists. The pattern for forest/shrub species was less pronounced, with a small increase in species richness, but a decrease in total cover through time. Large herbs and grasses were unaffected by time, both in terms of species richness and total cover (Table 3). Species richness of megaphanerophytes and large herbs and grasses furthermore differed among the treatments, with higher richness of megaphanerophytes, but lower richness of large herbs and grasses for the L treatment and higher richness of large herbs & grasses for the G treatment compared to the C treatment (Table 3; Fig. 2a). Most interestingly, these patterns remained significant after excluding the treatment species from the dataset (graminoids and legumes) (Table 3; Fig. 2b). Total cover of large herbs and grasses was lower for the G treatment compared to the C and L treatment in the second and third year following vegetation manipulation when including the treatment species (significant interaction term, post hoc results not shown). Total cover of sedges and shallow soil specialists was also significantly lower for the G treatment compared to the C and L treatment (significant treatment effect, Table 3). However, these differences disappeared when excluding the treatment species (Table 3).

Species replacement of graminoids, based on both species presence-absence and cover, during the first and last year of the experiment was not significantly different among treatments (Table 4). Similarly, species replacement of legumes between the first and last year of the experiment was similar among the three treatments (Table 4).

## Discussion

### General assembly patterns

Changes in species richness and functional group composition through time observed in the C treatment can be interpreted as the natural assembly patterns in the studied grasslands. This natural assembly process is characterized by the replacement of generalist by specialist species and an increase in total vegetation cover through time. These results largely confirm the assembly patterns

**Table 3 Parameter estimates of the performed repeated measures linear mixed models for emergent groups species richness and cover**

| | Time | | | | | | Treatment | | | | | Interaction | |
|---|---|---|---|---|---|---|---|---|---|---|---|---|---|
| | F | $R_\beta^2$ | β T0 | β T1 | β T2 | β T3 | F | $R_\beta^2$ | β C | β L | β G | F | $R_\beta^2$ |
| S EG 1 | 9.16*** | 0.17 | 1.75 | 1.63 | 1.57 | 1.19 | 6.05** | 0.13 | 1.19 | 1.63 | 1.19 | 0.93 | 0.02 |
| S EG 2 | 3.93* | 0.08 | 6.13 | 6.32 | 6.69 | 6.63 | 0.59 | 0.01 | 6.51 | 6.88 | 6.63 | 0.39 | 0.01 |
| S EG 4 | 36.40***/28.02*** | 0.45/0.38 | 8.43/5.13 | 9.25/5.94 | 10.68/7.07 | 11.31/7.38 | 0.37/0.09 | 0.01/<0.01 | 12.12/7.63 | 12.06/7.88 | 11.31/7.38 | 0.26/0.18 | 0.01/<0.01 |
| S EG 5 | 0.37/0.38 | 0.01/0.01 | 12.19/8.69 | 11.32/8.69 | 12.01/8.88 | 12.13/8.38 | 4.45*/3.72* | 0.09/0.08 | 11.82/8.19 | 10.57/7.73 | 12.13/8.88 | 0.89/0.50 | 0.02/0.01 |
| S EG 6 | 6.73***/8.03*** | 0.13/0.15 | 2.81/1.25 | 2.88/1.50 | 3.13/1.75 | 3.44/1.75 | 0.57/0.55 | 0.01/0.01 | 3.25/1.94 | 3.38/1.88 | 3.44/1.75 | 0.90/1.32 | 0.02/0.03 |
| S EG 7 | 10.61***/7.93*** | 0.19/0.15 | 2.94/2.37 | 2.19/1.69 | 2.44/1.94 | 1.44/1.06 | 0.22/0.03 | 0.01/<0.01 | 1.69/1.25 | 1.63/1.44 | 1.44/1.06 | 1.14/1.14 | 0.02/0.02 |
| cover EG 1[c] | 9.69*** | 0.18 | 0.60 | 0.58 | 0.50 | 0.44 | 1.01 | 0.02 | 0.35 | 0.46 | 0.44 | 0.93 | 0.02 |
| cover EG 2[b] | 5.14** | 0.10 | 6.87 | 6.62 | 6.27 | 6.37 | 0.37 | 0.01 | 5.71 | 5.76 | 6.37 | 0.41 | 0.01 |
| cover EG 4[b] | 24.19***/24.79*** | 0.35/0.36 | 5.81/4.59 | 6.78/5.69 | 8.27/7.06 | 9.31/7.69 | 0.11/0.10 | <0.01/<0.01 | 8.64/7.07 | 8.63/7.30 | 9.31/7.68 | 2.02°/0.89 | 0.04/0.02 |
| cover EG 5[b] | 0.53/2.91* | 0.01/0.06 | 8.20/6.62 | 7.49/6.86 | 7.58/6.88 | 8.07/6.57 | 0.44/2.18 | 0.01/0.05 | 8.15/6.00 | 7.71/5.67 | 8.07/6.57 | 2.71*/0.49 | 0.06/0.01 |
| cover EG 6[b] | 11.72***/5.36** | 0.21/0.14 | 3.78/1.74 | 3.59/1.85 | 4.43/1.82 | 5.54/2.11 | 6.54**/0.48 | 0.14/0.01 | 6.52/2.25 | 6.28/2.16 | 5.54/2.11 | 2.31°/2.30° | 0.05/0.07 |
| cover EG 7[b] | 3.53*/1.88 | 0.07/0.04 | 2.22/1.91 | 2.27/1.78 | 2.01/1.76 | 1.63/1.26 | 0.03/0.07 | <0.01/<0.01 | 1.99/1.70 | 2.07/1.94 | 1.63/1.26 | 0.90/1.24 | 0.02/0.03 |

Beta-coefficient, test statistic and semi-partial $R_\beta^2$ given for time, treatment and the interaction term

For EG 4–7 analyses were performed first for all species in the respective EGs (results before slash) and second after excluding treatment species (graminoids and legumes) from the EGs (results after slash)

*S* species richness, *EG* emergent group (see Table 1 for EGs names and content), *T* time since treatment (year), *C* control treatment, *L* legumes treatment, *G* graminoids treatment

° $0.10 \geq P > 0.05$ * $0.05 \geq P > 0.01$; ** $0.01 \geq P > 0.001$; *** $0.001 \geq P$

[b] Square root transformation

[c] Logarithm transformation

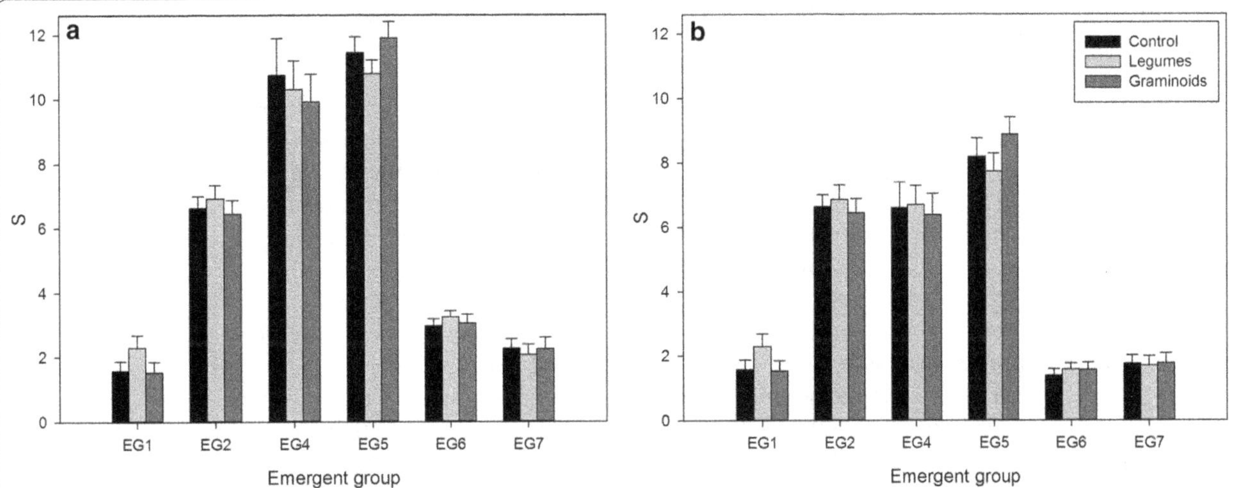

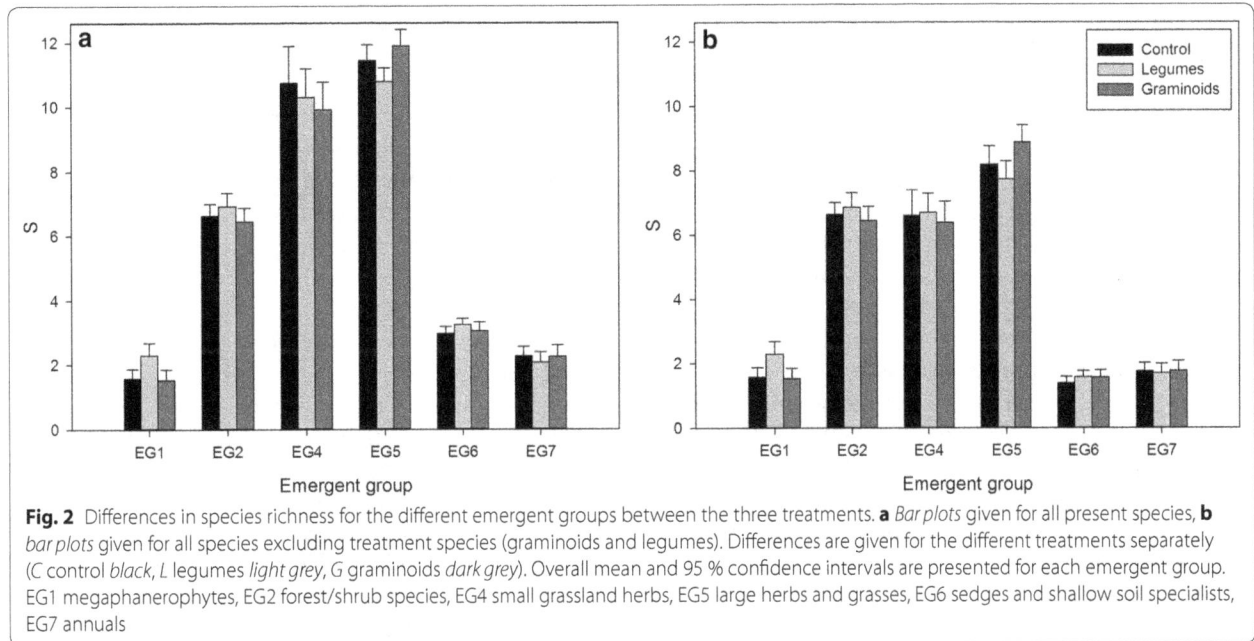

**Fig. 2** Differences in species richness for the different emergent groups between the three treatments. **a** *Bar plots* given for all present species, **b** *bar plots* given for all species excluding treatment species (graminoids and legumes). Differences are given for the different treatments separately (*C* control *black*, *L* legumes *light grey*, *G* graminoids *dark grey*). Overall mean and 95 % confidence intervals are presented for each emergent group. EG1 megaphanerophytes, EG2 forest/shrub species, EG4 small grassland herbs, EG5 large herbs and grasses, EG6 sedges and shallow soil specialists, EG7 annuals

**Table 4 Parameter estimates of the performed linear mixed models for species replacement**

| | Treatment | | | | |
|---|---|---|---|---|---|
| | F | $R^2_\beta$ | β C | β L | β G |
| G $^aR_{rel}$ (cover) | 0.12 | 0.003 | 0.36 | 0.40 | 0.38 |
| G $R_{rel}$ (presabs) | 0.25 | 0.006 | 0.41 | 0.44 | 0.39 |
| L $^aR_{rel}$ (cover) | 0.01 | 0.001 | 0.17 | 0.16 | 0.17 |
| L $R_{rel}$ (presabs) | 0.25 | 0.006 | 0.17 | 0.13 | 0.17 |

Beta-coefficient, test statistic and semi-partial $R^2_\beta$ given for treatment

*G* graminoids, *L* legumes, $^aR_{rel}$ relativized abundance replacement, $R_{rel}$ relativized species replacement, *C* control treatment, *presabs* presence–absence

* 0.05 ≥ *P* > 0.01; ** 0.01 ≥ *P* > 0.001; *** 0.001 ≥ *P*

previously observed through a chronosequence approach in similar dry semi-natural grasslands [39]. At the functional trait level, the assembly patterns also partly confirmed the previous results of Helsen et al. [39]. However, opposite to the chronosequence study, the number of forest/shrub species showed a small increase, and annuals a strong decrease in species numbers through time in this study. Large herbs and grasses were found to remain relatively constant through time in this study, while a decrease in richness was found for these groups in the chronosequence study [39]. These differences might be caused by the smaller time scale in this study.

### Species removal effects on non-treatment species

The initial treatments resulted in changes in the functional trait set of the grassland communities in the first years following manipulation, with, for the G treatment, a reduction of total cover, graminoid species richness and associated cover of the emergent groups that mainly consist of graminoid species. In the L treatment, manipulation initially resulted in a reduction of N fixating species, but had no significant effect on total cover.

Interestingly, species richness was not affected by treatment, with similar levels of both specialist and generalist species across treatments. This is in accordance with other priority effect experiments, where species richness was found to converge, independent of initial differences in species richness and treatment [10, 32]. Both the G and L treatments nevertheless affected species composition. More specifically, the removal of graminoids resulted in a small, but nonetheless significant increase in the number of species in EG 5 (large herbs and grasses). This is in accordance with our (inhibitory) niche preemption hypothesis, with species with similar niches as the removed graminoids benefitting from the treatment [26, 28]. In other words, the removal of large competitive grasses resulted in a small increase of large competitive herbs, likely through colonization. The absence of a similar pattern for EGs 4 and 6, which also contain many graminoids, is likely caused by the fast recovery of graminoids species in these grasslands, well before other species can colonize due to reduced within-niche competition (see the 'species removal effects on treatment species' discussion section further). Alternatively, it could be argued that the inhibitory competitive priority effects are less pronounced within these EGs, which are characteristic of high stress-low competition communities. It has indeed been suggested that the strength of direct (inhibitory niche preemption) priority effects are dependent on soil nutrient levels, implying that priority effects in experimental studies (using optimal nutrient concentrations) are likely much stronger than those occurring in natural (nutrient poor) communities [10, 36]. Since EG 5 mainly consists of relatively competitive species, this could explain the stronger effect of species removal for this specific group. Indeed, all else being equal, direct inhibitory niche preemption priority effects are expected to be more pronounced for competitive species that produce much biomass [24, 36]. Since soil legacies are strongly species-specific, we also cannot exclude the possibility of differential facilitative or inhibitory soil legacy effects of graminoid species, indirectly promoting the establishment of species of EG 5 [21, 22].

In accordance with our hypothesis, legume removal did not affect the species richness of EG 4 (or 7), which contain all legume species, but resulted in changes in unrelated functional groups (decrease of the number of large herbs and grasses, and an increase of megaphanerophytes). This suggests the occurrence of niche modification effects of legumes after their removal. As discussed earlier, the low overall competitive abilities of the legumes present in these grasslands (Additional file 2) likely explain why competitive exclusion (niche preemption) is limited within this functional group. Previous research has also shown that legumes do often not exert persistent inhibitory priority effects through size-asymmetric competition, and often facilitate higher biomass production of functionally different co-occurring plant species through nitrogen enrichment of the soil (facilitation) [19, 49]. This facilitative niche modification effect can be especially effective for plants growing in nutrient poor grasslands, as is the case in this study. The removal of legumes in the L treatment likely resulted in open patches with increased nitrogen availability. Megaphanoreophyte seedlings seem to be better at establishing at these former legume sites, suggesting facilitative soil legacies through nitrogen enrichment [21]. In this scenario, the observed decrease in species of EG 5 might be partly caused by the decreased competitive success of large (herbs and grasses) against megaphanerophytes. Alternatively, the absence of legumes might have resulted in a lower overall availability of nitrogen in the community, an effect that will most strongly inhibit the growth of species that are not adapted to nutrient poor conditions, such as those of EG 5. Indeed, this EG mainly contains generalist species adapted to fast growth and relatively nutrient rich soils (Table 1; [39]). The positive effect on megaphanerophytes on the other hand might then suggest that legumes have a negative effect on tree and shrub seedlings through inhibitory niche modification effects, independent of their effect on nitrogen availability.

**Species removal effects on treatment species**

In this experiment, both graminoid and legume removal was after 4 years almost completely compensated by the recolonization of graminoids and legumes, respectively, strongly suggesting that niche processes shape community assembly and priority effects in certain semi-natural grassland systems, as previously argued by Helsen et al. [31]. Although this was largely expected for graminoids, we did not expect similar patterns to occur for legumes. More surprisingly, the species replacement rates among the treatments show that this recolonization is effectuated by largely the same set of species as those that were removed. This suggests that niche preemption through size-asymmetric competition is likely only partly driving these patterns, since we would have expected some levels of species replacement (within functional groups) in this case. Likely, the observed patterns are also partly driven by localized soil legacies that promote the colonization of the same species (facilitative), or prevent the colonization of other species (inhibitory legacy effects acting within a functional group). Although some studies demonstrated that within species plant-soil feedbacks can be inhibitory [21, 35], other studies have indeed shown that many species exhibit weaker inhibitory, or even facilitative plant-soil feedbacks upon conspecifics compared to plant-soil feedbacks upon other species [23, 34].

The observed patterns can, however, also be at least partly explained by other confounding factors. Since (dead) belowground biomass of the treatment species was not actively removed, possible priority effects of these species might have been much stronger than would have been the case after complete removal of the species. Indeed, inhibitory size-asymmetric competitive priority effects are not solely driven by aboveground biomass, but can also remain strong when aboveground biomass is periodically removed through mowing [19]. Furthermore, since all treatments were performed in relatively small plots within a larger grassland, the removed species are also present in the direct vicinity of the treatment plot, enhancing the chances of recolonization of the plot by the same species set, thus deflating replacement rates. This effect might have been especially strong for legumes, since only a relatively small number of species was present in these grasslands. Most of the graminoids were furthermore strongly clonal (Fig. 1e, f), also allowing quick clonal recolonization of the plot by ramets present at the vicinity of the plot border. In conclusion, we believe that soil legacies likely resulted in reduced levels of species replacement, but that this affect was likely not as strong as suggest by the species replacement results.

### Treatment effects through time

Although changes in the EG compositions across the different treatments persisted after 4 years (no significant interaction between time and treatment), we did also observe a fast recovery of the number and composition of both legumes and graminoids during the same time span. Contrary to our predictions, this suggests that the effect of specific functional group removal during grassland assembly does not result in alternative assembly pathways, through newly enforced priority effects of the secondary colonized species. These results more likely suggest that soil legacies result in, at least partial, maintenance of initial priority effects after species removal. This, in turn, allows the fast recolonization of the removed species, with only limited changes in overall species composition. These results are partly in agreement with the study of Plückers et al. [37], where initial differences in species richness and functional composition (forbs, grasses and legumes) through differential seeding, became very small after 4 years, with communities seemingly converging toward similar species richness and functional composition.

## Conclusions

In this study, we explored how priority effects within and among functional groups affect community assembly during natural plant community assembly. More specifically, our results show that, in a low nutrient, (semi-)natural grassland system, inhibitory priority effects acting through niche preemption can be slightly relieved by species removal. However, this relief depended on the competitive ability of the removed species, with relief only observed for more competitive grasses and herbs, but not for smaller grassland specialists. Although competition effects acting within functional groups (niche preemption) were observed for graminoids, they do not seem to apply to legumes. Indeed, the removal of legumes mainly affected functionally unrelated generalist species and megaphanerophytes, likely through the facilitative niche modification effects of nitrogen fixation after legume removal [14].

On the other hand, species removal was, contrary to our expectations, almost completely compensated by recolonization of the same species set, suggesting that the net community composition effects of species (group) removal is rather limited in this natural system. This additionally suggests that soil legacies are, at least up to a certain extent, important drivers of assembly patterns during natural grassland assembly. We can expect that, in the context of ecological restoration, if unwanted species get established, it might prove difficult to neutralize their effect on the community assembly outcome, since their imposed priority effects might act long after their removal through imposed soil legacies [16].

## Additional files

**Additional file 1.** Overview of the selected traits, used for the delineation of the emergent groups. Description, scale and main data sources are given for every trait.

**Additional file 2.** Overview of the species list. Species are defined as generalist (g) of specialist (s), and subdivided in seven emergent groups (EGs). Group numbers correspond to the emergent groups described in Table 1. The treatment column indicates what species are graminoids (G) and legumes (L).

### Abbreviations

$^{a}R_{rel}$: relativized abundance replacement; C: control treatment; CWM: community weighted mean; E: Pielou's evenness; EG: emergent group; G: graminoids treatment; L: legumes treatment; LMM: linear mixed model; REML: restricted maximum likelihood; RMLMM: repeated measures linear mixed model; $R_{rel}$: relativized species replacement; S: species richness.

### Authors' contributions

KH performed fieldwork and statistical analyses and drafted the manuscript. OH and MH participated in the design of the study and helped to draft the manuscript. All authors read and approved the final manuscript.

### Author details

[1] Plant Conservation and Population Biology, Department of Biology, University of Leuven, Arenbergpark 31, 3001 Heverlee, Belgium. [2] Department of Biology, Norwegian University of Science and Technology, Høgskoleringen 5, 7034 Trondheim, Norway. [3] Division Forest, Nature and Landscape Research, Department Earth and Environmental Sciences, University of Leuven, Celestijnenlaan 200E, 3001 Heverlee, Belgium.

### Acknowledgements

Thanks goes out to Stijn Cornelis, Pieter Gijbels, Krista Takkis and Kasper Van Acker for assistance in the field.

### Competing interests

The authors declare that they have no competing interests.

### Funding

This research was funded by the Flemish Fund for Scientific Research (FWO). However, the manuscript was written while K.H. held a research grant of the University of Leuven (PDMK/13/093).

### References

1. Clements FE. Plant succession: an analysis of the development of vegetation. Carnegie Washington: Institution of Washington; 1916.
2. Chase JM. Community assembly: when should history matter? Oecologia. 2003;136:489–98.
3. Brudvig LA. The restoration of biodiversity: where has research been and where does it need to go? Am J Bot. 2011;98:549–58.
4. McCune B, Allen TFH. Will similar forests develop on similar sites? Can J Bot. 1985;63:367–76.
5. Fastie CL. Causes and ecosystem consequences of multiple pathways of primary succession at Glacier Bay, Alaska. Ecology. 1995;76:1899–916.
6. Honnay O, Verhaeghe W, Hermy M. Plant community assembly along dendritic networks of small forest streams. Ecology. 2001;82:1691–702.

7. Gleason HA. Further views on the succession-concept. Ecology. 1927;8:299–326.
8. Diamond JM. Assembly of species communities. In: Cody ML, Diamond JM, editors. Ecology and evolution of communities. Cambridge: Harvard University Press; 1975.
9. Young TP, Chase JM, Huddleston RT. Community succession and assembly Comparing, contrasting and combining paradigms in the context of ecological restoration. Ecol Restor. 2001;19:5–18.
10. Ejrnæs R, Bruun HH, Graae BJ. Community assembly in experimental grasslands: suitable environment or timely arrival? Ecology. 2006;87:1225–33.
11. Baeten L, Hermy M, Verheyen K. Environmental limitation contributes to the differential colonization capacity of two forest herbs. J Veg Sci. 2009;20:209–23.
12. Van der Putten WH, Bardgett RD, Bever JD, Bezemer TM, Casper BB, Fukami T, et al. Plant-soil feedbacks: the past, the present and future challenges. J Ecol. 2013;101:265–76.
13. Fukami T, Nakajima M. Complex plant-soil interactions enhance plant species diversity by delaying community convergence. J Ecol. 2013;101:316–24.
14. Fukami T. Historical contingency in community assembly: integrating niches, species pools, and priority effects. Annu Rev Ecol Evol Syst. 2015;46:1–23.
15. Weiner J. Asymmetric competition in plant populations. Trends Ecol Evol. 1990;5:360–4.
16. Grman E, Suding KN. Within-year soil legacies contribute to strong priority effects of exotics on native California grassland communities. Restor Ecol. 2010;18:664–70.
17. Cuddington K. Legacy effects: the persistent impact of ecological interactions. Biol Theory. 2011;6:203–10.
18. Hubbell SP. The unified neutral theory of biodiversity and biogeography. Princeton: Princeton University Press; 2001.
19. Körner C, Stöcklin J, Reuther-Thiébaud L, Pelaez-Riedl S. Small differences in arrival time influence composition and productivity of plant communities. New Phytol. 2008;177:698–705.
20. Collinge SK, Ray C. Transient patterns in the assembly of vernal pool plant communities. Ecology. 2009;90:3313–23.
21. van de Voorde TFJ, van der Putten WH, Bezemer MT. Intra- and interspecific plant-soil interactions, soil legacies and priority effects during old-field succession. J Ecol. 2011;99:945–53.
22. Hawkes CV, Kivlin SN, Du J, Eviner VT. The temporal development and additivity of plant-soil feedback in perennial grasses. Plant Soil. 2013;369:141–50.
23. Baxendale C, Orwin KH, Poly F, Pommier T, Bardgett RD. Are plant-soil feedback responses explained by plant traits? New Phytol. 2014;204:408–23.
24. Cleland EE, Esch E, Mckinney J. Priority effects vary with species identity and origin in an experiment varying the timing of seed arrival. Oikos. 2015;124:33–40.
25. Tan J, Pu Z, Ryberg WA, Jiang L. Species phylogenetic relatedness, priority effects, and ecosystem functioning. Ecology. 2012;93:1164–72.
26. Vannette RL, Fukami T. Historical contingency in species interactions: towards niche-based predictions. Ecol Lett. 2014;17:115–24.
27. Fargione J, Brown CD, Tilman D. Community assembly and invasion: an experimental test of neutral versus niche processes. Proc Natl Acad Sci USA. 2003;100:8916–20.
28. Mwangi PN, Schmitz M, Scherber C, Roscher C, Schumacher J, Scherer-Lorenzen M, et al. Niche pre-emption increases with species richness in experimental plant communities. J Ecol. 2007;95:65–78.
29. Price JN, Pärtel M. Can limiting similarity increase invasion resistance? A meta-analysis of experimental studies. Oikos. 2013;122:649–56.
30. Fukami T, Bezemer MT, Mortimer SR, Putten WH. Species divergence and trait convergence in experimental plant community assembly. Ecol Lett. 2005;8:1283–90.
31. Helsen K, Hermy M, Honnay O. Trait but not species convergence during plant community assembly in restored semi-natural grasslands. Oikos. 2012;121:2121–30.
32. Roscher C, Schumacher J, Gerighausen U, Schmid B. Different assembly processes drive shifts in species and functional composition in experimental grasslands varying in sown diversity and community history. PLoS One. 2014;9:e101928.
33. Grime JP. Trait convergence and trait divergence in herbaceous plant communities: mechanisms and consequences. J Veg Sci. 2006;17:255–60.
34. Kardol P, Cornips NJ, van Kempen MML, Bakx-Schotman JMT, van der Putten WH. Microbe-mediated plant–soil feedback causes historical contingency effects in plant community assembly. Ecol Monogr. 2007;77:147–62.
35. Harrison KA, Bardgett RD. Influence of plant species and soil conditions on plant-soil feedback in mixed grassland communities. J Ecol. 2010;98:384–95.
36. Kardol P, Souza L, Classen AT. Resource availability mediates the importance of priority effects in plant community assembly and ecosystem function. Oikos. 2013;122:84–94.
37. Plückers C, Rascher U, Scharr H, Von Gillhaussen P, Beierkuhnlein C, Temperton VM. Sowing different mixtures in dry acidic grassland produced priority effects of varying strength. Acta Oecol. 2013;53:110–6.
38. Helsen K, Hermy M, Honnay O. Data from: a test of priority effect persistence in semi-natural grasslands through the removal of plant functional groups during community assembly. Dryad Digit Repos. 2016. doi:10.5061/dryad.7s5s4.
39. Helsen K, Hermy M, Honnay O. Spatial isolation slows down directional plant functional group assembly in restored semi-natural grasslands. J Appl Ecol. 2013;50:404–13.
40. Weiher E, Werf A, Thompson K, Roderick M, Garnier E, Eriksson O. Challenging Theophrastus: a common core list of plant traits for functional ecology. J Veg Sci. 1999;10:609–20.
41. Díaz S, Lavorel S, de Bello F, Quétier F, Grigulis K, Robson TM. Incorporating plant functional diversity effects in ecosystem service assessments. Proc Natl Acad Sci USA. 2007;104:20684–9.
42. Fitter A, Peat H. The ecological flora database. J Ecol. 1994;82:415–25.
43. Klotz S, Kühn I, Durka W. BIOLFLOR—Eine Datenbank zu biologisch-ökologischen Merkmalen der Gefäßpflanzen in Deutschland. Bonn: Bundesamt für Naturschutz, Schriftenreihe für Vegetationskunde 38; 2002.
44. Lambinon J, De Langhe J, Delvosalle L, Duvigneaud J. Flora van België, het Groothertogdom Luxemburg, Noord-Frankrijk en de aangrenzende gebieden (Pteridofyten en Spermatofyten). Meise: Nationale plantentuin van België; 1998.
45. van Landuyt W, Hoste I, Vanhecke L, Van den Bremt P, Vercruysse W, De Beer D. Atlas van de flora van Vlaanderen en het Brussels Gewest. Brussel: Instituut voor natuur- en bosonderzoek, Nationale Plantentuin van België & Flo.Wer; 2006.
46. Podani J, Schmera D. A new conceptual and methodological framework for exploring and explaining pattern in presence—absence data. Oikos. 2011;120:1625–38.
47. Podani J, Ricotta C, Schmera D. A general framework for analyzing beta diversity, nestedness and related community-level phenomena based on abundance data. Ecol Complex. 2013;15:52–61.
48. Edwards LJ, Muller KE, Wolfinger RD, Qaqish BF, Schabenberger O. An $R^2$ statistic for fixed effects in the linear mixed model. Stat Med. 2008;27:6137–57.
49. Von Gillhaussen P, Rascher U, Jablonowski ND, Plückers C, Beierkuhnlein C, Temperton VM. Priority effects of time of arrival of plant functional groups override sowing interval or density effects: a grassland experiment. PLoS One. 2014;9:e86906.

9

# Genetic diversity of calcareous grassland plant species depends on historical landscape configuration

Christoph Reisch[1*], Sonja Schmidkonz[1], Katrin Meier[1], Quirin Schöpplein[1], Carina Meyer[1], Christian Hums[1], Christina Putz[1] and Christoph Schmid[2]

**Abstract**

**Background:** Habitat fragmentation is considered to be a main reason for decreasing genetic diversity of plant species. However, the results of many fragmentation studies are inconsistent. This may be due to the influence of habitat conditions, having an indirect effect on genetic variation via reproduction. Consequently we took a comparative approach to analyse the impact of habitat fragmentation and habitat conditions on the genetic diversity of calcareous grassland species in this study. We selected five typical grassland species (*Primula veris*, *Dianthus carthusianorum*, *Medicago falcata*, *Polygala comosa* and *Salvia pratensis*) occurring in 18 fragments of calcareous grasslands in south eastern Germany. We sampled 1286 individuals in 87 populations and analysed genetic diversity using amplified fragment length polymorphisms. Additionally, we collected data concerning habitat fragmentation (historical and present landscape structure) and habitat conditions (vegetation structure, soil conditions) of the selected study sites. The whole data set was analysed using Bayesian multiple regressions.

**Results:** Our investigation indicated a habitat loss of nearly 80% and increasing isolation between grasslands since 1830. Bayesian analysis revealed a significant impact of the historical landscape structure, whereas habitat conditions played no important role for the present-day genetic variation of the studied plant species.

**Conclusions:** Our study indicates that the historical landscape structure may be more important for genetic diversity than present habitat conditions. Populations persisting in abandoned grassland fragments may contribute significantly to the species' variability even under deteriorating habitat conditions. Therefore, these populations should be included in approaches to preserve the genetic variation of calcareous grassland species.

**Keywords:** AFLP, Dry grasslands, Habitat fragmentation, Genetic diversity, Grazing, Land use, Litter, Soil analysis

## Background

Calcareous grasslands are important hotspots of plant species diversity in central Europe. They contain many rare and endangered plant species and are of strong conservation interest [1–3]. However, due to land use changes, calcareous grasslands declined significantly in Europe during the last 150 years [4]. Agricultural intensification, increased fertilization and afforestation caused a drastic loss of grasslands. In some regions, up to 90% of the grasslands disappeared [5]. Today calcareous grasslands are thus often highly fragmented, with the area of the grassland patches continuously decreasing while their spatial isolation increases [6]. This process of habitat fragmentation is a general threat to biodiversity, reducing species richness within small and isolated habitat patches [7].

However, fragmentation has not only an impact on biodiversity at the species level, but also on the genetic diversity, due to geographic isolation of smaller populations [8]. In particular formerly widespread species are more susceptible to the effects of fragmentation than naturally rare and isolated species [9]. Genetic variation

*Correspondence: christoph.reisch@biologie.uni-regensburg.de
[1] Institute of Plant Sciences, University of Regensburg, 93040 Regensburg, Germany
Full list of author information is available at the end of the article

is directly related to population size [10] and often decreases during the process of fragmentation. Moreover, the exchange of pollen and seeds between populations is impeded by fragmentation, which decreases gene flow [11] and increases genetic drift. This results in a loss of genetic diversity and may lead to reduced generative [12] and vegetative performance [13]. As a consequence, there is an increase in the susceptibility to pathogens and herbivores in the short term [14] and the probability of extinction in the long term [8]. Finally, the loss of genetic diversity may explain the observed loss of species diversity in numerous studies analysing fragmentation effects [6].

However, the results of many studies dealing with the impact of fragmentation on genetic diversity are inconsistent. Some of these studies support the assumptions derived from the theory of island biogeography [15], whereas others do not [16]. This dilemma has recently been illustrated by a review of 259 fragmentation studies, which concluded that the broad generalisations on the effects of fragmentation are problematic [17].

One of the most important challenges of fragmentation studies is that habitat conditions may differ between remnant grassland patches [17]. Many calcareous grasslands have been abandoned because sheep grazing is no longer economical [1, 2]. In the absence of grazing, habitat conditions continuously deteriorate and typical open short-grass conditions get lost [18]. This process is enhanced by the spill-over of fertiliser from adjacent agricultural areas [18]. Absence of grazing and increased nitrogen deposition change the vegetation structure of calcareous grasslands [19], resulting in a decrease in species richness [20] and shifts in species composition [21]. Due to the lack of biomass removal and increasing productivity litter accumulates while gaps of open soil, which are indispensable for the germination of many dry grassland species, become rare [22]. Furthermore, litter acts as a trap, further reducing the number of germinating seeds [23]. Particularly for species which require light to germinate, ground shadowing caused by increasing vegetation height, dominance of grasses and litter accumulation leads to the regression those of species [24]. Therefore, successful reproduction is impeded when habitat conditions deteriorate, which may subsequently affect genetic diversity. This process is intensified by the loss of suitable dispersal vectors, such as migrating sheep, which also reduces the exchange of seeds [2] and subsequently, the immigration of genetically deviant individuals.

Hence, genetic diversity may be affected by the interfering effects of both fragmentation and habitat conditions, which might explain the inconsistent results of many fragmentation studies. Moreover, due to different biological traits, patterns of genetic variation differ between plant species [25] and these may react completely different to fragmentation, again explaining unclear results of many genetic fragmentation surveys.

In this study, we analysed the impact of fragmentation on genetic diversity of several calcareous grassland species in a comparative approach, including both habitat fragmentation and habitat conditions. We used the same analytical approach for all species and applied Bayesian multiple regressions, which enables a detailed interpretation of the data, while being more flexibly adaptable to the data structure than traditional frequentist methods. More specifically we ask the following questions: (i) has fragmentation of calcareous grasslands a significant impact on genetic diversity of plant populations? (ii) is historical landscape configuration more important for genetic diversity than present landscape configuration? (ii) is genetic diversity of grassland plant populations affected by habitat conditions?

## Methods

### Study sites, habitat fragmentation and habitat conditions

For our study, we randomly selected 18 remnant calcareous grasslands in the valleys of the rivers Naab and Laber on the Franconian Alb in south eastern Germany near Regensburg (Fig. 1). Within the study region, calcareous grasslands have been subjected to fragmentation due to afforestation, intensification and abandonment since the nineteenth century. The study sites and all other calcareous grasslands occurring within a radius of 3 km around these sites were vectorised using a Geographic Information System (Arc Info 10.0, Esri) based upon corrected aerial photos (orthophotos) from August to October 2013 to study the effects of this fragmentation process. Vectorised data was used to calculate the current area ($HA_{2013}$) and perimeter (P) of each grassland fragment as well as the distance ($D_{2013}$) to the nearest calcareous grassland within the 3 km radius. The shape of the study sites was characterised by the ratio of habitat area to perimeter ($HA/P_{2013}$), which was small for narrow and elongated grasslands and large for round and compact grasslands.

Using historical cadastral maps, which were available from local land surveying offices, we determined the area of the study sites ($HA_{1830}$) and the distance ($D_{1830}$) to the nearest calcareous grassland within the 3 km radius in 1830. The maps include a detailed legend, which allows the identification of calcareous grasslands. We compared the area covered by calcareous grasslands in 1830 and 2013 and calculated the habitat loss (HL) within the 3 km radius around each of our study sites as a percentage of the grassland area lost since 1830. Additionally, connectivity of each fragment to all other fragments within the 3 km radius in 2013 and 1830 was calculated according to Hanski [26] as $Si = \sum_{j \neq i} \exp\left(-\alpha d_{ij}\right) A_j$ where $Si$ is

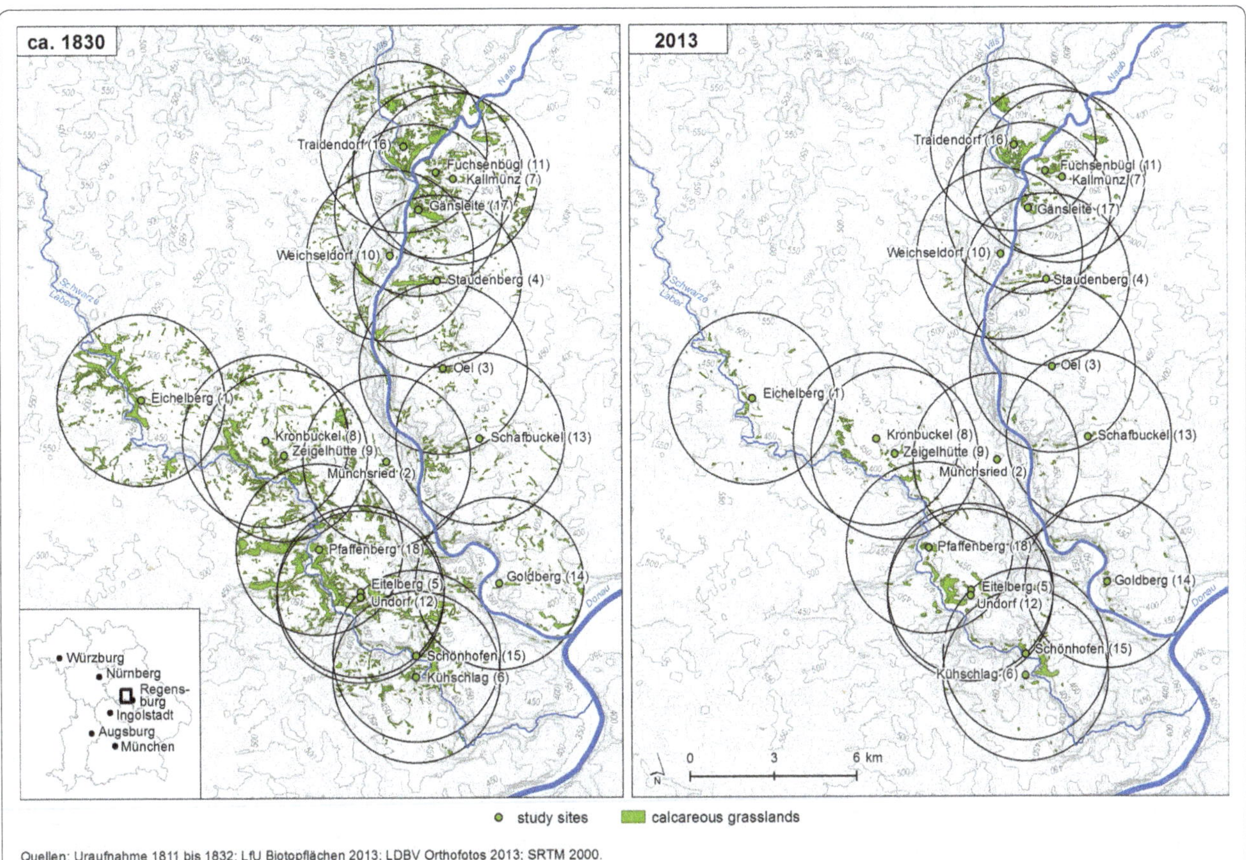


**Fig. 1** Geographic location of the 18 selected study sites (*labelled dots*) in the valleys of Naab and Laber on the Franconian Alb in south eastern Germany near Regensburg and all other calcareous grasslands (*grey areas*) within a radius of 3 km around the study sites in 1830 and 2013

the connectivity of the patch $i$, $d_{ij}$ is the distance between patches $i$ and $j$, and $Aj$ is the area of the patch $j$ [27]. Based upon the historical maps, grasslands were classified for further analyses as historically old, when they were already grassland in 1830, and historically young grasslands, when they originated after 1830.

The selected grasslands date back at least to the period of the Roman Empire [28] and have been grazed frequently until the 1960s, as have most other grasslands in central Europe [4]. Today they are abandoned or infrequently grazed. However, detailed information about the grazing history since the 1960s is not available. In 2014, we established ten study plots with the size of 2 × 2 m at each of the selected grasslands to analyse the impact of the habitat conditions on genetic diversity of our study species. In each plot we determined the vegetation height (VH) as well as the cover of grass (CG), litter (CL) and bare soil (BS). Furthermore, we took five soil samples at each study site with a core sampler, which were then pooled, in order to analyse the nutrient content of the soil. Pooled samples were dried in a heating cabinet at 50 °C for several days, cleaned by sieving with 2 mm mesh size and then stored at 4 °C until they were subjected to a soil chemical analysis following the procedures described by Bassler et al. [29]. We determined the phosphorous (P) and potassium (K) content, as well as the carbon to nitrogen ratio (C/N) as described previously [30].

### Study species and genetic variation

For our study we selected five typical and widespread calcareous grassland species (*Primula veris* L., *Dianthus carthusianorum* L., *Medicago falcata* (L.) Arc., *Polygala comosa* Schkuhr and *Salvia pratensis* L.), frequently occurring in calcareous grasslands in south eastern Germany. In the field we assessed population size (NI) by counting the number of present individuals (Table 1) at each site. For the analysis of genetic variation with a few exceptions (at three sites *P. comosa* could not be sampled), leaf material of 15 individuals per population and species was collected (Table 5). In total 1286 individuals from 87 populations were analysed. Leaf material was placed in plastic bags in the field and stored in a lab freezer at −20 °C until molecular analysis.

**Table 1 Size of the studied populations**

| St. | Name | *P.v.* | *D.c.* | *M.f.* | *P.c.* | *S.p.* |
|---|---|---|---|---|---|---|
| 01 | Eichelberg | 138 | 24 | 238 | 138 | 433 |
| 02 | Münchsried | 782 | 266 | 67 | 782 | 458 |
| 03 | Oel | 110 | 196 | 110 | 110 | 1057 |
| 04 | Staudenberg | 30 | 13 | 23 | 30 | 41 |
| 05 | Eitelberg | 145 | 302 | 415 | 145 | 287 |
| 06 | Kühschlag | 429 | 265 | 57 | 429 | 844 |
| 07 | Kallmünz | 120 | 741 | 351 | 120 | 487 |
| 08 | Kronbuckel | 1288 | 146 | 166 | 1288 | 1251 |
| 09 | Ziegelhütte | 1173 | 972 | 246 | 1173 | 1565 |
| 10 | Weichseldorf | 421 | 1790 | 905 | 421 | 740 |
| 11 | Fuchsenbügl | 61 | 7115 | 504 | 61 | 1140 |
| 12 | Undorf | 630 | 69 | 1348 | 630 | 735 |
| 13 | Schafbuckel | 7690 | 2757 | 784 | 7690 | 2312 |
| 14 | Goldberg | 143 | 2391 | 1809 | 143 | 3319 |
| 15 | Schönhofen | 380 | 368 | 702 | 380 | 3735 |
| 16 | Traidendorf | 2694 | 4097 | 401 | 2694 | 8410 |
| 17 | Gänsleite | 2707 | 4310 | 1174 | 2707 | 18,760 |
| 18 | Pfaffenberg | 12,938 | 8125 | 9457 | 12,938 | 53,585 |

Population size of *Primula veris* (*P.v.*), *Dianthus carthusianorum* (*D.c.*), *Medicago falcata* (*M.f.*), *Polygala comosa* (*P.c.*) and *Salvia pratensis* (*S.p.*) at the study sites determined as the number of occurring individuals

Genomic DNA was isolated from dry leaf material using the CTAB-based method [31] as described before [32]. Concentrations of the DNA extracts were measured photometrically. DNA solutions were diluted with water to 7.8 ng/μL and used for the analysis of Amplified Fragment Length Polymorphisms (AFLPs), which were conducted concordant with the protocol from Beckmann Coulter as described previously [33, 34].

DNA adapters were prepared by adding equal volumes of both single strands of *Eco*RI (4 μM) and *Mse*I (40 μM) adaptors (MWG Biotech), following a 5 min heating at 95 °C with a final 10 min step at 25 °C. DNA restriction and adapter ligation were performed in one step by adding a 3.6 μL mixture per reaction containing 2.5 U *Eco*RI (MBI Fermentas), 2.5 U *Mse*I (MWG Biotech), 0.1 μM *Eco*RI and 1 μM *Mse*I adapter pair, 0.5 U T4 Ligase with 0.1× of its corresponding buffer (MBI Fermentas), 0.05 M NaCl and 0.5 μg BSA (New England BioLabs) to 6.4 μL of genomic DNA in a concentration of 7.8 ng/μL. Following an incubation at 37 °C for 2 h with a final enzyme denaturation step at 70 °C for 15 min, the restriction-ligation products were diluted tenfold with 1× TE buffer for DNA (20 mM Tris–HCl, pH 8.0; 0.1 mM EDTA, pH 8.0).

For preselective DNA amplification, 1 μL diluted DNA restriction-ligation product, 0.25 μM preselective *Eco*RI and *Mse*I primers (MWG Biotech) were added to an AFLP Core Mix (PeqLab, Germany) containing 1x Buffer S, 0.4 mM dNTPs and 1.25 U/μL Taq-Polymerase. In a 5 μL reaction volume PCR was performed on at 94 °C for 2 min then 30 cycles of 20 s denaturation at 94 °C, 30 s annealing at 56 °C and 2 min elongation at 72 °C, a final 2 min 72 °C and 30 min 60 °C step for complete extension ending with a final cool down to 4 °C. After PCR, products were diluted 20 fold with 1× TE buffer for DNA.

After an extensive screening of 30 primer combinations, three primer combinations per species were chosen for a subsequent selective PCR reaction. For detection, *Eco*RI primers labelled with different fluorescent dyes (Beckman Coulter) were used (Table 2). Selective PCR was carried out in a total reaction volume of 5 μL containing an AFLP Core Mix (1× Buffer S, 0.4 mM dNTP's, 1.25 U/μL Taq-Polymerase, PeqLab, Germany), 0.05 μM selective *Eco*RI (Proligo, France), 0.25 μM *Mse*I (MWG Biotech) primers and 0.75 μL diluted preselecive

**Table 2 Selective primer pairs used for AFLP analysis of the study species**

| Species | D2 | D3 | D4 |
|---|---|---|---|
| *Primula veris* | CAA-AAC | CAA-ACG | CAG-ACA |
| *Dianthus carthusianorum* | CTC-AGC | CAA-AAG | CTG-ACT |
| *Medicago falcata* | CAC-ACC | CTA-ACG | CTT-ACA |
| *Polygala comosa* | CAA-AAC | CAT-ACG | CTA-ACA |
| *Salvia pratensis* | CTT-AGC | CTA-AGG | CTT-ACA |

amplification product. For detection, *Eco*RI primers labelled with different fluorescent dyes (D2, D3, D4) were used. PCR parameters used were treated 2 min at 94 °C, 10 cycles 20 s denaturation at 94 °C, annealing 30 s at 66 °C and 2 min elongation at 72 °C, where annealing temperature was reduced every subsequent step by 1 °C, additional 25 cycles of 20 s denaturation at 94 °C, 30 s annealing at 56 °C and 2 min elongation at 72 °C completed by a following 30 min step at 60 °C and a cool down to 4 °C.

Selective PCR products were diluted with 1× $TE_{0.1}$ buffer for AFLP and pooled. After pooling 5 μL of each selective PCR product of a given sample and adding them to a mixture of 2 μL sodium acetate (3 M, pH 5.2), 2 μL $Na_2EDTA$ (100 mM, pH 8) and 1 μL glycogen (20 mg/mL; Roche), DNA was precipitated in a 1.5 mL tube by adding 60 μL of 96% ethanol (−20 °C) and an immediate shaking. DNA was pelleted by 20 min centrifugation at 14,000*g* at 4 °C, the supernatant was poured off and the pellet was washed once by adding 200 μL 70% ethanol (−20 °C) and centrifugation at the latter conditions and was subsequently vacuum dried in a concentrator [33, 34].

After redissolving the pelleted DNA in a mixture of 24.8 μL Sample Loading Solution (SLS, Beckman Coulter) and 0.2 μL CEQ Size Standard 400 (Beckman Coulter), selective PCR products were separated by capillary gel electrophoresis on an automated sequencer (GeXP, Beckmann Coulter). Results were examined using the GeXP software (Beckman Coulter) and analysed using the software Bionumerics 4.6 (Applied Maths, Kortrijk, Belgium). From the computed gels, only those fragments that showed intense and articulate bands were taken into account for further analyses. Samples yielding no clear banding pattern or obviously representing PCR artefacts were repeated. Finally, 19 individuals were excluded from the analysis due to the lack of a clear banding pattern. Reproducibility of molecular analyses was investigated with 10% of all analysed samples by means of estimating the genotyping error rate [35], which was 3.8%.

From the AFLP bands, a binary (0/1) matrix was created for each species. Based upon this matrix, we calculated the genetic diversity of each population as Nei's Gene Diversity [36] using the program AFLP SURV [37].

### Bayesian multiple regressions

The impact of habitat fragmentation and habitat conditions on diversity was analysed using a robust hierarchical Bayesian multiple regression approach with regression parameters for the species level (not shown) and an overarching set of hyperparameters for the species-independent estimation of regression parameters. Predictor variables were grassland fragmentation parameters (fragment area, area/perimeter ratio as well as distance to the nearest calcareous grassland and habitat connectivity) and habitat condition parameters (vegetation height, cover of grass, litter and bare soil, contents of P, K and C/N ratio). This approach is equivalent, though not identical, to a linear mixed model with random slopes and random intercepts using species as a random effect. The overarching hyperparameters ensure a transfer of information between the species' parameters that shrinks outliers on species-level towards the main trend. The hyperparameters themselves are therefore suitable to depict the species-independent trends in the data.

There was no credible influence of habitat age detectable by the model and, hence, impeded accurate parameter estimation of the remaining parameters. Accordingly, habitat age as a parameter hampered model interpretation, and therefore was excluded from the final analysis. All predictor variables were checked for strong correlations (Pearson's correlation coefficient >0.8; observed maximum 0.7) to check for multicollinearity. Additionally, multicollinearity can be recognised in Bayesian models by extremely broad posterior distributions of correlated parameters. However, this was not the case in the presented analysis. Model stability was verified by re-running the analysis with and without the most strongly correlated predictors. Accordingly, habitat age as a parameter hampered model interpretation and was, therefore, excluded from the final analysis.

A Bayesian approach was chosen for being flexibly adjustable to the situation at hand, e.g. it can be easily modified to reduce false positives in parameter estimation or improved to accommodate outliers in the data. In the situation at hand, the applied model could easily be tailored to reflect the hierarchical structure of our data (Additional file 1). Furthermore, results from Bayesian models have a higher informative value than classical NHST methods as they provide full probability distributions on the estimated parameters. Modelling and interpretation were carried out using the software packages R 3.2.1 [38] and JAGS 3.2.0 for Markov Chain Monte Carlo (MCMC) sampling [39] as well as utility functions provided by Kruschke [40]. Errors were modelled as being t-distributed in order to accommodate outliers and conduct a robust regression. Regression parameters were regularised using mildly informed, double-exponential prior distributions with location parameter set to 0 and a fixed precision parameter set to 0.1, thereby reducing chances for false positive regression parameters. These settings, known as the Bayesian Lasso [41] avoid overfitting in complex models and reduce the overestimation of effects that can happen in AIC-based model selection procedures.

Sampling was carried out with four MCMC chains with 300,000 steps in total with thinning set to every 10th step, a burn-in period of 2000 steps and 1000 steps for adaption. All parameters were checked for chain convergence. Autocorrelation in the MCMC chains was assessed as the effective sample size (ESS) aiming at a lower limit of 10 k for the relevant parameters. Highest density intervals (HDIs) were computed for the regression coefficients to check if coefficients were credibly non-zero. The predictors' relative influences were assessed using standardised regression coefficients. A graph was produced by fixing all but the predictors of credible influence to their mean, resulting in a two dimensional scatter plot.

## Results

### Habitat fragmentation and habitat conditions

Our results indicated a strong decline of calcareous grasslands within the study region. The mean size of the selected grasslands patches decreased from 115,045 $m^2$ in 1830–14,881 $m^2$ in 2013 (Table 3). In contrast, the mean distance to the nearest grassland increased from 110 m in 1830–210 m in 2013. Confirming this observation, mean connectivity among grasslands decreased from 74.81 in 1830–27.90 in 2013. Mean loss of calcareous grasslands within the 3 km radius around each of our study sites was 78.62%.

Habitat conditions strongly differed between study sites. Vegetation height ranged from 0.51 to 1.18 m with an average of 0.93 m (Table 4). Large differences could also be observed for the cover of grass, which varied between 48.0 and 90.0% with a mean of 76.5%. The cover of litter ranged from 7.7 to 38.0% with a mean of 20.4%, whereas the proportion of bare soil was minimum 0% and maximum 5.5% with an average value of 0.7%.

The content of phosphorous also varied between sites and ranged from 8.04 to 53.76 mg/kg soil with a mean of 26.47 mg/kg soil (Table 4). Similarly, the content of potassium varied between 101.22 and 319.02 mg/kg soil. On average we observed a potassium content of 211.06 mg/kg soil. Finally, we determined the carbon to nitrogen ratio, which ranged from 10.9 to 42.0 with a mean of 19.6 (Table 4).

### Genetic variation and hierarchical Bayesian multiple regression

AFLP analysis resulted in 120 fragments for *Primula veris*, 148 fragments for *Dianthus carthusianorum*, 285 fragments for *Medicago falcata*, 166 fragments for *Polygala comosa* and 192 fragments for *Salvia pratensis*. The proportion of polymorphic bands per species was 88.3% (*P. veris*), 97.4% (*D. carthusianorum*), 97.8% (*M. falcata*), 98.7% (*P. comosa*) and 94.3% (*S. pratensis*). Genetic

**Table 3 Habitat fragmentation data**

| St. | Name | $HA_{2013}$ | $HA_{1830}$ | HA/P | $D_{2013}$ | $D_{1830}$ | $CO_{1830}$ | $CO_{2013}$ | HL |
|---|---|---|---|---|---|---|---|---|---|
| 01 | Eichelberg | 445 | 715,410 | 2.41 | 58 | 129 | 100.90 | 8.35 | 83.01 |
| 02 | Münchsried | 631 | 0 | 3.95 | 980 | 133 | 23.89 | 2.34 | 79.12 |
| 03 | Oel | 763 | 6725 | 3.44 | 391 | 81 | 9.76 | 1.96 | 61.88 |
| 04 | Staudenberg | 1020 | 0 | 5.31 | 97 | 70 | 37.16 | 11.65 | 67.04 |
| 05 | Eitelberg | 1399 | 0 | 8.56 | 117 | 97 | 130.80 | 44.65 | 84.25 |
| 06 | Kühschlag | 1440 | 3308 | 8.25 | 340 | 98 | 62.28 | 26.52 | 82.43 |
| 07 | Kallmünz | 1546 | 11,072 | 6.67 | 175 | 168 | 82.16 | 48.29 | 83.75 |
| 08 | Kronbuckel | 1695 | 1176 | 4.79 | 382 | 62 | 59.55 | 13.94 | 80.55 |
| 09 | Ziegelhütte | 2495 | 0 | 4.69 | 41 | 24 | 70.72 | 15.55 | 78.33 |
| 10 | Weichseldorf | 5659 | 37,519 | 12.26 | 290 | 273 | 43.21 | 9.24 | 71.19 |
| 11 | Fuchsenbügl | 6211 | 13,243 | 11.22 | 60 | 15 | 101.79 | 66.19 | 83.21 |
| 12 | Undorf | 8009 | 0 | 20.89 | 91 | 59 | 132.94 | 50.75 | 84.49 |
| 13 | Schafbuckel | 12,033 | 17,338 | 17.12 | 150 | 192 | 13.19 | 3.60 | 77.90 |
| 14 | Goldberg | 22,160 | 0 | 12.31 | 32 | 121 | 21.15 | 7.19 | 66.83 |
| 15 | Schönhofen | 21,894 | 58,015 | 23.84 | 211 | 251 | 78.99 | 32.96 | 81.99 |
| 16 | Traidendorf | 24,405 | 134,710 | 15.02 | 58 | 44 | 97.91 | 71.44 | 84.95 |
| 17 | Gänsleite | 64,984 | 440,768 | 23.17 | 222 | 63 | 97.09 | 48.36 | 78.88 |
| 18 | Pfaffenberg | 91,067 | 631,523 | 24.87 | 87 | 94 | 183.16 | 38.90 | 85.32 |
|  | Mean | 14,881 | 115,045 | 11.60 | 210 | 110 | 74.81 | 27.90 | 78.62 |
|  | SE | ±5828 | ±53,938 | ±1.79 | ±53 | ±17 | ±11.01 | ±5.38 | ±1.67 |

Area of the selected study sites in $m^2$ in 1830 and 2013 ($HA_{1830}$ and $HA_{2013}$), the area/perimeter ratio (HA/P) in 2013, the distance to the nearest calcareous grassland in meter ($D_{1830}$ and $D_{2013}$) and the connectivity of the grasslands ($CO_{1839}$ and $CO_{2013}$) within a radius of 3 km in 1830 and 2013, and the loss of calcareous grasslands within this radius since 1830 in % (HL)

**Table 4 Habitat condition data**

| St. | Name | VH | CG | CL | BS | P | K | C/N |
|---|---|---|---|---|---|---|---|---|
| 01 | Eichelberg | 1.18 | 92.8 | 23.0 | 0.0 | 14.70 | 369.53 | 18.3 |
| 02 | Münchsried | 0.95 | 62.5 | 19.0 | 0.3 | 15.66 | 101.22 | 13.7 |
| 03 | Oel | 0.94 | 90.0 | 24.0 | 0.0 | 36.48 | 232.81 | 15.6 |
| 04 | Staudenberg | 1.54 | 67.0 | 29.0 | 0.0 | 53.76 | 272.77 | 16.0 |
| 05 | Eitelberg | 1.08 | 88.2 | 16.6 | 0.0 | 14.18 | 130.26 | 22.6 |
| 06 | Kühschlag | 0.91 | 87.0 | 30.5 | 0.3 | 26.63 | 192.97 | 42.0 |
| 07 | Kallmünz | 0.77 | 82.5 | 15.0 | 5.5 | 12.70 | 195.62 | 17.6 |
| 08 | Kronbuckel | 1.13 | 88.8 | 10.3 | 0.4 | 23.85 | 220.63 | 16.6 |
| 09 | Ziegelhütte | 0.93 | 84.5 | 17.5 | 0.8 | 37.63 | 169.02 | 21.8 |
| 10 | Weichseldorf | 0.51 | 63.0 | 14.5 | 0.8 | 16.25 | 135.42 | 20.1 |
| 11 | Fuchsenbügl | 1.15 | 74.5 | 19.0 | 0.1 | 31.92 | 249.64 | 19.0 |
| 12 | Undorf | 1.13 | 90.3 | 25.5 | 0.1 | 41.19 | 173.90 | 37.4 |
| 13 | Schafbuckel | 1.01 | 78.0 | 07.7 | 0.0 | 37.90 | 240.98 | 18.1 |
| 14 | Goldberg | 1.13 | 62.0 | 29.0 | 0.2 | 37.57 | 127.73 | 19.9 |
| 15 | Schönhofen | 0.43 | 73.0 | 20.5 | 0.2 | 37.63 | 247.30 | 17.6 |
| 16 | Traidendorf | 0.31 | 48.0 | 38.0 | 1.5 | 09.62 | 319.02 | 13.9 |
| 17 | Gänsleite | 0.98 | 66.0 | 17.0 | 1.8 | 20.67 | 294.17 | 10.9 |
| 18 | Pfaffenberg | 0.65 | 78.0 | 10.4 | 0.5 | 08.04 | 126.00 | 11.1 |
|  | Mean | 0.93 | 76.5 | 20.4 | 0.7 | 26.47 | 211.06 | 19.6 |
|  | SE | ±0.1 | ±3.0 | ±1.9 | ±0.3 | ±3.11 | ±17.52 | ±1.9 |

Habitat conditions of the selected study sites, described by the height of the vegetation in meter (VH), the cover of litter in % (CL), the cover of grass in % (CG), the proportion of bare soil in % (BS) as well as the content of phosphorous in mg/kg soil (P), potassium in mg/kg soil (K) and the ratio of carbon and nitrogen (C/N)

diversity varied between species and populations. Mean genetic diversity of populations (Table 5) was highest in *M. falcata* (0.29), followed by *S. pratensis* (0.26), *P. comosa* (0.25) and *D. carthusianorum* (0.23). The lowest level of genetic diversity was observed in *P. veris* (0.18).

Considering each species separately and all species together in the hierarchical Bayesian multiple regressions, we observed a credible impact of the distance to the nearest calcareous grassland in 1830 on the genetic variation within populations of the studied grassland species (Fig. 2, Tables 6, 7). The distance to the nearest calcareous grassland in 1830 was negatively correlated to the genetic variation within populations. This means that high levels of genetic variation have been detected at study sites which were closely located to other fragments in 1830. However, habitat area, today's distance to the nearest calcareous grassland, habitat shape, habitat conditions and population size had no impact on the genetic diversity of the species.

## Discussion

### Impact of habitat fragmentation on genetic diversity

In our study, we observed an impact of the historical landscape configuration on the genetic diversity of the study species, since it depended on the distance to the nearest calcareous grassland in 1830 in the hierarchical Bayesian regressions. During the process of fragmentation the area of the habitat patches usually decreases [8]. Consequently, the size of plant populations occurring in remnant calcareous grasslands also declines, which results in a decline in genetic diversity [10]. In our study, we also observed a strong reduction of the habitat size. The contemporary area of the study sites was on average only 10% of the area in 1830. However, we observed no significant relationship between genetic diversity of the five study species and population size. Neither the current habitat area, nor the actual number of individuals per grassland fragment had an impact on genetic diversity. Similar findings have been reported in other fragmentation studies [12, 42]. Lag effects, a delayed reaction of genetic diversity on the reduction of population size [43], which is comparable to the extinction debt reported for species diversity [44, 45], could be a reason for the observed lack of a relationship between genetic diversity and population size [46]. In this case, genetic diversity should then be related to the historical area of the study sites. However, our study provided no evidence for such a relationship, which means that genetic diversity may generally be determined by factors other than habitat area or population size. Stochastic gene flow and long term survival under highly fragmented conditions are often considered as reasons for this observation [9, 47]. It has also been stated that

**Table 5 Genetic diversity of the study species**

| St. | Name | *P.v.* | n | *D.c.* | n | *M.f.* | n | *P.c.* | n | *S.p.* | n |
|---|---|---|---|---|---|---|---|---|---|---|---|
| 01 | Eichelberg | 0.23 | 13 | 0.26 | 15 | 0.39 | 15 | 0.34 | 15 | 0.35 | 13 |
| 02 | Münchsried | 0.30 | 14 | 0.35 | 15 | 0.38 | 15 | 0.33 | 15 | 0.36 | 15 |
| 03 | Oel | 0.20 | 15 | 0.26 | 15 | 0.38 | 15 | – | – | 0.35 | 15 |
| 04 | Staudenberg | 0.20 | 15 | 0.25 | 15 | 0.36 | 15 | – | – | 0.37 | 14 |
| 05 | Eitelberg | 0.25 | 15 | 0.32 | 15 | 0.36 | 15 | 0.27 | 15 | 0.34 | 15 |
| 06 | Kühschlag | 0.27 | 15 | 0.32 | 15 | 0.36 | 15 | 0.31 | 15 | 0.33 | 15 |
| 07 | Kallmünz | 0.30 | 15 | 0.28 | 15 | 0.35 | 15 | 0.29 | 15 | 0.33 | 15 |
| 08 | Kronbuckel | 0.29 | 12 | 0.30 | 15 | 0.37 | 15 | – | – | 0.37 | 15 |
| 09 | Ziegelhütte | 0.30 | 15 | 0.31 | 15 | 0.38 | 15 | 0.36 | 11 | 0.36 | 15 |
| 10 | Weichseldorf | 0.22 | 15 | 0.29 | 15 | 0.37 | 15 | 0.33 | 15 | 0.34 | 15 |
| 11 | Fuchsenbügl | 0.22 | 15 | 0.29 | 15 | 0.37 | 15 | 0.31 | 15 | 0.35 | 15 |
| 12 | Undorf | 0.22 | 15 | 0.32 | 15 | 0.37 | 15 | 0.28 | 15 | 0.33 | 15 |
| 13 | Schafbuckel | 0.30 | 15 | 0.34 | 15 | 0.38 | 15 | 0.31 | 14 | 0.35 | 15 |
| 14 | Goldberg | 0.23 | 15 | 0.35 | 15 | 0.39 | 15 | 0.35 | 15 | 0.36 | 15 |
| 15 | Schönhofen | 0.26 | 15 | 0.31 | 15 | 0.36 | 15 | 0.29 | 15 | 0.33 | 15 |
| 16 | Traidendorf | 0.24 | 15 | 0.31 | 15 | 0.36 | 15 | 0.31 | 15 | 0.32 | 15 |
| 17 | Gänsleite | 0.27 | 15 | 0.31 | 15 | 0.38 | 15 | 0.31 | 15 | 0.34 | 15 |
| 18 | Pfaffenberg | 0.32 | 12 | 0.29 | 13 | 0.37 | 15 | 0.37 | 15 | 0.36 | 15 |
| | Mean/total | 0.18 | 261 | 0.23 | 268 | 0.29 | 270 | 0.25 | 220 | 0.26 | 267 |
| | SE | ±0.01 | | ±0.01 | | ±0 | | ±0.01 | | ±0.00 | |

Nei's Gene Diversity of *Primula veris* (*P.v.*), *Dianthus carthusianorum* (*D.c.*), *Medicago falcata* (*M.f.*), *Polygala comosa* (*P.c.*) and *Salvia pratensis* (*S.p.*) and the respective sample sizes

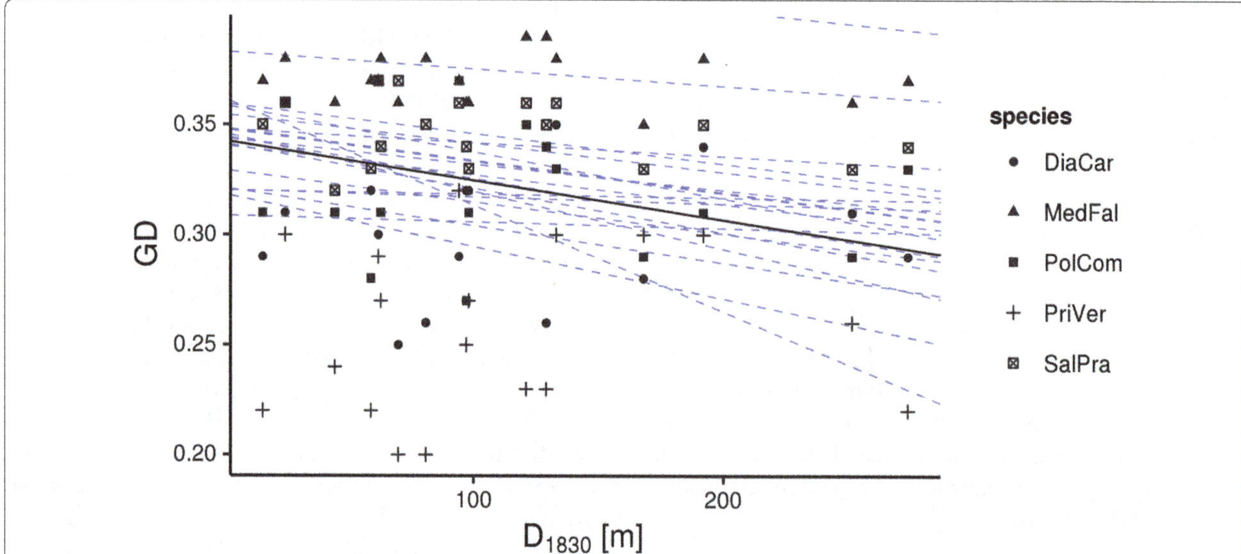

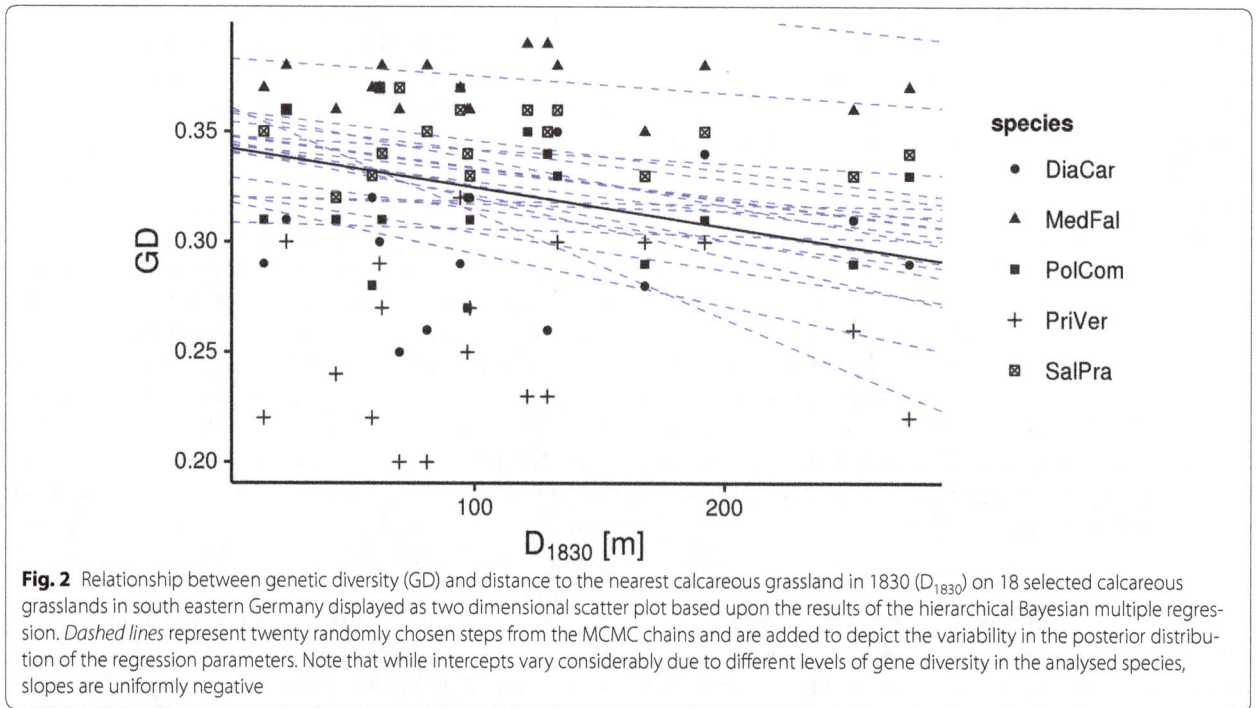

**Fig. 2** Relationship between genetic diversity (GD) and distance to the nearest calcareous grassland in 1830 ($D_{1830}$) on 18 selected calcareous grasslands in south eastern Germany displayed as two dimensional scatter plot based upon the results of the hierarchical Bayesian multiple regression. *Dashed lines* represent twenty randomly chosen steps from the MCMC chains and are added to depict the variability in the posterior distribution of the regression parameters. Note that while intercepts vary considerably due to different levels of gene diversity in the analysed species, slopes are uniformly negative

**Table 6 Bayesian multiple regressions for each of the analysed plant species**

| $GV_{within}$ | RC | ESS | Lower HDI limit | Upper HDI limit |
|---|---|---|---|---|
| *Primula veris* | | | | |
| Intercept | −1.24 | 170.597 | −1.44 | −1.04 |
| $HA_{1830}$ | 0.16 | 12.534 | −0.28 | 0.58 |
| $HA_{2013}$ | −0.14 | 17.675 | −0.62 | 0.31 |
| HA/P | 0.14 | 17.476 | −0.29 | 0.58 |
| *$D_{1830}$* | *−0.28* | *22.780* | *−0.53* | *−0.04* |
| $D_{2013}$ | 0.02 | 114.451 | −0.14 | 0.25 |
| $CO_{1830}$ | −0.08 | 11.826 | −0.50 | 0.32 |
| $CO_{2013}$ | −0.20 | 11.931 | −0.63 | 0.22 |
| P | −0.07 | 27.770 | −0.40 | 0.19 |
| K | −0.17 | 12.490 | −0.43 | 0.12 |
| C/N | 0.09 | 35.921 | −0.14 | 0.34 |
| VH | −0.13 | 49.200 | −0.43 | 0.08 |
| CG | −0.11 | 35.207 | −0.36 | 0.20 |
| CL | −0.25 | 50.816 | −0.59 | −0.02 |
| BS | 0.16 | 57.745 | −0.05 | 0.43 |
| NI | 0.26 | 104.064 | −0.13 | 0.99 |
| *Dianthus carthusianorum* | | | | |
| Intercept | −0.31 | 233.219 | −0.50 | −0.12 |
| $HA_{1830}$ | 0.05 | 13.246 | −0.38 | 0.52 |
| $HA_{2013}$ | −0.20 | 18.615 | −0.69 | 0.26 |
| HA/P | 0.31 | 19.995 | −0.11 | 0.81 |
| *$D_{1830}$* | *−0.27* | *23.119* | *−0.52* | *−0.03* |
| $D_{2013}$ | 0.00 | 119.211 | −0.17 | 0.20 |
| $CO_{1830}$ | −0.14 | 12.332 | −0.54 | 0.28 |
| $CO_{2013}$ | −0.18 | 12.178 | −0.62 | 0.27 |
| P | −0.07 | 29.190 | −0.38 | 0.21 |
| K | −0.18 | 13.066 | −0.48 | 0.09 |
| C/N | 0.12 | 38.994 | −0.10 | 0.40 |
| VH | −0.09 | 53.816 | −0.31 | 0.12 |
| CG | −0.19 | 36.515 | −0.48 | 0.08 |
| CL | −0.14 | 50.200 | −0.37 | 0.10 |
| BS | 0.01 | 43.275 | −0.19 | 0.20 |
| NI | 0.02 | 87.303 | −0.73 | 0.51 |
| *Medicago falcata* | | | | |
| Intercept | 1.06 | 209.393 | 0.85 | 1.28 |
| $HA_{1830}$ | 0.26 | 12.321 | −0.18 | 0.69 |
| $HA_{2013}$ | −0.18 | 17.310 | −0.66 | 0.27 |
| HA/P | 0.18 | 16.981 | −0.23 | 0.60 |
| *$D_{1830}$* | *−0.28* | *21.349* | *−0.51* | *−0.04* |
| $D_{2013}$ | −0.03 | 114.837 | −0.21 | 0.13 |
| $CO_{1830}$ | −0.10 | 11.349 | −0.51 | 0.29 |
| $CO_{2013}$ | −0.19 | 11.427 | −0.63 | 0.21 |
| P | −0.02 | 25.143 | −0.29 | 0.26 |
| K | −0.13 | 13.169 | −0.39 | 0.17 |
| C/N | 0.06 | 36.149 | −0.16 | 0.29 |
| VH | −0.07 | 55.816 | −0.27 | 0.13 |
| CG | −0.12 | 33.158 | −0.36 | 0.12 |
| CL | −0.13 | 49.603 | −0.34 | 0.09 |

**Table 6 continued**

| $GV_{within}$ | RC | ESS | Lower HDI limit | Upper HDI limit |
|---|---|---|---|---|
| BS | 0.01 | 42.573 | −0.17 | 0.20 |
| NI | 0.01 | 131.943 | −0.91 | 0.55 |
| *Polygala comosa* | | | | |
| Intercept | −0.02 | 99.030 | −0.49 | 0.35 |
| $HA_{1830}$ | 0.30 | 13.267 | −0.14 | 0.80 |
| $HA_{2013}$ | −0.08 | 18.712 | −0.53 | 0.42 |
| HA/P | 0.05 | 18.994 | −0.45 | 0.48 |
| *$D_{1830}$* | *−0.29* | *22.677* | *−0.55* | *−0.06* |
| $D_{2013}$ | −0.06 | 120.552 | −0.25 | 0.11 |
| $CO_{1830}$ | −0.10 | 11.951 | −0.53 | 0.30 |
| $CO_{2013}$ | −0.28 | 12.171 | −0.72 | 0.15 |
| P | −0.02 | 29.128 | −0.31 | 0.31 |
| K | −0.14 | 12.706 | −0.45 | 0.13 |
| C/N | 0.04 | 42.379 | −0.19 | 0.28 |
| VH | −0.12 | 45.828 | −0.38 | 0.09 |
| CG | −0.16 | 34.579 | −0.45 | 0.10 |
| CL | −0.14 | 56.817 | −0.36 | 0.10 |
| BS | 0.02 | 46.885 | −0.18 | 0.20 |
| NI | 0.04 | 90.504 | −1.45 | 1.05 |
| *Salvia pratensis* | | | | |
| Intercept | 0.55 | 171.962 | 0.34 | 0.76 |
| $HA_{1830}$ | 0.20 | 12.550 | −0.24 | 0.62 |
| $HA_{2013}$ | −0.20 | 20.218 | −0.75 | 0.33 |
| HA/P | 0.12 | 17.543 | −0.34 | 0.51 |
| *$D_{1830}$* | *−0.29* | *22.443* | *−0.52* | *−0.05* |
| $D_{2013}$ | −0.03 | 111.415 | −0.20 | 0.13 |
| $CO_{1830}$ | −0.10 | 11.783 | −0.50 | 0.31 |
| $CO_{2013}$ | −0.19 | 11.411 | −0.63 | 0.21 |
| P | 0.02 | 27.334 | −0.27 | 0.31 |
| K | −0.13 | 12.997 | −0.40 | 0.15 |
| C/N | 0.03 | 40.560 | −0.20 | 0.26 |
| VH | −0.04 | 63.303 | −0.24 | 0.20 |
| CG | −0.14 | 32.465 | −0.37 | 0.11 |
| CL | −0.16 | 49.189 | −0.38 | 0.05 |
| BS | 0.01 | 44.441 | −0.17 | 0.20 |
| NI | 0.02 | 90.645 | −0.18 | 0.24 |

Results of the Bayesian multiple regressions on genetic variation within populations ($GV_{within}$) calculated on species-dependent level. Modal values of marginal distributions of each standardised regression coefficient are given together with the effective sample size (ESS) of all parameters. A 90% highest density interval (HDI) was computed for each model parameter. The distance to the next calcareous grassland in 1830 ($D_{1830}$) exhibits a credible impact on the genetic variation of the selected species (in italic letters) as its HDI excludes zero (*RC* standardised regression coefficient)

the absence of this relationship may occur when the habitat area rapidly changes relative to the generation time of the study species [15]. Indeed, the calcareous grasslands in this study were formerly widely distributed and may have exhibited more or less erratic gene flow due to grazing. In combination with the long term

**Table 7 Hierarchical Bayesian multiple regression**

| $GV_{within}$ | RC | ESS | Lower HDI limit | Upper HDI limit |
|---|---|---|---|---|
| Intercept | 0.01 | 300.000 | −1.47 | 1.46 |
| $HA_{1830}$ | 0.20 | 13.665 | −0.34 | 0.75 |
| $HA_{2013}$ | −0.16 | 18.771 | −0.74 | 0.40 |
| HA/P | 0.15 | 20.222 | −0.39 | 0.70 |
| $D_{1830}$ | *−0.29* | *22.242* | *−0.57* | *<0.00* |
| $D_{2013}$ | −0.02 | 130.088 | −0.23 | 0.20 |
| $CO_{1830}$ | −0.09 | 11.912 | −0.60 | 0.37 |
| $CO_{2013}$ | −0.21 | 11.813 | −0.73 | 0.30 |
| P | −0.03 | 29.774 | −0.38 | 0.31 |
| K | −0.15 | 12.880 | −0.48 | 0.19 |
| C/N | 0.07 | 40.678 | −0.20 | 0.36 |
| VH | −0.10 | 58.448 | −0.37 | 0.17 |
| CG | −0.15 | 36.036 | −0.46 | 0.17 |
| CL | −0.16 | 62.511 | −0.46 | 0.12 |
| BS | 0.04 | 59.163 | −0.21 | 0.31 |
| NI | 0.02 | 164.573 | −0.96 | 0.87 |
| Variance parameter | 0.44 | 35.412 | 0.24 | 0.58 |
| Normality parameter | 4.45 | 7044 | 1.00 | 82.89 |

Results of the hierarchical Bayesian multiple regression on genetic variation within populations ($GV_{within}$) calculated on species-independent level. Modal values of marginal distributions of each standardised regression coefficient are given together with the effective sample size (ESS) of all parameters. A 95% highest density interval (HDI) was computed for each model parameter. The distance to the next calcareous grassland in 1830 ($D_{1830}$) exhibits a credible impact on the genetic variation of all species at the selected study sites (in italic letters) as HDI <0 (*RC* standardised regression coefficient)

persistence of the grassland species [48] these factors may be the most likely explanation for the lack of relationship between habitat area or population size and genetic diversity in this study.

Aside from decreasing habitat area, isolation of calcareous grasslands is also an important consequence of habitat fragmentation [7, 8]. The continuous process of increasing isolation affects genetic variation both between and within populations since gene flow by seeds and pollen declines with increasing isolation [8]. As gene flow decreases, the effects of genetic drift and inbreeding are intensified [49]. This results in an increased level of genetic variation between populations and a progressive loss of genetic variation within populations [50].

Gene flow by pollen is normally restricted to the nearest vicinity of plant populations to distances of less than 1 km [51, 52]. However, rare pollination events may also allow gene flow over larger distances [53]. Therefore, gene flow among fragmented calcareous grasslands at a larger scale is mainly caused by endo- and ectozoochorous seed dispersal, especially from migration of sheep flocks [2]. It has been shown, that the genetic structure of plant populations depends on present landscape connectivity [54] and that genetic variation between populations is affected by geographic distance between populations [8]. Moreover, it has also been demonstrated that genetic diversity may depend on habitat age both in natural [55] and semi-natural [56] habitats. In contrast to previous studies, which reported higher levels of genetic diversity in populations from historically older habitat fragments of forests [57] or calcareous grasslands [58], we observed no impact of habitat age on genetic diversity in our analysis.

Surprisingly, neither historical nor present habitat connectivity had an impact on genetic diversity in our study. This may be traced back to the fact that the calculated connectivity reflects the spatial structure, but not the real gene flow, which may be strongly affected by the migration of sheep flocks [46]. However, we observed a relationship between genetic diversity of the grassland species and the distance to the nearest grassland patch in 1830. Therefore, the historical landscape configuration is more important for the genetic diversity of calcareous grassland species than the present landscape structure. The effect of historical landscape configuration on species diversity has so far been demonstrated in several studies [59, 60], whereas the impact on genetic diversity has scarcely been shown [27]. However, it has recently been reported that the genetic diversity of the grassland species *Succisa pratensis* depends on the historical landscape structure of the habitat [15], which supports the results of our study. Moreover, it has been shown that the genetic diversity of *Dianthus carthusianorum* depends on patch connectivity by shepherding [61] and that population disconnection can create a genetic bottleneck, even in the absence of a demographic collapse [62], which underlines the importance of historical gene flow for the level of current genetic diversity.

#### Impact of habitat conditions on genetic variation

The results of our analyses indicated no impact of habitat conditions on genetic diversity of the studied calcareous grassland species. It has been demonstrated that, alongside habitat fragmentation, changes in habitat conditions have a strong impact on the species richness and composition of remnant calcareous grasslands fragments [18, 21]. Lack of grazing and the accumulation of soil nutrients lead to the loss of the typical open short-grass vegetation structure, allowing for the existence of many less competitive herbs. Under the conditions of abandonment and due to increased levels of nutrients, grasses such as *Brachypodium pinnatum* become increasingly dominant [19] and litter accumulates [63]. As a consequence, species requiring light for germination decline due to the effects of ground shadowing [24]. For the calcareous grasslands studied here, it has been demonstrated, that species diversity strongly depends on vegetation height and litter cover. Lack of grazing is therefore the most important reason for the declining species diversity of the grasslands, whereas fragmentation aspects play no significant role.

In contrast to the impact of land use on species diversity, the relationship between land use and genetic diversity is much less clear. However, it has already been demonstrated, that seedling recruitment and establishment in grasslands are positively affected by grazing and the removal of litter [64]. Schleuning et al. [65] even stated that the grassland species *Trifolium montanum* is more threatened by the effects of habitat degradation in the short term, than by the effects of fragmentation. Since genetic diversity depends explicitly on the degree of sexual reproduction [66], it appears possible that habitat conditions may also have an effect on genetic diversity. For the grassland species *Dianthus seguieri*, it has recently been shown that increased vegetation height and coverage as well as a high proportion of graminoids due to land use abandonment reduce genetic diversity and seed set [67]. As previously reported, land use by grazing generally has a positive impact on genetic diversity [68] since it reduces the cover of litter, and therefore may stimulate sexual reproduction.

However, in this study genetic diversity depended neither on vegetation structure, nor on soil nutrient levels. One reason for this observation may lie in the life span of the investigated plant species. It has already been reported that the frequency of plant species in remnant calcareous grasslands depends mainly on their persistence [48]. All species included in this study are long-lived perennials [69] and since the process of habitat deterioration due to abandonment goes back only about 50 years, many individuals we analysed may have been established before litter accumulation reached a critical level. This means that impaired habitat conditions may not yet have resulted in decreased levels of genetic diversity. Another reason for the lack of relationship between habitat conditions and genetic diversity could be the persistence of seeds in the soil seed bank, which may have contributed to the regeneration of the populations and to the maintenance of genetic diversity within the studied populations [70].

## Conclusions

The results of our study provide evidence that the genetic diversity of calcareous grassland plant species depends on historical landscape configuration, rather than on the present population size or habitat conditions. In practice, efforts to preserve calcareous grasslands mainly concentrate on large fragments exhibiting the typical open-shortgrass habitat conditions and high species diversity. However, a comprehensive conservation approach should also consider the genetic diversity of calcareous grassland plant species. From our results it can be concluded that populations in smaller grassland fragments may, depending on historical landscape configuration, substantially contribute to the genetic variation of the plant species even under conditions of habitat deterioration. Preferably, these populations should therefore be included in strategies to preserve calcareous grasslands as the local biodiversity hotspots they are.

#### Authors' contributions

CR conceived and designed the study. SS, KM, QS, CM, CH and CP collected data. CS conducted the statistical analyses, CS and CR wrote the manuscript. All authors read and approved the final manuscript.

#### Author details

[1] Institute of Plant Sciences, University of Regensburg, 93040 Regensburg, Germany. [2] German Research Center for Environmental Health, Research Group Comparative Microbiome Analysis, Ingolstädter Landstr. 1, 85764 Neuherberg, Germany.

#### Acknowledgements

Special thanks go to Christine Kammel, Annika Sezi and Mathias Rass for molecular analyses, to Petra Schitko for technical assistance in the lab, to Sabine Fischer for her help with GIS, to Günther Kolb for technical assistance during soil analyses, to Ellen Pagel, Jose Valdez and Sharon Wirth for improving the language and to Peter Poschlod for his generous support.

#### Competing interests

The authors declare that they have no competing interests.

#### References

1. Wallis De Vries MF, Poschlod P, Willems JH. Challenges for the conservation of calcareous grasslands in northwestern Europe: integrating the requirements of flora and fauna. Biol Conserv. 2002;104:265–73.
2. Poschlod P, Wallis De Vries MF. The historical and socioeconomic perspective of calcareous grasslands—lessons from the distant and recent past. Biol Conserv. 2002;104(3):361–76.
3. Kajtoch Ł, Cieslak E, Varga Z, Paul W, Mazur MA, Sramkó G, Kubisz D. Phylogeographic patterns of steppe species in Eastern Central Europe: a review and the implications for conservation. Biodivers Conserv. 2016;25:2309–39.
4. Poschlod P. Geschichte der Kulturlandschaft. Stuttgart: Eugen Ulmer; 2015.

5. Cousins SAO, Ohlson H, Eriksson O. Effects of historical and present fragmentation on plant species diversity in semi-natural grasslands in Swedish rural landscapes. Landsc Ecol. 2007;22:723–30.
6. Krauss J, Klein A-M, Steffan-Dewenter I, Tscharntke T. Effects of habitat area, isolation, and landscape diversity on plant species richness of calcareous grasslands. Biodivers Conserv. 2004;13:1427–39.
7. Fahrig L. Effects of habitat fragmentation on biodiversity. Annu Rev Ecol Syst. 2003;34:487–515.
8. Ouborg NJ, Vergeer P, Mix C. The rough edges of the conservation genetics paradigm. J Ecol. 2006;94:1233–48.
9. Honnay O, Jacquemyn H. Susceptibility of common and rare species to the genetic consequences of habitat fragmentation. Conserv Biol. 2007;21(3):823–31.
10. Leimu R, Mutikainen P, Koricheva J, Fischer M. How general are positive relationships between plant population size, fitness and genetic variation. J Ecol. 2006;94:942–52.
11. Listl D, Reisch C. Genetic variation of *Sherardia arvensis* L.—How land use and fragmentation affect an arable weed. Biochem Syst Ecol. 2014;55:164–9.
12. Schmidt K, Jensen K. Genetic structure and AFLP variation of remnant populations in the rare plant *Pedicularis palustris* (Scrophulariaceae) and its relation to population size and reproductive components. Am J Bot. 2000;87:678–89.
13. de Jong T, Klinkhamer PGL. Plant size and reproductive success through female and male function. J Ecol. 1994;82:399–402.
14. Ellstrand NC, Elam DR. Population genetic consequences of small population size: implications for plant conservation. Annu Rev Ecol Syst. 1993;24:217–42.
15. Münzbergová Z, Cousins SAO, Herben T, Plačkova I, Mildén M, Ehrlén J. Historical habitat connectivity affects current genetic structure in a grassland species. Plant Biol. 2013;15:195–202.
16. Heelemann S, Krug CB, Esler KJ, Poschlod P, Reisch C. Low impact of fragmentation on genetic variation within and between remnant populations of the typical renosterveld species *Nemesia barbata* in South Africa. Biochem Syst Ecol. 2014;54:59–64.
17. Ibánez I, Katz DSW, Peltier D, Wolf SM, Connor Barrie BT. Assessing the integrated effects of landscape fragmentation on plants and plant communities: the challenge of multiprocess-multiresponse dynamics. J Ecol. 2014;102:882–95.
18. Zulka KP, Abensberg-Traun M, Milasowszky N, Bieringer G, Gereben-Krenn B-A, Holzinger W, Hölzler G, Rabitsch W, Reischütz A, Querner P, et al. Species richness in dry grassland patches of eastern Austria: a multi-taxon study on the role of local, landscape and habitat quality variables. Agric Ecosyst Environ. 2014;182:25–36.
19. Bobbink R, Willems JH. Increasing dominance of *Brachypodium pinnatum* (L.) Beauv. in chalk grasslands: a threat to a species-rich grassland. Biol Conserv. 1987;40:301–14.
20. Bobbink R, Hicks K, Galloway J, Spranger T, Alkemade R, Ashmore M, Bustamante M, Cinderby S, Davidson E, Dentener F, et al. Global assessment of nitrogen deposition effects on terrestrial plant diversity: a synthesis. Ecol Appl. 2010;20:30–59.
21. Diekmann M, Jandt U, Alard D, Bleeker A, Corcket E, Gowing DJG, Stevens CJ, Dupré C. Long-term changes in calcareous grassland vegetation in North-western Germany—no decline in species richness, but a shift in species composition. Biol Conserv. 2014;172:170–9.
22. Ruprecht E, Enyedi M, Eckstein RL, Donath TW. Restorative removal of plant litter and vegetation 40 years after abandonment enhances re-emergence of steppe grassland vegetation. Biol Conserv. 2010;143:449–56.
23. Ruprecht E, Szabó A. Grass litter is a natural seed trap in long-term undisturbed grassland. J Veg Sci. 2012;23:495–504.
24. Jensen K, Gutekunst K. Effects of litter on establishment of grassland plant species: the role of seed size and successional status. Basic Appl Ecol. 2003;4:579–87.
25. Reisch C, Bernhardt-Römermann M. The impact of study design and life history traits on genetic variation of plants determined with AFLPs. Plant Ecol. 2014;215:1493–511.
26. Hanski I. Patch occupancy dynamics in fragmented landscapes. Trend Ecol Evol. 1994;9:131–5.
27. Vandepitte K, Jacquemyn H, Roldán-Ruiz I, Honnay O. Landscape genetics of the self-compatible forest herb *Geum urbanum*: effects of habitat age, fragmentation and local environment. Mol Ecol. 2007;16:4171–9.
28. Poschlod P, Baumann A. The historical dynamics of calcareous grasslands in the Central and Southern Franconian jurassic mountains—a comparative pedoanthracological and pollen analytical study. Holocene. 2010;20:13–23.
29. Bassler R, Schmitt L, Siegel O: Methodenbuch/Verband Deutscher Landwirtschaftlicher Untersuchungsanstalten, vol. Teillieferung 3, 4. Auflage edn. Darmstadt: VDLUFA-Verlag; 2003.
30. Karlík P, Poschlod P. History of abiotic filter: which is more important in determining the species composition of calcareous grasslands. Preslia. 2009;81:321–40.
31. Rogers SO, Bendich AJ. Extraction of total cellular DNA from plants, algae and fungi. In: Gelvin SB, Schilperoort RA, editors. Plant molecular biology manual. Dordrecht: Kluwer Academic Press; 1994. p. 1–8.
32. Reisch C, Kellermeier J. Microscale variation in alpine grasslands: aFLPs reveal a high level of genotypic diversity in *Primula minima* (Primulaceae). Bot J Linn Soc. 2007;155:549–56.
33. Bylebyl K, Poschlod P, Reisch C. Genetic variation of *Eryngium campestre* L. (Apiaceae) in Central Europe. Mol Ecol. 2008;17:3379–88.
34. Reisch C. Glacial history of *Saxifraga paniculata* (Saxifragaceae)—molecular biogeography of a disjunct arctic-alpine species in Europe and North America. Biol J Linn Soc. 2008;93:385–98.
35. Bonin A, Belleman E, Eidesen PB, Pompanon F, Brochmann C, Taberlet P. How to track and assess genotyping errors in population genetic studies. Mol Ecol. 2004;13:3261–73.
36. Nei M. Analysis of gene diversity in subdivided populations. Proc Natl Acad Sci USA. 1973;70:3321–3.
37. Vekemans X. Aflp-Surv Version 1.0. In. Belgium: Distributed by the Author. Laboratoire De Génétique Et Ecologie Végétale, Université Libre De Bruxelles; 2002.
38. R-Core-Team. R: a language and Environment for Statistical Computing. In: Vienna, Austria, http://www.R-project.org/. R foundation for statistical computing; 2013.
39. Plummer M. JAGS: a program for analysis of Bayesian graphical models using Gibbs sampling; 2003.
40. Kruschke JK. Doing Bayesian Data analysis—a tutorial with R, JAGS, and Stan. Waltham: Academic Press; 2015.
41. Park T, Casella G. The Bayesian lasso. J Am Stat Assoc. 2008;103:681–6.
42. Lauterbach D, Ristow M, Gemeinholzer B. Genetic population structure, fitness variation and the importance of population history in remnant populations of the endangered plant *Silene chlorantha* (Willd.) Ehrh. (Caryophyllaceae). Plant Biol. 2011;13:667–77.
43. Tilman D, May RM, Lehman CL, Novak MA. Habitat destruction and the extinction debt. Nature. 1994;371:65–6.
44. Helm A, Hanski I, Pärtel M. Slow response of plant species richness to habitat loss and fragmentation. Ecol Lett. 2006;9:72–7.
45. Krauss J, Bommarco R, Guardiola M, Heikkinen RK, Helm A, Kuusaari M, Lindborg R, Öckinger E, Pärtel M, Pino J, et al. Habitat fragmentation causes immediate and time-delayed biodiversity loss at different trophic levels. Ecol Lett. 2010;13:597–605.
46. Honnay O, Coart E, Butaye J, Adriaens D, Van Glabeke S, Roldán-Ruiz I. Low impact of present and historical landscape configuration on the genetics of fragmented *Anthyllis vulneraria* populations. Biol Conserv. 2006;127:411–9.
47. Kuss P, Pluess AR, Ægisdóttir HH, Stöcklin J. Spatial isolation and genetic differentiation in naturally fragmented plant populations of the Swiss Alps. J Plant Ecol. 2008;1:149–59.
48. Maurer K, Durka W, Stöcklin J. Frequency of plant species in remnants of calcareous grassland and their dispersal and persistence characteristics. Basic Appl Ecol. 2003;4:307–16.
49. Slatkin M. Gene flow in natural populations. Annu Rev Ecol Syst. 1985;16:393–430.
50. Widén B, Svensson L. Conservation of genetic variation in plants: The importance of population size and gene flow. In: Hansson L, editor. Ecological principles of nature conservation: applications in temperate and boreal environments. New York: Springer; 1992. p. 113–61.
51. Aavik T, Holderegger R, Bolliger J. The structural and functional connectivity of the grassland plant *Lychnis flos-cuculi*. Heredity. 2014;112:471–8.
52. Kwak MM, Velterop O, van Andel J. Pollen and gene flow in fragmented habitats. Appl Veg Sci. 1998;1:37–54.
53. Ellstrand NC. Current knowledge of gene flow in plants: implications for transgene flow. Philos Trans R Soc Lond B Biol Sci. 2003;358:1163–70.

54. Helm A, Oja T, Saar L, Takkis K, Talve T, Pärtel M. Human influence lowers plant genetic diversity in communities with extinction debt. J Ecol. 2009;97:1329–36.
55. Powolny M, Poschlod P, Reisch C. Genetic variation in *Silene acaulis* increases with population age. Botany. 2016. doi:10.1139/cjb-2015-0195.
56. Vellend M. Parallel effects of land-use history on species diversity and genetic diversity of forest herbs. Ecology. 2004;85(11):3043–55.
57. Jacquemyn H, Honnay O, Galbusera P, Roldán-Ruiz I. Genetic structure of the forest herb *Primula elatior* in a changing landscape. Mol Ecol. 2004;13:211–9.
58. Prentice HC, Lönn M, Rosquist G, Ihse M, Kindström M. Gene diversity in a fragmented population of *Briza media*: grassland continuity in a landscape context. J Ecol. 2006;94:87–97.
59. Lindborg R, Eriksson O. Historical landscape connectivity affects present plant species diversity. Ecology. 2004;85:1840–5.
60. Reitalu T, Johannson LJ, Sykes MT, Hall K, Prentice HC. History matters: village distances, grazing and grassland species diversity. J Appl Ecol. 2010;47:1216–24.
61. Rico Y, Boehmer HJ, Wagner HH. Effect of rotational shepherding on demographic and genetic connectivity of calcareous grasslands. Conserv Biol. 2013;28:467–77.
62. Broquet T, Angelone S, Jaquiery J, Joly P, Lena J-P, Lengagne T, Plenet S, Luquet E, Perrin N. Genetic bottlenecks driven by population disconnection. Conserv Biol. 2010;24:1596–605.
63. Kelemen A, Török P, Valkó O, Miglécz T, Tóthmérész B. Mechanisms shaping plant biomass and species richness: plant strategies and litter effect in alkali and loess grasslands. J Veg Sci. 2013;24:1195–203.
64. Brys R, Jacquemyn H, Endels P, De Blust G, Hermy M. Effect of habitat deterioration on population dynamics and extinction risks in a previously common perennial. Conserv Biol. 2005;19:1633–43.
65. Schleuning M, Niggemann M, Becker U, Matthies D. Negative effects of habitat degradation and fragmentation on the declining grassland plant *Trifolium montanum*. Basic Appl Ecol. 2009;10:61–9.
66. Jongejans E, Soons MB, De Kroon H. Bottlenecks and spatiotemporal variation in the sexual reproduction pathway of perennial meadow plants. Basic Appl Ecol. 2006;7:71–81.
67. Busch V, Reisch C. Population size and land use affect the genetic variation and performance of the endangered plant species *Dianthus seguieri* ssp. *glaber*. Conserv Genet. 2015. doi:10.1007/s10592-10015-10794-10591.
68. Rudmann-Maurer K, Weyand A, Fischer M, Stöcklin J. Microsatellite diversity of the agriculturally important alpine grass *Poa alpina* in relation to land use and natural environment. Ann Bot. 2007;100:1249–58.
69. Schweingruber FH, Poschlod P. Growth rings in herbs and shrubs: life span, age determination and stem anatomy. For Snow Landsc Res. 2005;79(3):195–415.
70. Honnay O, Bossuyt B, Jacquemyn H, Shimono A, Uchiyama K. Can a seed bank maintain the genetic variation in the above ground plant population? Oikos. 2008;117:1–5.

# 10

# To have your citizen science cake and eat it? Delivering research and outreach through Open Air Laboratories (OPAL)

Poppy Lakeman-Fraser[1*], Laura Gosling[1], Andy J. Moffat[2], Sarah E. West[3], Roger Fradera[1], Linda Davies[1], Maxwell A. Ayamba[4] and René van der Wal[5]

**Abstract**

**Background:** The vast array of citizen science projects which have blossomed over the last decade span a spectrum of objectives from research to outreach. While some focus primarily on the collection of rigorous scientific data and others are positioned towards the public engagement end of the gradient, the majority of initiatives attempt to balance the two. Although meeting multiple aims can be seen as a 'win–win' situation, it can also yield significant challenges as allocating resources to one element means that they may be diverted away from the other. Here we analyse one such programme which set out to find an effective equilibrium between these arguably polarised goals. Through the lens of the Open Air Laboratories (OPAL) programme we explore the inherent trade-offs encountered under four indicators derived from an independent citizen science evaluation framework. Assimilating experience from the OPAL network we investigate practical approaches taken to tackle arising tensions.

**Results:** Working backwards from project delivery to design, we found the following elements to be important: ensuring outputs are fit for purpose, developing strong internal and external collaborations, building a sufficiently diverse partnership and considering target audiences. We combine these 'operational indicators' with four pre-existing 'outcome indicators' to create a model which can be used to shape the planning and delivery of a citizen science project.

**Conclusions:** Our findings suggest that whether the proverb in the title rings true will largely depend on the identification of challenges along the way and the ability to address these conflicts throughout the citizen science project.

**Keywords:** Citizen science, Evaluation framework, Lessons learned, OPAL, Outputs, Outreach, Public participation in scientific research, Research, Trade-off, Volunteers

## Background

Citizen science, in all its diverse manifestations, is a burgeoning field of scientific endeavour. Considered by some to be part of 'public participation in scientific research' (PPSR) [1], it is a branch of contemporary science which is used to describe a vast array of activities. It spans subjects from identifying simple morphological classifications of galaxy shapes [2]; to competing in a multiplayer online game to discover protein structure models [3]; to field-based monitoring of commercial poachers in the Congo basin rainforest by Mbendjele hunter-gather communities [4]. As such, the umbrella term has come to mean different things to different people, but is now defined in the Oxford English Dictionary as "scientific work undertaken by members of the general public, often in collaboration with or under the direction of professional scientists and scientific institutions" [5].

Data collection by amateurs has, in many cases, predated paid scientific professions [6, 7]; however, the modern movement of citizen science is still in its infancy. It is a term used to describe a new approach to scientific investigation that, riding on the wave of technology,

*Correspondence: p.lakeman-fraser@imperial.ac.uk
[1] Centre for Environmental Policy, Imperial College London, South Kensington, London SW7 1NA, UK
Full list of author information is available at the end of the article

is open to a broad audience, rather than a wealthy few 'gentleman scientists' [8]. The term 'citizen science' was coined independently in the mid-1990s by Rick Bonney in the US [9] and Alan Irwin in the UK [10]. Citizen science to Bonney was concerned with science communication and public participation in science; whereas for Irwin, the focus was to enhance the accessibility of science policy processes to the public [11]. These descriptions broadly align with two academic movements that have influenced and sculpted the discipline of citizen science: 'public understanding of science and technology' (PUST) which enhances public knowledge and acceptance of science; and 'public engagement in science' (PES) which draws on participatory democratic ideals in scientific research, practice and policy [12]. Paralleling drivers of PUST and PES, investigations into citizen science project goals [13], reasons for participants to become involved [14] and benefits that projects yield [15] reveal two broad themes—outreach and research. 'Outreach' (i.e. an effort to bring services or information to people [16]) includes potential benefits to individuals [through providing learning and training opportunities (e.g. about the natural world)]; benefits to the scientific community [such as promoting science as a worthy cause or expanding awareness of new application areas (e.g. astronomy)]; or benefits to society [such as changing public behaviour (e.g. to prevent spread of invasive species)]. 'Research' (i.e. detailed study of a subject, especially in order to discover (new) information or reach (new) understanding [17]) not only includes potential benefits for scientists (in gathering, analysing and interpreting large data sets); but also benefits for policy makers or for society as a whole (via collectively gathering evidence and acquiring knowledge from non-traditional sources).

Some suggest [14] that citizen science "must place equal emphasis on scientific outcomes and learning outcomes" (p. 313) and many see striving to obtain both goals as a 'win–win' situation [18, 19], where increased participation in science yields enhanced learning opportunities and advanced research outcomes [14]. Trade-offs can however be experienced. For example, creating a project which yields rigorous data sets through complex protocols can be a barrier, potentially limiting the number and retention of participants [20, 21]; or alternatively striving for strong outreach benefits whilst paying little attention to accuracy can potentially lead to datasets of unknown quality and limit their value [22, 23]. Dickinson and Bonney [13] studied 80 projects and asked developers to assign a weight to the goals of the project. They found a significant negative relationship between the goals of education and scientific research, suggesting that investment in one compromised investment in the other. In a similar vein, Zoellick et al. [14] found that the more the students in a classroom benefited from the citizen science experience the less the scientist benefited and vice versa. Given this recognised trade-off, is it possible for citizen science projects to successfully achieve both aims, and if so how? Heeding the advice of one citizen science practitioner [11]—"next to all the enthusiastic endorsements of the many undoubted positive aspects of CS [citizen science] we also keep in mind the limits of what CS can realistically achieve, and keep up a conversation about how to address the limitations of CS" (p.118)—we investigate a programme that aims to balance these goals at a broad scale, Open Air Laboratories (OPAL).

## Methods

### Case study

OPAL is a UK-based public engagement in science programme which utilises citizen science to deliver both outreach and research. It does so from local to national scales, aiming to create 'citizen science for everyone' regardless of age, background or ability. Initiated in 2007 by Imperial College London, the programme was funded with £11.7 million (with three later awards in 2010, 2011 and 2013 increasing this total to £17.4 million) by the Big Lottery Fund (BLF). The original phase of OPAL represented a network of 15 organisations including: ten universities, one natural history museum, one educational organisation, one biological recording organisation, a parks consortium and an environmental government department. Initially operating across England (the period on which this review focusses) and in 2014 expanding to Scotland, Wales and Northern Ireland, the consortium aimed to place scientists into communities to share knowledge and engage individuals in field-based research [7]. It did this through a network of Community Scientists (science engagement staff), project leaders (academics based in each of the institutions), PhD students (based in nine geographically designated regions across the country) and external organisations (who provided an advisory role for specific activities). Five 'research centres' (academic research consortia) were assimilated from individuals from partner organisations on the topics of Air, Water, Climate, Soil and Biodiversity. OPAL's operations traversed the gradient from research approaches to outreach approaches involving professional researchers and citizen scientists to varied extents (Fig. 1). The network took a number of approaches, from delivering citizen science through online tools [24–26] to local co-created citizen science projects. However, the primary mechanism used was the series of seven environmental national surveys led by each of the aforementioned research centres and shaped by other relevant external organisations. These were the:

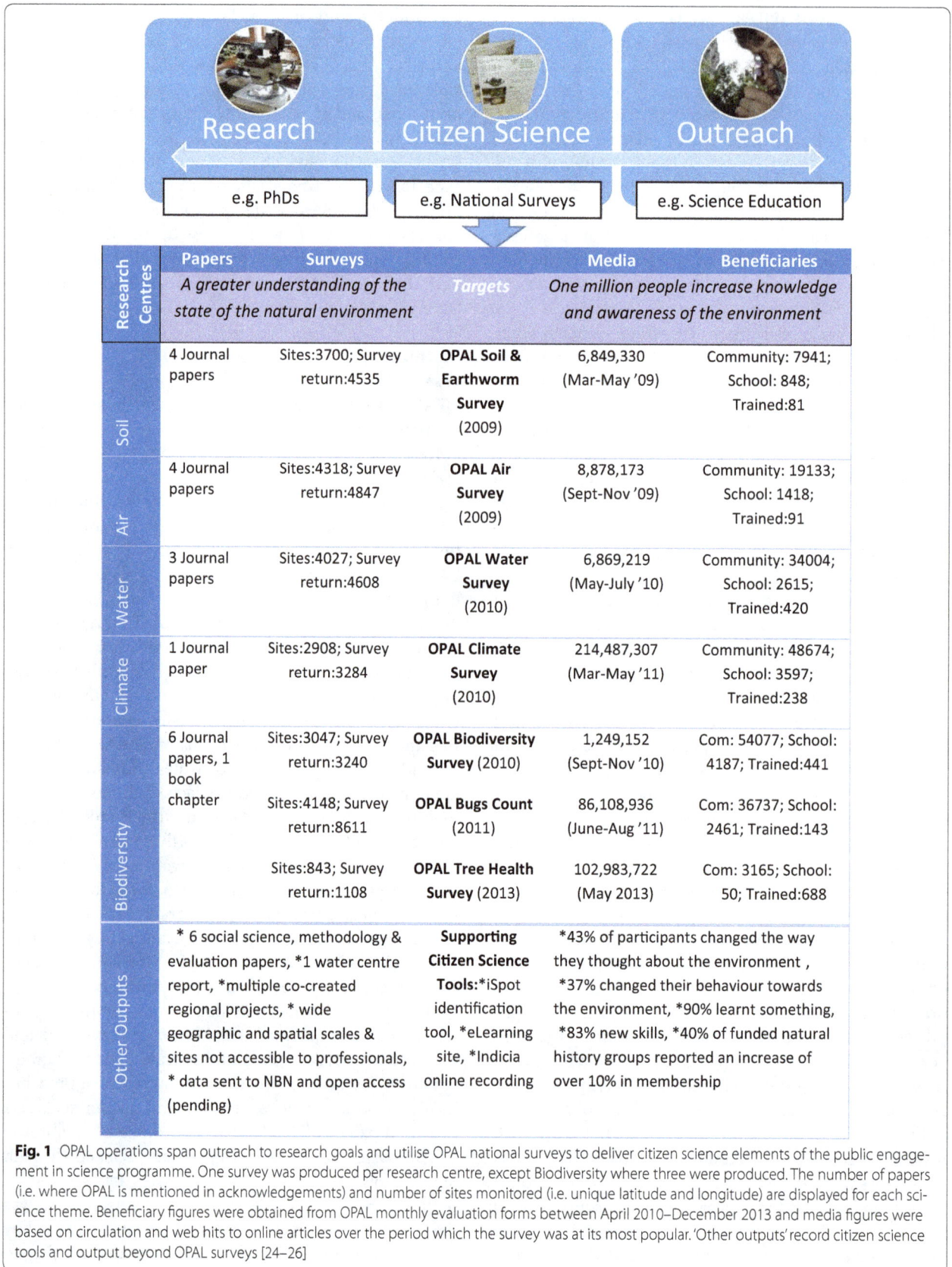

| Research Centres | Papers | Surveys | Targets | Media | Beneficiaries |
|---|---|---|---|---|---|
| | *A greater understanding of the state of the natural environment* | | | *One million people increase knowledge and awareness of the environment* | |
| Soil | 4 Journal papers | Sites:3700; Survey return:4535 | **OPAL Soil & Earthworm Survey** (2009) | 6,849,330 (Mar-May '09) | Community: 7941; School: 848; Trained:81 |
| Air | 4 Journal papers | Sites:4318; Survey return:4847 | **OPAL Air Survey** (2009) | 8,878,173 (Sept-Nov '09) | Community: 19133; School: 1418; Trained:91 |
| Water | 3 Journal papers | Sites:4027; Survey return:4608 | **OPAL Water Survey** (2010) | 6,869,219 (May-July '10) | Community: 34004; School: 2615; Trained:420 |
| Climate | 1 Journal paper | Sites:2908; Survey return:3284 | **OPAL Climate Survey** (2010) | 214,487,307 (Mar-May '11) | Community: 48674; School: 3597; Trained:238 |
| Biodiversity | 6 Journal papers, 1 book chapter | Sites:3047; Survey return:3240 | **OPAL Biodiversity Survey** (2010) | 1,249,152 (Sept-Nov '10) | Com: 54077; School: 4187; Trained:441 |
| | | Sites:4148; Survey return:8611 | **OPAL Bugs Count** (2011) | 86,108,936 (June-Aug '11) | Com: 36737; School: 2461; Trained:143 |
| | | Sites:843; Survey return:1108 | **OPAL Tree Health Survey** (2013) | 102,983,722 (May 2013) | Com: 3165; School: 50; Trained:688 |
| Other Outputs | * 6 social science, methodology & evaluation papers, *1 water centre report, *multiple co-created regional projects, * wide geographic and spatial scales & sites not accessible to professionals, * data sent to NBN and open access (pending) | | **Supporting Citizen Science Tools:***iSpot identification tool, *eLearning site, *Indicia online recording | *43% of participants changed the way they thought about the environment , *37% changed their behaviour towards the environment, *90% learnt something, *83% new skills, *40% of funded natural history groups reported an increase of over 10% in membership | |

**Fig. 1** OPAL operations span outreach to research goals and utilise OPAL national surveys to deliver citizen science elements of the public engagement in science programme. One survey was produced per research centre, except Biodiversity where three were produced. The number of papers (i.e. where OPAL is mentioned in acknowledgements) and number of sites monitored (i.e. unique latitude and longitude) are displayed for each science theme. Beneficiary figures were obtained from OPAL monthly evaluation forms between April 2010–December 2013 and media figures were based on circulation and web hits to online articles over the period which the survey was at its most popular. 'Other outputs' record citizen science tools and output beyond OPAL surveys [24–26]

OPAL Soil and Earthworm Survey, OPAL Air Survey, OPAL Water Survey, OPAL Climate Survey, OPAL Biodiversity Survey, OPAL Bugs Count and OPAL Tree Health Survey. Each survey was made freely available to participants (either in hard copy format or through digital downloads) to ensure inclusivity and consisted of a pack which contained everything required to conduct the survey. For example, the OPAL Soil and Earthworm Survey pack contained a field notebook (rationale for conducting the research and recording sheets to collect data on site characteristics, soil properties and earthworms), a field guide (earthworm identification guide and survey steps) and equipment (pH strips, magnifier, vinegar and mustard). All surveys aimed to raise awareness about key environmental issues and scientific methodologies; and generate data on selected scientific questions, e.g. species distributions, changing environmental conditions or the impact of urbanisation on biodiversity.

The programme was designed with the dual purpose of "bringing scientists and communities together to deliver a research programme focused on three environmental themes: loss of biodiversity, environmental degradation and climate change" (research) and "motivating outdoor exploration and providing participants with the knowledge, skills and confidence needed to study nature" (outreach) [27]. The research aim was driven by the Conventions on Climate Change and Biodiversity [28] and the crisis in taxonomy [29]. OPAL's outreach objective was driven by a decline in outdoor learning in the UK [30] and a call for programmes addressing education and engagement of local communities by the BLF [31] who award money to projects that improve health, education and the environment [7]. Targets were set for both research and outreach, the former being driven by the broad aim to 'achieve a greater understanding of the state of the natural environment' and the latter having the specific aim for 'one million people to increase knowledge and awareness of the environment'. These targets were monitored throughout the programme (Fig. 1). For research outputs, over 230,000 packs were distributed to the public and surveys were submitted which translated to 25,000 field sites being sampled [27]. The aim was to send all appropriate species level data to the National Biodiversity Network with the eventual target to make all data open access. In addition, academic journal papers were produced reporting on ecological results from the national surveys [19, 32–35], specific regional research [36–39], methodological research [19, 32, 34, 40–42], effective working practices [43, 44], and perceptions of citizen science amongst scientists [11]. All outreach targets were reached and in many cases exceeded with a total of over 850,000 direct beneficiaries (i.e. distinct learning experiences through events, lessons and community presentations) and almost 1.7 million website hits reached between February 2008 and November 2013 (Fig. 1). Online questionnaires filled out by a sample of participants following the data entry of national surveys revealed outcomes such as awareness raising, behaviour change, learning outcomes and impact on learning pathways (Fig. 1). Because of this broad approach and range of outputs, OPAL provides a useful case study through which to explore the trade-offs inherent in dual-aim programmes and investigate the factors which helped deliver the citizen science elements of the programme.

### Data

Monitoring and evaluation has been conducted since OPAL was initiated (2008–2013 reported here) in order gather evidence of the number of participants in OPAL activities (referred to as 'beneficiaries') for the funder (BLF) and to capture experiences of those involved in OPAL because of its large and distinct modus operandi so that lessons could be learned and communicated to benefit future citizen science endeavours. One study [15] suggested the importance of evaluating a citizen science programme not only in terms of its scientific outputs (which tend to be confined to hypothesis-led research and mainly positive data outcomes) but also from experiences of programme staff (who were well placed to comment upon the developmental process of the project and report on problems encountered). This investigation therefore not only employs quantitative information (i.e. beneficiary numbers) but also uses qualitative material on lessons learned from staff and participants.

Quantification of the extent to which people engaged with the OPAL programme was obtained from monitoring forms which the original 15 OPAL partners returned on a monthly basis to the programme management team. They primarily reported on the number of members of the public who participated in OPAL activities and also the schools and community, voluntary and statutory organisations they had worked with. A total of 1107 monitoring forms were collected from OPAL partners between 2008 and 2013, and these formed the basis of the quantitative data reported. At the end of every project year, partners were also required to report on their activities over the past 12 months. As part of these, partners were asked to answer the following questions: 'What are the five most important lessons learned since your project began?' (Year 4) and 'What are the five most important lessons learned from delivering your project?' (Year 5). A total of 60 annual reports formed the basis of our qualitative data. A log of 'lessons learned' was also maintained by OPAL management. This formed additional data that were qualitatively appraised. Qualitative data was used to deepen understanding gained from formal quantitative reporting to explore trade-offs and potential solutions for balanced operations. Key themes were drawn out of the evidence, and text fragments identified as belonging to each category were brought together and assimilated into the four themes presented in Table 1 and Fig. 2. In the presentation of our findings we use quotes to illustrate both widely shared and minority views, with explicit indication of the latter.

## Investigation and management of trade-offs across the outreach-research spectrum

Haywood and Besley [12, 45] developed a set of standards for PPSR projects (of which citizen science is a part), deeming four key dimensions central to the integration of 'education outreach' (broadly aligned with our definition of outreach) and 'participatory engagement' (broadly aligned with our definition of research) traditions. The findings suggested that a successful project should: A. *meet the needs* of all stakeholders, B. *foster trust and confidence* among stakeholders and in science, C. *broaden scope and influence*, and D. build *social capacity* to respond to ecological challenges. We use their conceptualisation as a theoretical model to: (1) reveal the key trade-offs between delivering research and outreach within the large-scale OPAL programme; and (2) discuss operational considerations to manage the identified trade-offs. Based on this exploration, we propose an evaluation framework (Fig. 2), which builds on Haywood and Besley's [45] standards ('outcome indicators') and pairs these with effective working practices ('operational indicators') that could address trade-off challenges.

### Revealing key trade-offs within OPAL

#### *A. Needs met*

The first (outcome) indicator which Haywood and Besley [45] identify in their assessment framework is the "degree to which the products generated (intellectual or material) meet the legitimate needs and expectations of participants [all stakeholders involved]" (p. 5). A key trade-off emerged from OPAL across the research outreach spectrum in this context of 'needs met': *outreach gets in the way of science* (i.e. science needs being compromised) *and science gets in the way of outreach* (i.e. outreach needs being compromised) (Table 1). Firstly, regarding outreach getting in the way of science, one scientist collaborator commented that "experience from OPAL [Tree Health Survey] and Sylva Tree Watch surveys suggest that lay involvement in tree health surveillance is at best only partially successful from a scientific perspective." While every effort was made to ensure scientific thoroughness, data acquisition could be compromised because "groups adopted a pick-and-mix approach to the different tasks and activities in the survey. This is driven by the fact that every group has their own range of ages and abilities, levels of scientific knowledge, plus unique time and logistical constraints." Next, the return rates for the number of OPAL surveys distributed compared to completed surveys submitted was around 10 %, partially because "after a field session people did not always want to sit down and upload all the data at a computer". While anecdotally this may be a good level compared to industry standards, such return rates may not be acceptable in terms of the efficient utilisation of scientific resources (depending upon whether hard copies are used or packs downloaded). Perhaps one of the most fundamental debates in citizen science is the preconception that it is either possible to collect large quantities of data which is of questionable accuracy or small quantities of highly accurate data. The use of photographs and apps for mass collection of verifiable data (for example the UK Ladybird Survey and OPAL Species Quest [46, 47]) can help here, but where this (or another quality monitoring technique) was not effectively utilised within OPAL, scientists found that "data are patchy and of poor quality". The scientists were not the only ones who expressed concern about generating usable data: "it would appear that most participants will try to undertake the survey to the best of their ability [...and] generally appear to be concerned about data quality and some even decline to submit their data as a result." Rigorous, complete data sets are required for science so are outputs being compromised with the citizen science approach?

Secondly, OPAL staff regularly commented that "science (activity) may interfere or get in the way of actual involvement [i.e. learning by doing]". OPAL's focus on fixed packs, developed to simultaneously generate outreach and data to address specific questions, meant that outreach opportunities were sometimes missed because the focal landscape components of some packs (e.g. ponds, hedges) could be in short supply: "while the survey was written for anyone to be able to participate, one of the biggest hurdles has been for people to find a publically accessible pond." The one-size-fits-all approach for generating useful data created further challenges: "OPAL designed each survey to appeal to an age range of 13–14 years. We felt this would also provide a valuable experience for newcomers and those without previous knowledge of the topics. Many people used the surveys with young children and they were not suitable, particularly without high levels of adult supervision". While it may seem that research may initially reduce the impact of outreach, "there is a need to be realistic in expectations of immediate payback - perhaps the greater return (in terms of developing scientific interest) will come much later if scientific interest is sustained in those who take part". Is it therefore possible to maintain high quality outreach outputs, and if so how?

**Table 1 Textual data obtained through monthly monitoring by OPAL staff which inform operational indicators**

| | Outcome indicator & trade-off | Operational indicator | Key questions | Example quotations providing evidence for operational indicators |
|---|---|---|---|---|
| A | *Needs met*<br>Trade-off: *Outreach gets in the way of research and research gets in the way of outreach?* | Ensure outputs are *fit for purpose* | Is the project *designed* and *monitored* appropriately to ensure outreach and research of an appropriate quality? | *"Activities are best split into bite-sized chunks that can be done singly with less interested individuals and in multiples with the more interested." *"resources ...have to be concise, visually interesting, and different." *"important not to ask the public to run before they can walk" *"focus on single or small set of bio-indicator species addressed the challenge of identification expertise, whilst proving less daunting and more empowering for the volunteer" *"citizen science projects must (1) provide sufficient training to ensure data are collected accurately, and (2) regularly monitor and screen incoming data to ensure continued accuracy." *"Planning a programme of evaluation from the outset is very useful and ensures that it is ingrained in everybody's thinking form the outset." *"Qualitative and quantitative evaluation are both valuable" |
| B | *Trust*<br>Trade-off: *Build a reputation with partners or participants?* | Develop *strong collaborations* | Is there adequate *buy-in* from partners and is *feedback* maintained? | *"Important to get appropriate buy-in from scientists who should/will be involved... especially in scientific disciplines where citizen science is new, novel or perceived as threatening." *"the key to making links with existing community groups is finding and highlighting ways in which it is possible to work together".*"The willingness of people to initially engage with the OPAL project appears to be enhanced when they are introduced to the project through face-to-face contact. Once initial engagement is made, many continue to request survey packs and information about events etc. in a more remote manner (telephone, email)." *"The surveys need to give instant results that people can relate to the quality of their local environment. The water survey was particularly good for this as the Pond Health score gave people a measure of their pond's water quality" |
| C | *Scope*<br>Trade-off: *Jack of all trades, master of none?* | Build a sufficiently *diverse partnership* | Is there appropriate *expertise* within the programme? | *"My main take away lessons from my time as a Community Scientist are that you need to share your passion for the natural world." *"Community Scientists were involved with the development of all 7 national surveys, e.g. from testing the survey with local communities to providing feedback on the final survey materials...This resulted in the development of surveys which the general public could participate in/contribute successfully to, as well as generating meaningful scientific data for OPAL." *"it is vital to obtain formal agreement from senior managers of partner organisations to provide the resources (especially time) to fulfil their obligations." *"Initially it was difficult to build up relationships with schools or to have anything other than one-off interactions." *"Academics are not usually involved in this scale of public outreach. It proved to be very rewarding on many fronts" |
| D | *Social capacity*<br>Trade-off: *Who is contributing and to what extent?* | *Target audience* | Which *sectors of society* are considered and is *technology* integrated appropriately? | *"Genuinely hard to reach community groups require large commitments of time and energy to build up relationships to the level where outreach can be delivered successfully." *"because once relocated to England, and especially for second generation and the younger generation, human activities are no longer seen as part of any ecosystem function." *"Issues of inclusivity have to be faced professionally" *"Technical developments intended to be a major part of a public engagement project need to be carefully planned for, well in advance, to ensure that they can be taken up effectively by participants." *"People have also increasingly moved towards using mobile and tablet devices since OPAL first started and as this tech is now part of their everyday lives, we need to respond to this demand." *"Use of digital technology (e.g. social media) offers us a way to reach out to this audience in the spaces that they already frequent, at very little expense to us" |

These indicators inform the practical considerations when addressing trade-offs

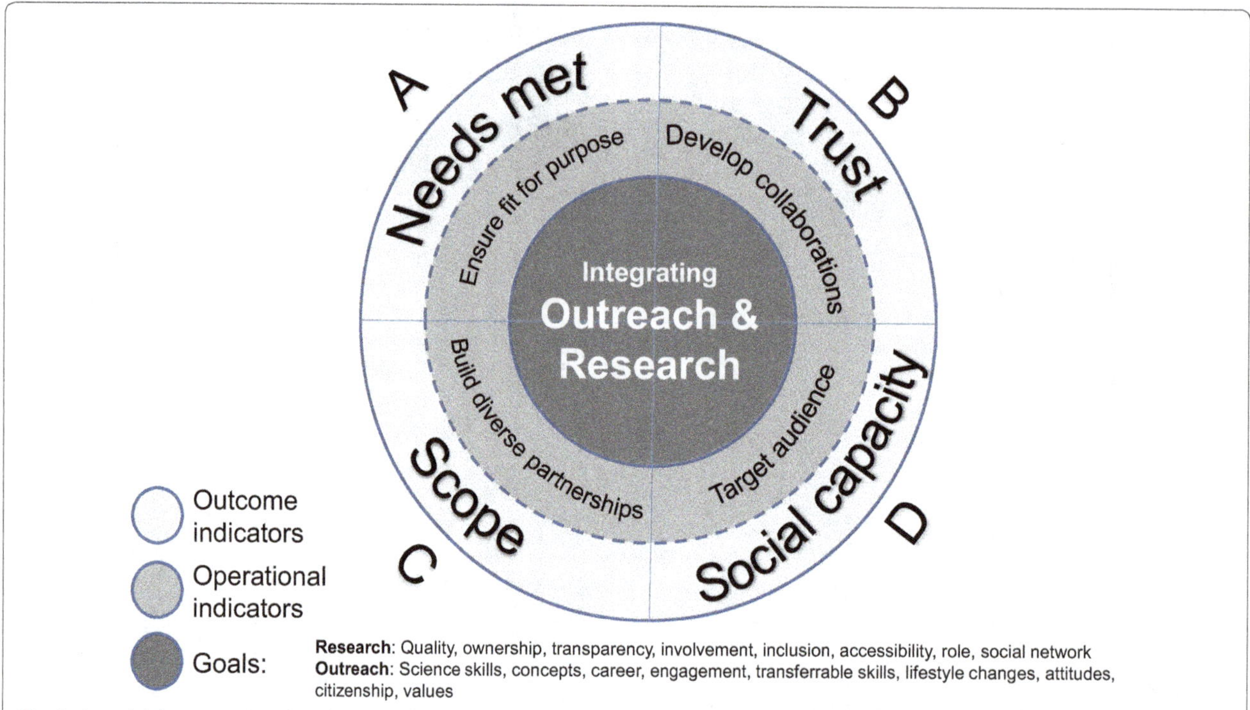

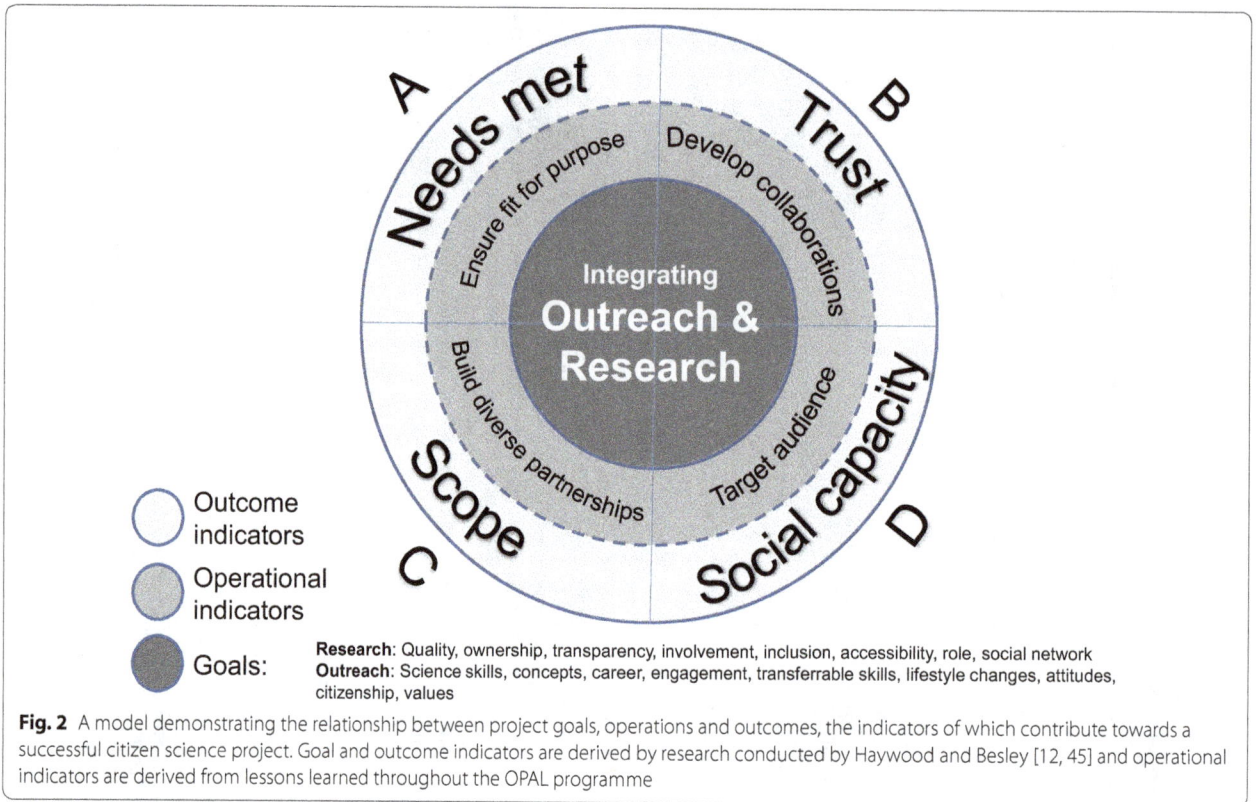

**Fig. 2** A model demonstrating the relationship between project goals, operations and outcomes, the indicators of which contribute towards a successful citizen science project. Goal and outcome indicators are derived by research conducted by Haywood and Besley [12, 45] and operational indicators are derived from lessons learned throughout the OPAL programme

***B. Trust and confidence***

A second outcome indicator on which citizen science programmes should be evaluated according to Haywood and Besley [45] is the "degree to which the project fosters general trust, confidence, and respect among project participants and in science" (p.5). A key trade-off emerged from OPAL across the research outreach spectrum in the context of 'trust and confidence': *Do you put resources into building a reputation with partners (research) or participants (outreach)?* (Table 1). As one partner commented, "developing collaboration requires more time and support than originally envisaged and has been a major challenge." Firstly, building trust and respect between organisations is draining on resources: "many such relationships, particularly between academia and local and national government and other organizations have developed but required considerable effort." It can be difficult to plan for when building relationships as "it was challenging to define an enabling project on the basis of the expected needs of a new and unknown partnership." Working closely with other organisations can also create challenges: "Elements of our project have been reliant upon OPAL partner organisations for their successful delivery, and in some cases unforeseen logistical problems have prevented them taking place or greatly increased the workload involved." A 'one-size-fits-all' approach may not work when building relationships with groups because organisations vary in the objectives they have to meet and their expectations from the working relationship. For example, "some who lent expertise to OPAL felt that their society should be paid at a consultancy rate for it. At the other end of the spectrum, others felt that claiming travel and subsistence funds when they attended events was wrong because they were doing it for the love of it and didn't feel comfortable reclaiming the money." Secondly, building the confidence of participants is integral to yielding social benefits as well as scientific outputs: "the real satisfaction of the [Community Scientist] role is watching others' skills and confidence grow" and "we have observed people gain the confidence to not only lead surveys in their own communities but to realise that resources such as iSpot [a species identification website] are accessible and engaging." On top of this, building trust and confidence between practitioners and participants is fundamentally important: "if survey coordinators do not find it possible to trust potential participants to undertake the surveys in the spirit they are intended then the voluntary participant approach is probably not the right approach for the task." This raises an important point, namely that citizen science is not always appropriate for all types of research questions (see [41] for a tool to help practitioners decide whether citizen science is appropriate). Is it an 'either/or' situation, or can resources, trust and confidence be built with both partners and participants, and if so, how?

### *C. Scope*

According to Haywood and Besley [45] projects should also be evaluated on the "degree to which products generated (intellectual or material) impact broader social, economic, or environmental systems and relevant policy (e.g. local laws and procedures, national standards, corporate practices)" (p. 5). A key trade-off emerged from OPAL across the research outreach spectrum in the context of *'scope': jack of all trades and master of none* (i.e. are both outreach and research needs being compromised)*?* (Table 1). OPAL aimed to create "science-society-policy interactions which would lead to a more democratic research, based on evidence-informed decision making", but in doing so does the programme and its staff become generally effective across multiple sectors but not outstanding in any of them? For example in policy, OPAL Tree Health Survey is referenced in a parliamentary 'POSTNOTE' paper as an example of a project which contributes towards Defra's Tree Health and Biosecurity Action Plan [48]; in education, OPAL activities are thought to have had an influence on improving science GCSE grades [27]; and in research, the OPAL Air Survey found that citizens access new areas, previously underexplored by professional scientists [32]. The trade-off lies potentially in the progress in each of these areas. Have efforts into realising impact in society meant that not as much was being achieved in research? To consider this, Fig. 1 demonstrates the relative balance between the impact in each sector. One scientist collaborator who was sceptical about the usefulness of the scientific output generated commented that it is "important for national biosecurity that a formal assessment of citizen science impact is made sooner rather than later or not at all so that alternative strategies be rolled out—the subject is too important for prevarication, political correctness or fudging." Indeed, the influence of citizen science on academia for example can look minimal when assessed using recognised institutional indices (i.e. publication output) [43]. While staff may have been making inroads in enriching learning experience of participants and enhancing awareness, all that is recorded is the 'pounds per paper' so that a large investment of money looks squandered against academic output. This can put a strain on the staff delivering these projects "it's several roles rolled into one... it's essentially...two part time jobs..., they're not, they're two full time jobs to do them properly" [49]. This was also reflected in the feedback that was obtained suggesting that there was "an understandable reticence for some sectors of the scientific profession to engage in CS—'what will this add to my project,—or my career?'" Therefore when measured against traditional systems, whether that be outreach or research—the amount of impact may not be as large as concentrating on one discipline. This, of course, is not accounting for cumulative impacts of the two, measurement scales for which may not fully exist yet. Paradigms are however shifting as the Research Excellence Framework (the system used to assess the quality of research in UK higher education institutions [50]) increasingly recognises *impact* ("'reach and significance' of impacts on the economy, society and/or culture that were underpinned by excellent research" [51]). While the majority is focussed on outputs (e.g. academic papers), impact now carries a weighting of 20 % and citizen science projects are now being recognised as ideal case studies and excellent ways to generate publicity for institutions. Is it therefore possible to have a recognised impact across a range of sectors (academia, education and policy) rather than just one, and if so, how?

### *D. Social capacity*

A final dimension Haywood and Besley [45] propose citizen science projects to be evaluated on is the "degree to which the project influences the capacity of communities/social groups to respond to social or ecological challenges, negotiate conflicts, and develop solutions" (p.5). A key trade-off emerged from OPAL across the research-outreach spectrum in the context of 'social capacity': *who is contributing* (well informed public-supporting research needs, or all sectors of society-supporting outreach) *and to what extent* (high level of involvement of scientists-potentially benefiting research, or high levels of involvement of the public-potentially benefiting outreach)? (Table 1). OPAL's remit was to work with hard to reach communities and staff commented: "the deepest interactions are undoubtedly the most rewarding, but they are also the most resource intensive". It was therefore a conundrum to many Community Scientists between reaching as many beneficiaries to support them in producing high quality science and reaching sectors of society that may not have the opportunity to access these opportunities. One of OPAL's collaborators suggested, perhaps controversially, that given the "data are not from a random sample anyway, does it matter scientifically if some sections of society are excluded?"

Within citizen science there is a spectrum of engagement from contributory (designed by scientists, public contribute data) to collaborative (designed by scientist, public contribute to design, data and analysis) to co-created (scientists and public work together on all aspects of process) projects [1], and typically a particular project will focus on one approach. For OPAL, the primary focus has been on allocating resources to the 'contributory' style national survey with far less emphasis being placed on the citizen- driven 'co-created' studies. Although at a regional level Community Scientists worked with local groups to develop research important to these groups (on topics ranging from invasive crayfish monitoring to hedgehog tracking), this was a small element of the programme. There was a call to move towards this end of the spectrum with one Community Scientist suggesting "to ensure successful delivery of the programme the community has to help shape and plan the project aims and objectives (i.e. bottom-up)". Those OPAL staff who operated that way suggested that through involving participants from the beginning on issues that are important to them, projects were more likely to recruit participants and collect data that would be used, if only locally. Others within the network held alternative opinions: "I have mixed views on co-creation. It seems to be the scientists who are saying that this is what people want. This may be true sometimes and in some situations but not always. Sometimes citizens are content to let the experts develop the project and to assist by data gathering. I have even heard negative attitudes to greater involvement: we do not have the skills/that is what experts are for/we are too busy with our own jobs etc." Is it therefore too challenging to take multiple approaches when planning the level at which participants get involved and if so what is the important factor to consider to support social capacity?

**Managing trade-offs in OPAL**

Is it possible to address these trade-offs and deliver outputs useful for both outreach and research? Qualitative data generated by OPAL staff and participants (Table 1) contribute to an understanding of how to practically confront each of the four challenges under the outcome indicators set out by Haywood and Besley [45]. These effective working practices were used to develop four operational indicators in order to provide a potential solution to a number of the trade-offs raised: ensuring outputs are fit for purpose, developing strong collaborations, building a sufficiently diverse partnership and targeting specific audiences.

***A. Fit for purpose (to ensure needs are met)***

Given that the first of the trade-offs was: *outreach gets in the way of science and science gets in the way of outreach,* there is a clear requirement for the products generated (resources, events, data etc.) to be *fit for purpose* for both research and outreach. Two essential points feed into this: the quality of the products need to be *designed* appropriately for their intended use and *monitoring* needs to be built in to track quality of the data and outreach (Table 1).

From the perspective of outreach, outputs that are fit for purpose attract participants, maintain their enthusiasm and ensure the science content is understandable. Resources (e.g. the OPAL packs) should clearly communicate an aim, be visually interesting, understandable and sufficiently flexible in their use (Table 1). To explore this theme in more detail, we focus on the creation of scientific outputs that are fit for purpose. To ensure that the data collected are of a quality which is usable for the intended purpose, methodological design is key, training should be considered, technology used appropriately, monitoring of accuracy implemented and analysis techniques tailored to the data utilised (Table 1). One contributor noted that "it was in fact these quite substantial worries about data quality that drove them [practitioners] to be methodologically innovative in their approach to interpreting, validating and manipulating their data and making sure that the science being produced was indeed new, important and worth everyone's time." In many cases, survey leaders thought carefully about balancing the needs of participants and data users. For example in the Bugs Count, the first activity asked the public to classify invertebrates into broad taxonomic groups (which were easier to identify than species) and the second activity asked participants to photograph just six easy-to-identify species. Participants therefore learned about what features differentiate different invertebrate groups whilst collecting valuable verifiable information on species distribution (e.g. resulting OPAL tree bumblebee data were used in a study comparing skilled naturalist and lay citizen science recording [52]). Data quality monitoring was conducted to varying degrees between surveys. The Water Survey [34] for example, integrated training by Community Scientists, identification quizzes, photographic verification, comparison to professional data and data cleaning techniques. Survey leads on the Air Survey [32] compared the identification accuracy of novice participants and expert lichenologists and found that for certain species of lichen, average accuracy of identification across novices was 90 % or more, however for others accuracy was as low as 26 %. Data with a high level of inaccuracy were excluded from analysis and "this, together with the high level of participation makes it likely that results are a good reflection of spatial patterns [of pollution] and abundances [of lichens] at a national [England-wide] scale" [32]. For the Bugs Count Survey, information on the accuracy of different groups of participants was built into the analysis as a weight, so that data from groups (age and experience) that were on average more accurate, contributed more towards the statistical model [19]. This exemplifies that if data quality is being tracked, and sampling is well understood, then a decision can be made by the end user about which datasets are suitable for which purpose.

***B. Develop strong collaborations (to build trust and confidence)***

To tackle the second key trade-off—*building a reputation with partners (research) or participants (outreach)?*—in order to build trust and confidence, effective collaborations (within practitioner organisations and between practitioners and participants) are imperative (Table 1). Being a programme delivered by a network of organisations and working with a range of audiences, this was essential to the functioning of OPAL. Indeed it is important for all citizen science projects as they require the input not only of both scientists and participants but often a wide array of other partners too.

Firstly, is there enough *buy-in* from partners? Receiving adequate buy-in from all organisations involved can require considerable effort, time and resources (Table 1) yet failing to get the support from either the experts informing the project, the data end users, the outreach staff or the participants can create difficult working relationships and inadequate outputs. This was highlighted by one external collaborator who sat on an advisory committee for the OPAL Tree Health Survey and felt that buy-in from professional scientists was particularly key in "scientific disciplines where citizen science is new, novel or perceived as threatening". It was also clear from the data that "the most effective projects are those that have a clear champion within the organisation taking part". Recognising that partners are often stretched for time, OPAL assigned funding towards resources to support an effective balance between research and outreach by for example, funding PhD students to join the research labs of academics with the aim of freeing up time for these professional scientists to contribute to citizen science. When there are no leading supporters, relationships can break down, for example one collaborator noted "in our case, no one person was designated as leading on the OPAL-related work we were carrying out, and this may have reduced the impact of our work to other OPAL partners." Champions are also very valuable in other parts of the project, from a strategic level (such as a member of a government department sitting on an advisory board) to a community level (such as an active member of the public coordinating group monitoring in a local park).

Secondly, how is *feedback* maintained between all stakeholders (partners, external organisations and participants)*?* The Community Scientist network was a key strength in the OPAL programme. These staff members provided an effective conduit between the in-house scientists and the community groups (Table 1). With diverse backgrounds in research, education and science communication, they were ideally placed for taking science into communities (as opposed to individuals coming to science centres like museums or universities). In addition to these staff, identifying key individuals who support the project's work was found to be an effective way to spread messages throughout a community. Teachers for example, are gatekeepers to young people's experience with science and as such OPAL created curriculum support for teachers by using the surveys to deliver their lessons. Participants could also undertake the surveys independently, and for this to be an effective experience for both learning and data collection in the absence of Community Scientist support, available resources needed to be clear, innovative and intellectually matched to the audience, and involve feedback to participants. In all surveys once participants had entered data, their results appeared on an interactive map so that they could compare their record to others and in some surveys participants could work out an instant score of environmental health (Table 1). This feedback benefit was clear as in the climate survey there was no such instant measure of quality so "although people liked the idea of contributing to a national data set many participants did not go away feeling they had learnt more about their local environment." Intermediary results were also uploaded to the website to provide clear infographics displaying findings once data had been processed, and then lay summaries of scientific papers were also posted on the website. Forging collaborations between practitioners and participants can therefore be gained through effective communication.

***C. Build a sufficiently diverse partnership (to widen scope)***

It is important to tackle the third key trade-off—*jack of all trades, master of none?*—if scope and influence is to be initiated and expanded upon. There are many dimensions to a citizen science project and many sectors on which they can have an impact. In order to reduce the likelihood of scope and influence being compromised, we suggest that it is important to *build a sufficiently diverse partnership* (Table 1). The key to this is having "appropriate staff members to deliver complex projects" and diverse *expertise* where projects have broad aims. The OPAL programme was planned to bring together staff across research and outreach sectors. On the research side, academics based at English universities formed the regional project leaders, PhD students were employed to support regional research, and external personnel from government departments and environmental organisations sat on strategy boards and working groups. For outreach, museums, environmental educators and other public facing organisations formed part of the core partnership; and Communication Officers and Web Editors were employed for remote engagement. Sitting between the two sectors, the management team coordinate balancing the two aims and Community Scientists take science to the public with their expertise in science and experience in science communication. Participants also generate impact, i.e. through promoting "knowledge exchange rather than using citizens as mere suppliers of information" alongside enabling wider geographic and potentially more diverse areas to be studied.

Each of these stakeholders bring unique expertise to the partnership which contributes to its success at commandeering influence. Take Community Scientists for example, they: share their passion and knowledge (one noted "excitement about a tiny parasitic wasp or how fungi reproduce is what draws others in and brings it alive for them"); they integrate experience into resource development (one partner commented that they "bring their knowledge and experience of working with people of all backgrounds, ages and abilities to each individual survey") and they understand about effective partnership working (one Community Scientist noted "working with amateur natural history societies requires a sensitive and responsive approach.") While distinct roles were carefully planned, in reality most stakeholders pitched into many activities which yielded "unforeseen benefits of public engagement: First, [academic] staff developed and widened their communication and teaching skills by having to engage with new audiences (some of whom may not initially be interested in the subject). Second, we benefited from discussions with members of the public and hearing a wide range of views. Third, feedback strongly suggests that the public enjoyed and benefited from meeting scientists and being able to ask them questions."

#### *D. Target audience (to build social capacity)*

In order to address the trade-off—*who is contributing and to what extent?*—within the field of social capacity it is helpful to understand who your *target audience* is and what their requirements are (Table 1). There is a spectrum of levels of involvement and sectors of society that citizen science projects work across and below we reflect on the latter.

Which *sectors of society* are considered? The OPAL portfolio aims to support 'citizen science for everyone' and has sought to provide multiple mechanisms for people to get involved, from taking a quick photograph of a Species Quest invertebrate through to undertaking a detailed hour long survey. Efforts were made by all staff to engage with groups that were classed as 'hard to reach.' More than 129,000 beneficiaries were from these communities which included for example, those in areas of deprivation (as identified by the Index of Multiple Deprivation) or from black and ethnic minority (BME) communities through organisations such as Sheffield Black and Ethnic Minority Environmental Network. The primary message from OPAL staff in working with hard to reach groups was "know your audience." For example, when working with BME communities (Table 1), one practitioner suggested that it may be useful to "identify BME community leaders and champions (e.g. from churches, mosques, temples and other places of worship) as an effective way to overcoming barriers to engagement and suspicion." Of broader significance, when engaging BME groups in environmental work, it is important to recognise that some BME groups may feel excluded from the natural environment and experience a cultural severance (Table 1). To OPAL, reaching these sectors of society was fundamental, not only for the social benefits but also because participants from socio-economically deprived areas are under-represented in monitoring schemes [42] and because evidence suggests that deprived areas experience higher levels of environmental degradation [53]. Whoever the target audience is, they need to be understood.

Secondly, the mechanism through which participation can occur is important and as such, is *technology* integrated appropriately? Technology is a huge driver behind the current wave of citizen science. Within OPAL, technology enabled data to be captured, verified, stored, shared, transferred, analysed interpreted and understood and had the capacity to motivate and innovate. For example, comments highlighted that reaching some audiences—in particular young adults—was made easier by the greater use of technology and digital media which offered an inexpensive route in for citizen science practitioners to reach spaces already frequented by this audience (Table 1). Tools in the OPAL programme (Fig. 1) ranged from: an online identification tool—iSpot—which taps into a community of enthusiasts who support each other in verifying species identifications from photographs (390,000 observations received until mid-2014 of which 94 % received a determination [54]); to a digital learning suite—the OPAL Learning Lab—which provides guidance and fun online activities which mimic the Community Scientists role; to recording software—Indicia and iRecord—which enables communities to create local databases of interest to them (such as seaweed recording forms by the British Phycological Society to the Sea Life Tracker app by the organisation Nature Locator [55]). The Species Quest mobile app allowed participants to send photographs of species along with location and date information (key components of a biological record), which allowed scientists to subsequently verify the species identification and add positive record data points with confidence. The benefits of apps for high quality, rapid response geo-referenced records are clear and recommendations for design of smartphone applications can be found elsewhere [56], however, not all audiences (such as parts of natural history societies and older generations) have access to a computer/smartphone, or if they do, some may not be 'tech-savvy'. Therefore digital communications were not solely relied upon. Indeed Mueller et al. [57] note that it is often those that may benefit most from new technologies that lose out. To provide a means for those without access to a computer to send in their results, a freepost address was set up shortly after the launch of the first OPAL survey [33]; this increased the amount of data returned and ensured that all those who wanted to participate (i.e. OPAL's target audience) could do so. It is therefore essential to assess how technology can be used to maximum effect and where more traditional methods could be maintained. These questions are key considerations in order to understand whether communities are effectively supported to respond to social or ecological challenges.

## Evaluation framework: Can a project have its citizen science cake and eat it?

Is the proverb correct—is it impossible to 'have it both ways' in order to achieve two apparently conflicting objectives? Being a portfolio of projects aiming to balance research and outreach objectives, OPAL provided a lens through which to identify trade-offs and investigate mechanisms to work around these challenges in order to address this question. While OPAL is only one programme in a sea of other citizen science approaches, and some findings will be specific to OPAL, many other projects also aim to balance outreach and research. We therefore attempted to capture key lessons learned in the OPAL programme in order to investigate the practical approaches for overcoming trade-offs.

To illustrate how practical approaches can lead to a successful balance between research and outreach objectives, we have illustrated the findings from OPAL by combining them with the evaluation framework of Haywood and Besley [12, 45] (Fig. 2). We have taken learning points from the daily operations of the OPAL programme (which we term 'Operational Indicators') and matched them (slice A–D) with the factors that Haywood and Belsey suggest lead to a successful programme outcome (which we term 'Outcome Indicators'). The first slice (A) proposes that through a project's operations the outreach and research should be *fit for purpose* (through designing products to ensure quality products and monitoring quality throughout) with the end result that the *needs are met* of both scientists and participants. The second slice (B) proposes that *working collaboratively* (by getting adequate buy-in and obtaining feedback) is important when delivering multi-aim projects in order to build *trust and confidence* between the partners. The third slice (C) proposes that *building a diverse partnership* (will ensure expertise are available) in order to widen and advance the *scope* of the project. The last slice (D) suggests that considering the *target audience* (which sectors of society should be targeted and how technology should support this) is imperative when maximising *social capacity.*

OPAL received a relatively substantial budget with which to carry out its national scale operations which provided the resources for the network to explore the full spectrum of citizen science from outreach to research and evaluate the challenges and potential solutions along the way. While every project will have differing objectives and different levels of financial support, a number of common challenges are likely to be encountered. The solution to these will of course vary in detail but we suggest four broad operational approaches which may help to alleviate trade-offs encountered with dual-aim projects. Given the evidence presented within this manuscript we therefore believe that it may indeed be possible to have your citizen science cake and eat it, given appropriate planning, monitoring and adaptation.

### Abbreviations

BLF: Big Lottery Fund; BME: black and ethnic minority communities; CS: citizen science; OPAL: Open Air Laboratories; PES: public engagement in science; PUST: public understanding of science and technology; PPSR: public participation in scientific research.

### Authors' contributions

PLF reviewed data to draw out key trade-offs and operational solutions to challenges, formulated the evaluation framework and drafted the manuscript. LG assimilated much of the data from OPAL evaluation records and provided prose for the methodological and quantitative sections of the paper. AJ, SW and MA contributed prose and insight to particular qualitative data sources. LD conceived the original design of the programme and along with RF refined the salient lessons learned and provided guidance during drafting. RvdW shaped the content and presentation of the information within the manuscript. All authors read and approved the final manuscript.

### Author details

[1] Centre for Environmental Policy, Imperial College London, South Kensington, London SW7 1NA, UK. [2] Forest Research, Alice Holt Lodge, Farnham, Surrey GU10 4LH, UK. [3] Stockholm Environment Institute, University of York, Heslington, York YO10 5DD, UK. [4] Department for the Natural and Built Environment, Faculty of Development and Society, Sheffield Hallam University, Sheffield S1 1WB, UK. [5] Aberdeen Centre for Environmental Sustainability, School of Biological Sciences, University of Aberdeen, Aberdeen AB24 3UU, UK.

### Acknowledgements

We would like to thank the Big Lottery Fund for financing this research through the OPAL programme and Defra for supporting the publication of this manuscript. David Slawson and Kate Martin provided guidance throughout the development of this manuscript. Lastly, we are grateful to the OPAL participants, without whom this citizen science programme would not be possible.

### Competing interests

The authors declare that they have no competing interests.

### Declarations

This article has been published as part of *BMC Ecology* Volume 16 Supplement 1, 2016: Citizen science through the OPAL lens. The full contents of the supplement are available online at http://bmcecol.biomedcentral.com/articles/supplements/volume-16-supplement-1. Publication of this supplement was supported by Defra.

**References**


1. Bonney R, Ballard H, Jordan R, McCallie E, Phillips T, Shirk J, Wilderman CC. Public Participation in Scientific Research: Defining the Field and Assessing Its Potential for Informal Science Education. A CAISE Inquiry Group Report. Washington: Center for Advancement of Informal Science Education (CAISE); 2009.
2. Lintott C, Schawinski K, Bamford S, Slosar A, Land K, Thomas D, Edmondson E, Masters K, Nichol RC, Raddick MJ, Szalay A, Andreescu D, Murray P, Vandenberg J. Galaxy Zoo 1: data release of morphological classifications for nearly 900,000 galaxies. Mon Not R Astron Soc. 2011;410:166–78.
3. Khatib F, Cooper S, Tyka MD, Xu K, Makedon I, Popovic Z, Baker D, Players F. Algorithm discovery by protein folding game players. Proc Natl Acad Sci. 2011;108:18949–53.
4. Stevens M, Vitos M, Lewis J, Haklay M. Participatory monitoring of poaching in the Congo basin. In: GISRUK 2013, 21st GIS Research UK conference, Liverpool. 2013.
5. Oxford English Dictionary Online: "citizen, n. and adj." 2014. http://www.oed.com/view/Entry/33513?redirectedFrom=citizen+science%20.
6. Miller-Rushing A, Primack R, Bonney R. The history of public participation in ecological research. Front Ecol Environ. 2012;10:285–90.
7. Davies L, Fradera R, Riesch H, Lakeman Fraser P. Surveying the citizen science landscape: an exploration of the design, delivery and impact of citizen science through the lens of the Open Air Laboratories (OPAL) programme. BMC Ecol 2016, 16(Suppl 1).
8. Silvertown J. A new dawn for citizen science. Trends Ecol Evol. 2009;24:467–71.
9. Bonney R. Citizen science: a lab tradition. Living Bird. 1996;15:7–15.
10. Irwin A. Citizen Science: a study of people, expertise and sustainable development. New York: Routledge; 1995.
11. Riesch H, Potter C. Citizen science as seen by scientists: methodological, epistemological and ethical dimensions. Public Underst Sci. 2014;23:107–20.
12. Haywood BK, Besley JC. Education, outreach, and inclusive engagement: towards integrated indicators of successful program outcomes in participatory science. Public Underst Sci. 2014;23:92–106.
13. Dickinson J, Bonney R: Why Citizen Science? In: Dickinson J, Bonney R, eds. Citiz Sci Public Particip Environ Res. Ithaca and London: Cornell University Press; 2012:1–14.
14. Zoellick B, Nelson SJ, Schauffler M. Participatory science and education: bringing both views into focus. Front Ecol Environ. 2012;10:310–3.
15. Freitag A, Pfeffer MJ. Process, not product: investigating recommendations for improving citizen science "success". PLoS One. 2013;8:1–5.
16. "Outreach". http://dictionary.cambridge.org/dictionary/british/outreach.
17. "Research". http://dictionary.cambridge.org/dictionary/british/research.
18. Toogood M. Engaging publics: biodiversity data collection and the geographies of citizen science. Geogr Compass. 2013;7:611–21.
19. Bates AJ, Lakeman Fraser P, Robinson L, Tweddle JC, Sadler JP, West SE, Norman S, Batson M, Davies L. The OPAL bugs count survey: exploring the effects of urbanisation and habitat characteristics using citizen science. Urban Ecosyst. 2015;18(4):1477–97.
20. Bates AJ, Sadler JP, Everett G, Grundy D, Lowe N, Davis G, Baker D, Bridge M, Clifton J, Freestone R, Gardner D, Gibson C, Hemming R, Howarth S, Orridge S, Shaw M, Tams T, Young H. Assessing the value of the Garden Moth Scheme citizen science dataset: how does light trap type affect catch? Entomol Exp Appl. 2013;146:386–97.
21. Thiel M, Penna-Díaz MA, Luna-Jorquera G, Salas S, Sellanes J, Stotz W. Citizen scientists and marine research: volunteer participants, their contributions and projections for the future. Oceanogr Mar Biol. 2014;52:257–314.
22. Bonney R, Cooper CB, Dickinson J, Kelling S, Phillips T, Rosenberg KV, Shirk J. Citizen science: a developing tool for expanding science knowledge and scientific literacy. Bioscience. 2009;59:977–84.
23. Wiggins A, Newman G, Stevenson RD, Crowston K. Mechanisms for data quality and validation in citizen science, In 2011 IEEE Seventh Int Conf e-Science Work. IEEE. 2011;2011:14–9.
24. iSpot. http://www.opalexplorenature.org/ispot.
25. The OPAL Learning Lab. http://www.opalexplorenature.org/learning-lab.
26. Idicia. http://www.opalexplorenature.org/Indicia.
27. Davies L, Gosling L, Bachariou C, Fradera R, Manomaiudom N, Robins S. OPAL Community Environment Report. 2013.
28. United Nations General Assembly: United Nations Framework Convention on Climate Change: Resolution. 1994.
29. Hopkins GW, Freckleton RP. Declines in the numbers of amateur and professional taxonomists: implications for conservation. Anim Conserv. 2002;5:245–9.
30. Waite S. Losing our way? The downward path for outdoor learning for children aged 2–11 years. J Adventure Educ Outdoor Learn. 2010;10:111–26.
31. Big Lottery Fund: Changing Spaces England Booklet. 2008.
32. Seed L, Wolseley P, Gosling L, Davies L. Power S a: modelling relationships between lichen bioindicators, air quality and climate on a national scale: results from the UK OPAL air survey. Environ Pollut. 2013;182:437–47.
33. Bone J, Archer M, Barraclough D, Eggleton P, Flight D, Head M, Jones DT, Scheib C, Voulvoulis N. Public participation in soil surveys: lessons from a pilot study in England. Environ Sci Technol. 2012;46:3687–96.
34. Rose NL, Turner SD, Goldsmith B, Gosling L, Davidson T. Quality control in public participation assessments of water quality: The OPAL Water Survey. BMC Ecol. 2016; 16(Suppl 1).
35. Gosling L, Sparks TH, Araya Y, Harvey M, Ansine J: Differences between urban and rural hedges in England revealed by a citizen science project. BMC Ecol. 2016; 16(Suppl 1).
36. Lee MA, Power SA. Direct and indirect effects of roads and road vehicles on the plant community composition of calcareous grasslands. Environ Pollut. 2013;176:106–13.
37. Lee MA, Davies L, Power SA. Effects of roads on adjacent plant community composition and ecosystem function: an example from three calcareous ecosystems. Environ Pollut. 2012;163:273–80.
38. OPAL: OPAL Water Centre Monitoring Report 2008–2012. 2013.
39. Bone J, Barraclough D, Eggleton P, Head M, Jones DT, Voulvoulis N. Prioritising soil quality assessment through the screening of sites: the use of publically collected data. L Degrad Dev. 2012;25:251–66.
40. Fowler A, Whyatt JD, Davies G, Ellis R. How reliable are citizen-derived scientific data? Assessing the quality of contrail observations made by the general public. Trans GIS. 2013;17:488–506.
41. Everett G, Geoghegan H: Initiating and continuing participation in citizen science for natural history. BMC Ecol. 2016, 16(Suppl 1).
42. Hobbs SJ, White PCL. Motivations and barriers in relation to community participation in biodiversity recording. J Nat Conserv. 2012;20:364–73.
43. Stagg B, Donkin M. Teaching botanical identification to adults: experiences of the UK participatory science project "Open Air Laboratories". J Biol Educ. 2013;47:104–10.
44. Davies L, Bell JNB, Bone J, Head M, Hill L, Howard C, Hobbs SJ, Jones DT, Power SA, Rose N, Ryder C, Seed L, Stevens G, Toumi R, Voulvoulis N, White PCL. Open Air Laboratories (OPAL): a community-driven research programme. Environ Pollut. 2011;159:2203–10.
45. Haywood BK, Besley JC. Online Appendix of Education, outreach, and inclusive engagement: towards integrated indicators of successful program outcomes in participatory science. Public Underst Sci. 2014;23:92–106.
46. UK Ladybird Survey. http://www.ladybird-survey.org/.
47. OPAL: OPAL Species Quest; 2015.
48. Parliamentary Office of Science and Technology: Environmental Citizen Science; 2014.
49. Riesch H, Potter C, Davies L: Combining citizen science and public engagement: the Open Air Laboratories Programme. J Sci Commun. 2013; 12.
50. Research Excellence Framework. http://www.ref.ac.uk/.
51. Research Excellence Framework: REF 2014: Assessment Framework and Guidance on Submissions; 2011.
52. Van der Wal R, Anderson H, Robinson A, Sharma N, Mellish C, Roberts S, Darvill B, Siddharthan A. Mapping species distributions: comparing the spread of UK bumblebees as recorded by the national depository and a citizen science approach. AMBIO Spec Issue Digit Conserv. 2015;44(Suppl 4):584–600.
53. Pye S, King K, Sturman J. Air quality and social deprivation in the UK: an environmental inequalities analysis—Final Report to Defra. Contract RMP/2035 AEA Technology Plc; 2006.
54. Silvertown J, Harvey M, Greenwood R, Dodd M, Rosewell J, Rebelo T, Ansine J, Mcconway K. Crowdsourcing the identification of organisms: a case-study of iSpot. Zookeys. 2015;146:125–46.
55. List of Indicia online recording sites and apps. http://www.nbn.org.uk/nbn_wide/media/Documents/Indicia/List-of-Indicia-forms.pdf.

56. Adriaens T, Sutton-croft M, Owen K, Brosens D, Van Valkenburg J, Kilbey D, Groom Q, Ehmig C, Thürkow F, Van Hende P, Schneider K. Trying to engage the crowd in recording invasive alien species in Europe: experiences from two smartphone applications in northwest Europe. Manag Biol Invasions. 2015;6:215–25.

57. Mueller M, Tippins D, Bryan L. The future of citizen science. Democr Educ. 2012;20:1–12.

# 11

# Floral volatiles interfere with plant attraction of parasitoids: ontogeny-dependent infochemical dynamics in *Brassica rapa*

Gaylord A Desurmont[1*], Diane Laplanche[1], Florian P Schiestl[2] and Ted C J Turlings[1]

**Abstract**

**Background:** The role of plant ontogeny on investment in direct defense against herbivores is well accepted, but the transition from the vegetative to the reproductive stage can also affect indirect resistance traits (i.e. attraction of the natural enemies of plant attackers). Here, we conducted behavioral bioassays in olfactometers to determine whether the developmental stage (vegetative, pre-flowering, and flowering) of *Brassica rapa* plants affects attraction of *Cotesia glomerata*, a parasitoid of the herbivore *Pieris brassicae*, and examined the blends of volatile compounds emitted by plants at each developmental stage.

**Results:** *Pieris*-infested plants were always more attractive to parasitoids than control plants and plants infested by a non-host herbivore, independently of plant developmental stage. On the other hand, the relative attractiveness of *Pieris*-infested plants was ontogeny dependent: *Pieris*-infested plants were more attractive at the pre-flowering stage than at the vegetative stage, and more attractive at the vegetative stage than at the flowering stage. Chemical analyses revealed that the induction of leaf volatiles after herbivory is strongly reduced in flowering plants. The addition of synthetic floral volatiles to infested vegetative plants decreased their attractiveness to parasitoids, suggesting a trade-off between signaling to pollinators and parasitoids.

**Conclusion:** Our results show that putative indirect resistance traits are affected by plant development, and are reduced during *B. rapa* reproductive stage. The effects of ontogenetic shifts in resource allocation on the behavior of members of the third trophic level may have important implications for the evolution of plant defense strategies against herbivores.

**Keywords:** Plant ontogeny, Herbivore-induced plant volatiles, Indirect defense, Host location, Plant signaling, Floral VOCs

## Background

The production of secondary metabolites (i.e. compounds that are not directly involved in the growth or reproduction of the plant) detrimental to herbivores is one of the main strategies plants have evolved to fend off their consumers. Secondary metabolites can directly deter or negatively affect the performance of herbivores (direct defense) [1], or attract and facilitate the action of natural enemies of herbivores (indirect resistance) [2].

Investment in secondary metabolism can vary tremendously through the lifetime of single plants [3]. Such temporal changes can be separated in two categories: changes driven by fluctuating environmental conditions are referred to as seasonal, and changes associated with the development of the plant are referred to as ontogenetic [4]. Both types of temporal variation can play an important role in plant defense and insect–plant interactions. For example, decrease in plant resistance due to suboptimal environmental conditions may create a window of vulnerability to herbivores [5]. Additionally, plants may not have the same needs for defenses or equal amounts of resources to invest in defensive compounds

*Correspondence: gaylord.desurmont@unine.ch
[1] Institute of Biology, University of Neuchâtel, Rue Emile-argand 11, 2000 Neuchâtel, Switzerland
Full list of author information is available at the end of the article

at all points of their development [3], and such variation may also lead to herbivore adaptation (e.g. herbivore adjusting its "offense" to ontogenetic shifts in plant defense) [6, 7].

The role of ontogeny on direct defenses has been thoroughly studied in the context of the general trade-off between defense and growth/reproduction in plants [8, 9]. It is generally admitted that plant tissues are more valuable, and herbivory more impactful, at the early stages of development [10], and thus that young plant tissues (e.g. seedlings and saplings for herbaceous plants, young expanding leaves for trees) should be more protected than older ones [11], as predicted by the optimal defense hypothesis [12, 13]. However, plants may not always have the capacity to protect young tissues optimally: resource limitations may constrain the production of defensive compounds until the plant has gained enough biomass to allocate resources for defenses [14]. As plants mature and reach the reproductive stage, the dynamics of defenses may change again: levels of defensive compounds may be reallocated to valuable reproductive tissues (e.g. flowers, seeds), or globally decrease if the production of secondary metabolites constrains the production of reproductive tissues [3, 15]. In the case of insect-pollinated plants, the production of defensive compounds may also diminish if those interfere with the action of mutualist pollinators [16]. In contrast to these studies on ontogeny-mediated variation in direct defenses, much less experimental work exists on the effects of plant development on indirect resistance.

Indirect resistance refers to plant adaptations encouraging the presence of natural enemies of insect herbivores, which in turn help plants by reducing herbivore damage. When the action of natural enemies consistently results in increased plant fitness, the term "indirect defense" may be used [2]. These adaptations fall into two main categories: rewards such as food or shelters for natural enemies (e.g. extrafloral nectaries, domatia) [17], and cues facilitating foraging behavior (e.g. herbivore-induced volatiles) [18, 19]. These adaptations have the potential to be ontogeny-dependent: organs rewarding natural enemies may not be developed in the early developmental stages [20], or the emission of plant volatiles may fluctuate over the plant's lifetime [21].

Several studies have shown that blends of herbivore-induced plant volatiles (HIPVs) can vary over a plant's lifetime [22], but how these ontogeny-driven changes in HIPVs affect the foraging behavior of natural enemies is vastly unknown [23], particularly during the transition between the vegetative stage and the reproductive stage [24, 25]. This transition is important for members of the third trophic level, because nectar and/or pollen are important food sources for numerous species of natural enemies, including parasitic wasps [26, 27]. Infested flowering plants may offer a "double reward" (host + food) to visiting parasitoids [28], and thus be more attractive than infested vegetative plants. Alternatively, floral volatiles may also interfere with attractive HIPVs in the headspace and reduce their attractiveness to parasitoids. The reliability of HIPVs as specific cues for natural enemies in search of host or prey has been the subject of much debate [29, 30], and ontogeny-driven variations in HIPVs may alter the quality and specificity of the cues natural enemies are looking for, which could have major consequences for the foraging behavior of specialized natural enemies such as parasitoids. Here, we use the plant *Brassica rapa* and one of its major herbivores, the butterfly *Pieris brassicae*, to test whether the attractiveness of herbivore-damaged plants to a specialized natural enemy of *P. brassicae*, the parasitoid *Cotesia glomerata*, changes through plant ontogeny. *Cotesia glomerata* is the main larval parasitoid of *P. brassicae* in Western Europe and can be very abundant in the field [31], but it remains unclear whether *Brassica* plants benefit from its attraction in terms of reduction of herbivore damage and fitness gain [32, 33]. Indeed *C. glomerata* kills its host at the end of its larval development [34], once most of the damage has been done to the plant. In order to test the specificity of the volatiles cues emitted by *B. rapa* in response to *P. brassicae*, we compared *C. glomerata* attraction toward *Pieris*-infested plants to plants infested by a non-host herbivore, *Spodoptera littoralis* through plant ontogeny. Specifically, we used behavioral bioassays to answer the following questions:

(1) Are pre-flowering and flowering plants infested by *Pieris* caterpillars attractive to parasitoids, and how does this attractiveness compare to infested vegetative plants and non-infested plants?

(2) Do *C. glomerata* wasps show preferences between volatiles emitted by plants infested by its host compared to volatiles emitted by plants damaged by a non-host herbivore, *S. littoralis*, at different plant developmental stages?

Because we saw a decrease in attractiveness to parasitoids in infested flowering plants, we then used manipulative experiments to test two potential proximal mechanisms explaining this result: (a) floral volatiles directly interfere with attractive leaf volatiles, and (b) a behavioral change from folivory to florivory (i.e. eating flowers) in *P. brassicae* caterpillars leads to reduced parasitoid attraction.

(3) How do herbivore-induced plant volatiles vary through plant ontogeny? In addition to the behavioral assays abovementioned, we conducted chemical analyses to measure quantitative and qualitative variation in volatile emissions between undamaged and *Pieris*-infested

plants at each of the developmental stages tested (vegetative, pre-flowering, and flowering)

## Methods

### Insect and plant material

Plants used in the study came from a wild accession of *B. rapa* whose seeds were collected in 2012 near Maarsen, the Netherlands. Plants were grown in controlled growth chambers under 16:8 L:D light regime at 25°C, in cylindrical plastic pots (4 × 10 cm) with fertilized commercial soil (Ricoter Aussaaterde, Aarberg, Switzerland). Plants were between 3 and 7 weeks old when used for the experiments: plant age could not be standardized because the time at which plants would enter the reproductive stage was unpredictable. Plants were considered in the vegetative stage when they only bore leaves, in the pre-flowering stage once they produced a 8–10 cm long flowering stalk (ca. 5 day before flowering), and in the flowering stage once they started producing open flowers (Figure 1). Plants in the flowering stage were used 2–15 days after the opening of the first flower.

*Pieris brassicae* (Lepidoptera: Pieridae) is a specialist herbivore of crucifers widespread in most of Eurasia. The larvae are typically leaf-feeders, but are known to switch from folivory to florivory after the second larval instar on *Brassica nigra* plants [35], as well as on our *B. rapa* plants (GAD personal observation). *P. brassicae* came from our laboratory rearing, started with field-collected individuals from the Zürich area (Switzerland), and were reared on *Brassica* plants. In order to prepare the *Pieris*-infested plants used for the behavioral bioassays and the chemistry analysis, 15 first instar (L1) larvae were randomly placed on the leaves of the treatment plant using a fine brush the day prior to the experiment. Herbivores were left feeding on the plant during the experiment. First instar larvae were used for infestation treatments because they are the most susceptible to parasitism by *Cotesia glomerata*.

**Figure 1** *Brassica rapa* plants at the different developmental stages used in the study: flowering stage (**a**), pre-flowering stage (**b**), and vegetative stage (**c**). The *arrow* indicates the flowering stalk, indicative that the plant has entered the pre-flowering stage.

The generalist herbivore *Spodoptera littoralis* is native to Africa and invasive in the southern part of Europe [36]. Individuals of *S. littoralis* used for the experiments came from eggs shipped weekly by Syngenta (Stein, Switzerland). In order to prepare *Spodoptera*-infested plants for the experiments, 25 first and second instar larvae were placed randomly on the leaves of the treatment plant using a fine brush 12–18 h prior to the experiment. Herbivores were left feeding on the plant during the experiment. We used more *S. littoralis* larvae than *P. brassicae* larvae for herbivore-infested treatments (25 vs. 15) because these numbers of larvae resulted in comparable amounts of damage on *B. rapa* leaves after 24 h during preliminary experiments.

The specialist natural enemy of *P. brassicae*, the parasitic wasp *C. glomerata* (Hymenoptera: Braconidae) is a major endoparasitoid of *P. brassicae* in temperate Western Europe, and parasitizes early instars of the caterpillar. Individuals of *C. glomerata* used for the behavioral bioassays came from our laboratory rearing, started with field-collected individuals from the Netherlands and later supplemented with field-collected individuals from the Neuchâtel area (Switzerland). Newly emerged wasps were left in rearing cages at ambient temperature with water and honey for feeding and mating for 48 h, a period that is typically sufficient to ensure successful mating [37]. Then, rearing cages were transferred in a growth chamber at 13°C (16/8 L:D light regime) with water and honey until parasitoids were needed for the experiments (1–6 weeks after emergence). Parasitoid age variability did not differ between the different olfactometer experiments. Parasitoids were removed from their growth chamber and left at ambient temperature for ~1 h before the bioassays. Only naive females (i.e. females that had never encountered a host prior to the experiment) were used in the bioassays, and insects were only used once for experimental purposes.

### Behavioral bioassays

The preferences of *C. glomerata* females toward certain odor blends were investigated using 4 and 6-arm olfactometer settings [38, 39]. In these settings, wasps were given the choices between 4 or 6 odor sources (=treatments) contained in separated glass bottles. Individual air flows were connected to each odor source, and all air flows converged to a central glass piece, where the wasps were released. After 30 min spent in the olfactometer, wasps were recollected and the treatment they chose was recorded. Wasps that did not make a choice were recorded as "no choice" in the analysis of the results. An olfactometer test (=1 replicate) consisted in five consecutive releases of five wasps (wasps were replaced between releases) for the 4-arm olfactometer tests, and in six consecutive releases of six wasps (wasps were replaced between releases) for the 6-arm olfactometer tests. We conducted a minimum of five replicates for each experiment. Plants were changed and glassware was cleaned between replicates. The cleaning process of the glassware consisted in rinsing the glassware sequentially with three solvents: water, acetone, and pentane, and putting the glassware in an oven at 250°C for a minimum of 3 h. A minimum of five replicates were conducted for each of the experiments described below.

#### Chemical analysis

To identify and quantify the blends of volatile organic compounds (VOCs) emitted by undamaged and infested *B. rapa* plants at the vegetative and flowering stage, plants (n = 14, 12, and 12 for the vegetative, pre-flowering, and flowering stage, respectively) were placed in a VOC collection setup [40] for 5 h. VOCs were collected using a trapping filter containing 25 mg of 80–100 mesh SuperQ absorbent. Before use, trapping filters were cleaned with 300 μL of methylene chloride. After each collection, VOCs were extracted from the filters with 150 μL of methylene chloride. The collection was performed first with undamaged plants then, using the same plants, 24 h after infestation by 15 L1 *P. brassicae* caterpillars. Caterpillars were not removed from the plants during the volatile collection.

The samples were stored at −80°C before analysis. Two internal standards (*n*-octane and nonyl acetate, each 200 ng in 10 mL methylene chloride) were added to each sample. VOCs were analysed with an Agilent 6850 gas chromatograph with a flame ionization detector. A 2 μL aliquot of each sample was injected in the pulsed splitless mode onto a non-polar column (HP-1 ms, 30 m, 0.25 mm ID, 0.25 μm film thickness, Agilent J&W Scientific, USA). Helium at constant flow (1.9 mL/min) was used as carrier gas. The quantities of the major components of the blends were estimated based on the peak areas of the compounds compared to the peak areas of the internal standards. Identities of the compounds were confirmed by mass spectrometry analysis whenever possible. Compounds were identified by comparing the spectra obtained from the samples with those from a reference database (NIST mass spectral library).

#### Experiments

(1) Are pre-flowering and flowering plants infested by *Pieris* caterpillars attractive to parasitoids, and how does this attractiveness compare to infested vegetative plants and non-infested plants?

To determine the attractiveness of *Pieris*-infested plants compared to control plants at each developmental stage, two separated choice-tests, one with vegetative plants and one with flowering plants, were conducted in 4-arm olfactometers (n = 5 replicates, 125 wasps tested). For these tests, wasps were given the choice between a *Pieris*-infested plant, a non-infested plant (control), and two empty odor sources (blanks). Then, to directly compare the attractiveness of *Pieris*-infested vegetative and flowering plants, a 6-arm olfactometer test was conducted, giving the wasps the following choices: a *Pieris*-infested vegetative plant, a *Pieris*-infested flowering plant, a non-infested vegetative plant, a non-infested flowering plant, and two empty odor sources (blanks) (n = 5, 180 wasps tested). The same experiment was repeated with pre-flowering plants instead of flowering plants (n = 5, 180 wasps tested).

(2) Do *C. glomerata* wasps show preferences between volatiles emitted by plants infested by its host compared to volatiles emitted by plants damaged by a non-host herbivore, *Spodoptera littoralis*, at different plant developmental stages?

In order to determine the specificity of HIPVs produced by infested plants at different developmental stages, two separate 4-arm olfactometer choice-tests were conducted, one with vegetative plants and one with flowering plants (n = 5, 125 wasps tested). For these tests, wasps were given the choice between a *Pieris*-infested plant, a *Spodoptera*-infested plant, a non-infested plant, and one empty odor source (blank).

(a) Do floral volatiles directly interfere with attractive HIPVs?

We used synthetic blends of floral volatiles and vegetative *Pieris*-infested plants to determine the influence of floral odors on the foraging behavior of parasitoids. Synthetic blends of volatiles were preferred to real inflorescences for this experiment to limit variability. Specifically we conducted a 4-arm olfactometer experiment, giving the wasps the following choices: *Pieris*-infested vegetative plant, *Pieris*-infested vegetative plant + synthetic floral blend, non-infested vegetative plant, and an empty odor source (blank) (n = 5, 125 wasps tested). Synthetic floral blends included six of the most abundant compounds found in the floral bouquet of our wild accession of *B. rapa* (Knauer and Schiestl unpublished): phenylacetaldehyde (≥90%, Sigma-Aldrich, Buchs, Switzerland) 3 μL/mL, nonanal (Givaudaudan, Dübendorf, Switherland) 9 μL/mL, decanal (Givaudaudan) 4 μL/mL, acetophenone (Givaudaudan) 24.5 μL/mL, *p*-Anisaldehyde (puriss. Sigma Aldrich) 27 μL/mL, and α-Farnesene (mixture of isomers, Sigma Aldrich) 492 μL/mL, diluted in methylene chloride (HPLC grade). Before olfactometer tests, rubber septa (GR-2, 5 mm Supelco, Bellefonte, PA,

USA) were soaked in the synthetic floral blend solution for 1 h, then were allowed to dry for 4 h. The concentration of each compound in the solution was adjusted so that the emission rate of each compound from a septa was comparable to one inflorescence (ca. 30 flowers) of *B. rapa* [41] Septa were then placed above the chosen treatment plant just prior to the tests, using a fine wire mesh to fix them at the desired location inside the olfactometer. Fine wire mesh was also added to the treatments without synthetic floral blends in order to account for the potential effects of wire odors. Preliminary trials were conducted before the experiments to show that rubber septa soaked only in solvent (methylene chloride) did not have an effect on parasitoid attraction under the same experimental conditions.

(b) Is the behavioral change from folivory to florivory in *P. brassicae* caterpillars linked to reduced attractiveness to parasitoids?

When they were placed randomly on flowering plants and let free for 24 h, *P. brassicae* first instar larvae generally disperse across the plant and damaged both leaves and flowers. Thus, to test the influence of florivory on parasitoid attraction, we gave the choice to parasitoids between *Pieris*-infested plants with larvae restrained to the leaves, and *Pieris*-infested plants with larvae let free to feed on both leaves and flowers. A 4-arm olfactometer test was conducted, giving the wasps the following choices: *Pieris*-infested flowering plant, *Pieris*-infested flowering plant whose flowers were bagged in a fine-mesh net to prevent larvae from reaching them, non-infested flowering plant, and an empty odor source (blank) ($n = 5$, 125 wasps tested). In order to avoid the potential effects of bag odors on parasitoid behavior, a fine-mesh net was added to all the treatments and placed next to the plants.

#### Statistical analysis

Preferences of *C. glomerata* females were analyzed for each test using a generalized linear model (GLM) fitted by maximum quasi-likelihood estimation according to Turlings et al. [38]. Means were compared using a Chi square test and a multiple comparison Wilcoxon test ($\alpha = 0.01$, JMP9). The number of wasps choosing the different branches (treatments) of the olfactometer constituted the dependent variable. Treatments were then compared using the all-pairwise Tukey–Kramer HSD procedure (JMP9). Results are presented as percentages in the figures illustrating olfactometer tests for easier comprehension: percentage attractiveness of a given treatment was calculated as the number of wasps that chose that particular treatment divided the total number of wasps that made a choice during the test *100 (wasps that did not make a choice were excluded from the calculations of percentage attractiveness). Results of the chemical analysis were analyzed in several ways. Firstly, the compounds emitted by vegetative plants were separated in two categories: leaf compounds and floral compounds. Floral compounds were compounds either only produced or significantly more produced by flowering plants compared to vegetative plants, regardless of infestation. In total, 64 compounds were isolated in *B. rapa* plants: 51 leaf compounds and 13 floral compounds. Secondly, the total volatile emission from plants at each developmental stage before and after infestation by *Pieris* caterpillars was compared for leaf and floral compounds using a paired t test ($\alpha = 0.05$, JMP9) ($n = 11$ pairs for both vegetative and flowering plants, $n = 12$ for pre-flowering plants). Thirdly, the complete blends of volatiles produced by plants at each developmental stage before and after infestation ($n = 25$ for vegetative plants, $n = 24$ for pre-flowering plants, $n = 23$ for flowering plants,) were compared using a principal component analysis (PCA) with the peak areas of the 15 most common compounds found in the blend of vegetative plants, the 25 most common compounds from the blends of pre-flowering plants, and the 27 most common compounds found in the blend of flowering plants. For our analysis of the blends of *B. rapa* plants, we used the two first principal components, accounting for 32 and 19% of the total variation in the dataset for vegetative plants, for 33 and 18% of the total variation in the dataset for pre-flowering plants, and for 22.2 and 16% of the total variation in the dataset for flowering plants. Clear separation between two points on the axes, representing projections of the principal components (Additional file 1: Figure S2), indicates divergence between the whole blends emitted by two plants.

### Results

Overall, the mean percentage of parasitoids that made a choice during the various olfactometer tests was $84.4 \pm 3.5$ (%, mean $\pm$ SE), ranging from 63.2 to 95.3%.

(1) Are pre-flowering and flowering plants infested by *Pieris* caterpillars attractive to parasitoids, and how does this attractiveness compare to infested vegetative plants and non-infested plants?

Tests conducted in 4-arm olfactometers with vegetative plants and flowering plants separately showed significant differences in attractiveness between the treatments ($\chi^2 = 42.2$, $P < 0.0001$, and $\chi^2 = 49.9$, $P < 0.0001$, respectively, $df = 3$). For both developmental stages, infested plants were more attractive than control plants: infested vegetative plants were five times more attractive than controls, and infested flowering plants twice as attractive as controls (Figure 2A, B). In addition, undamaged flowering plants were more attractive than empty odor sources, while control vegetative plants and empty odor

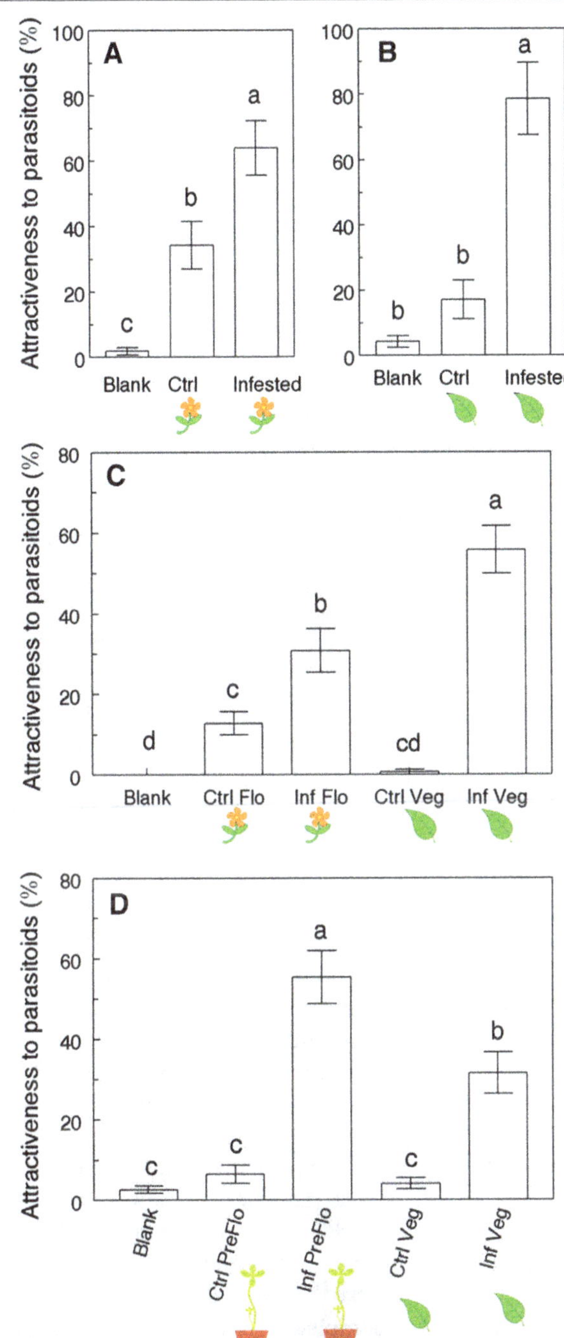

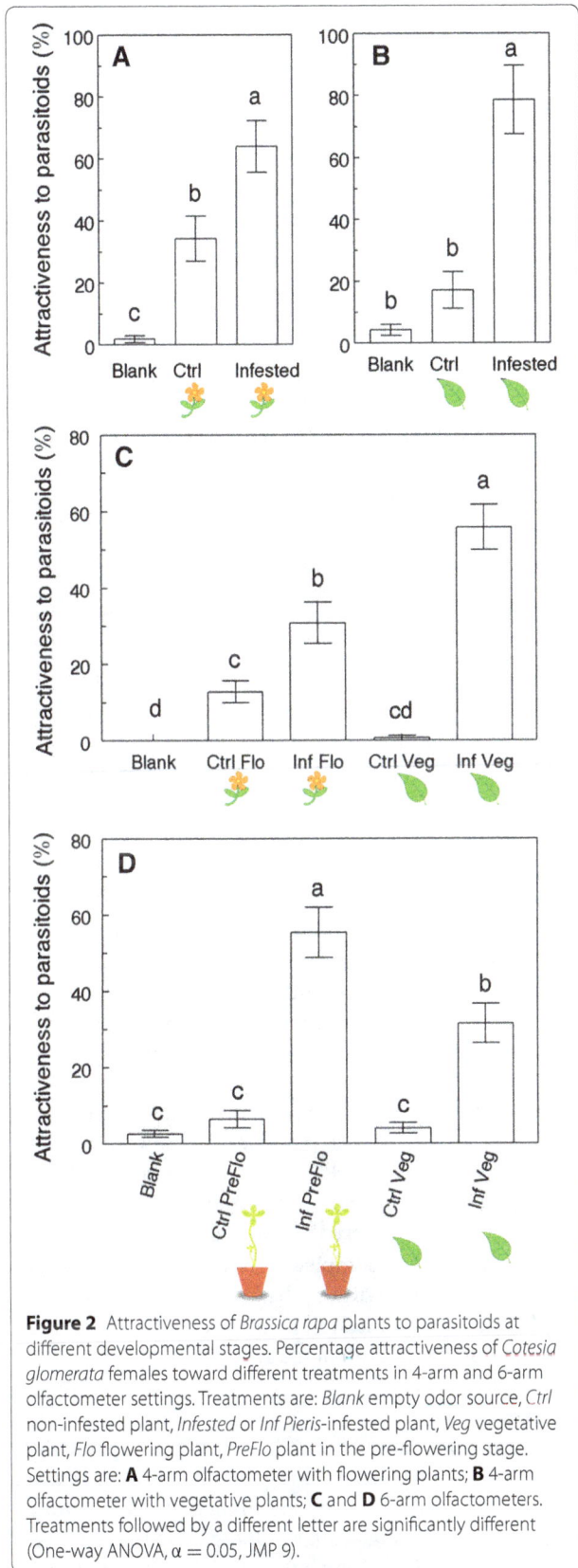

**Figure 2** Attractiveness of *Brassica rapa* plants to parasitoids at different developmental stages. Percentage attractiveness of *Cotesia glomerata* females toward different treatments in 4-arm and 6-arm olfactometer settings. Treatments are: *Blank* empty odor source, *Ctrl* non-infested plant, *Infested* or *Inf Pieris*-infested plant, *Veg* vegetative plant, *Flo* flowering plant, *PreFlo* plant in the pre-flowering stage. Settings are: **A** 4-arm olfactometer with flowering plants; **B** 4-arm olfactometer with vegetative plants; **C** and **D** 6-arm olfactometers. Treatments followed by a different letter are significantly different (One-way ANOVA, $\alpha = 0.05$, JMP 9).

sources had comparably low attractiveness (Figure 2A). When plants from both developmental stages were compared in a 6-arm olfactometer setting, there were significant differences between the treatments ($\chi^2 = 103.8$, $P < 0.0001$, df = 5): most importantly, infested vegetative plants were significantly more attractive to parasitoids than infested flowering plants (55.8 vs 30.8% attractiveness, respectively) (Figure 2C). On the other hand, when pre-flowering plants were compared to vegetative plants in a 6-arm olfactometer setting, infested pre-flowering plants were 75% more attractive than infested vegetative plants (55.4 vs 31.5% attractiveness, respectively, $\chi^2 = 82.1$, $P < 0.0001$, df = 3) (Figure 2D).

(2) Do *C. glomerata* show preferences between volatiles emitted by plants infested by its host compared to volatiles emitted by plants damaged by a non-host herbivore, *Spodoptera littoralis*, at different plant developmental stages?

Tests conducted in 4-arm olfactometers with vegetative plants and flowering plants separately showed significant differences between the treatments ($\chi^2 = 59.4$, $P < 0.0001$, and $\chi^2 = 25.6$, $P < 0.0001$, respectively, df = 3). For both developmental stages, the patterns of attractiveness observed were the same: *Pieris*-infested plants were the most attractive treatment, and *Spodoptera*-infested plants were as unattractive as control plants and empty odor sources (Figure 3A, B).

(a) Do floral volatiles directly interfere with attractive HIPVs?

The hypothesis that floral volatiles interfere with parasitoid foraging behavior was supported by our results. In a 4-arm olfactometer setting with infested vegetative plants with and without floral odors, we saw the following pattern of preferences ($\chi^2 = 23.6$, $P < 0.0001$, df = 3): infested plants without synthetic floral blends were the most attractive treatment followed by infested plants with synthetic floral blends, and control plants and empty odor sources were the least attractive treatments. The presence of floral volatiles decreased the attractiveness of infested vegetative plants by approximately 35% (Figure 4).

(b) Is the behavioral change from folivory to florivory in *P. brassicae* caterpillars linked to reduced attractiveness to parasitoids?

The hypothesis that florivory by *P. brassicae* caterpillars reduces the production of attractive HIPVs was not supported by our results. In a 4-arm olfactometer setting with infested flowering plants where caterpillars had access to the flowers or were denied access to the flowers (i.e. flowers were enclosed in bags preventing florivory), the following preferences were observed ($\chi^2 = 43.2$, $P < 0.0001$): infested plants with access to the flowers

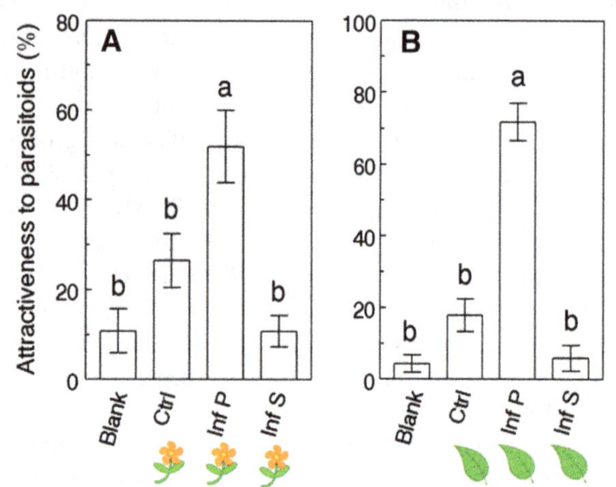


**Figure 3** Specificity of plant volatiles emitted by *Brassica rapa* at different developmental stages. Percentage attractiveness of *Cotesia glomerata* females toward different treatments in olfactometer settings. Treatments are: *Blank* empty odor source, *Ctrl* non-infested plant, *Inf P Pieris*-infested plant, *Inf S Spodoptera*-infested plant. Settings are: **A** 4-arm olfactometer with flowering plants; **B** 4-arm olfactometer with vegetative plants. Treatments followed by a different letter are significantly different (One-way ANOVA, $\alpha = 0.05$, JMP 9).

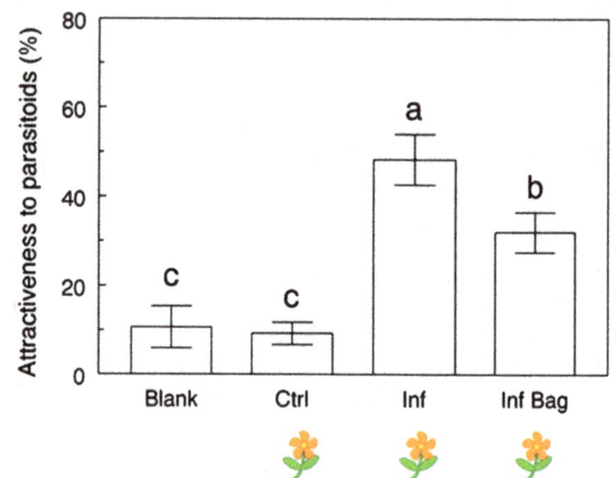


**Figure 5** Effect of florivory on parasitoid attraction. Percentage attractiveness of *Cotesia glomerata* females toward different treatments in a 4-arm olfactometer setting. Treatments are: *Blank* empty odor source, *Ctrl* non-infested flowering plant, *Inf Pieris*-infested flowering plant, *Inf Bag Pieris*-infested plant whose flowers were bagged to prevent florivory. Treatments followed by a different letter are significantly different (One-way ANOVA, $\alpha = 0.05$, JMP 9).

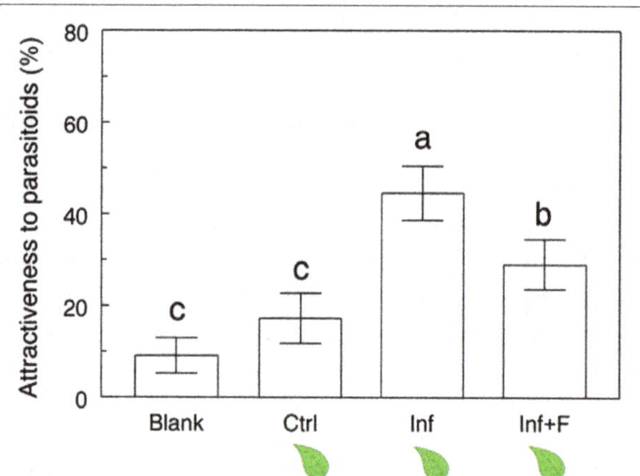


**Figure 4** Direct impact of floral odors on parasitoid attraction. Percentage attractiveness of *Cotesia glomerata* females toward different treatments in a 4-arm olfactometer setting. Treatments are: *Blank* empty odor source, *Ctrl* non-infested vegetative plant, *Inf Pieris*-infested vegetative plant, *Inf + F Pieris*-infested vegetative plant + synthetic floral blend. Treatments followed by a different letter are significantly different (One-way ANOVA, $\alpha = 0.05$, JMP 9).

were the most attractive treatment, followed by infested plants with enclosed flowers. Infested flowering plants with access to the flowers were 50% more attractive than infested flowering plants with enclosed flowers (48.3 vs. 32.0%, respectively). Both these treatments were more attractive than control plants and empty odor sources (Figure 5).

In all olfactometer tests containing more than one herbivore-infested treatment, the surface of leaf area damaged by herbivores did not significantly differ between treatments ($Ps > 0.05$, Additional file 1: Figure S1).

(3) How do herbivore-induced plant volatiles vary through plant ontogeny?

Analysis of the complete volatile emission of vegetative, pre-flowering, and flowering plants before and after infestation by *Pieris* caterpillars revealed that the total amount of leaf volatiles emitted was higher after infestation in vegetative and pre-flowering plants (i.e. there was significant induction of these volatiles) (paired t test, $t = 2.4$, $P = 0.03$ for vegetative plants, and $t = 3.83$, $P = 0.003$ for pre-flowering plants), and there was a non-significant trend toward induction in flowering plants ($t = 1.93$, $P = 0.08$). In addition, the total amount of floral volatiles emitted was higher after infestation in flowering plants ($t = 3.17$, $P = 0.01$) (Figure 6). Principal component analysis of the complete blends of volatiles showed divergence and little overlap between the blends emitted by undamaged and infested plants for each developmental stage tested (Additional file 1: Figure S2).

## Discussion

Plant development and secondary metabolism are tightly linked: the composition and abundance of defensive compounds present in different organs can vary through the lifetime of a single plant. Ontogeny-driven changes in plant metabolism may also affect indirect resistance and attraction of natural enemies [42], through variations

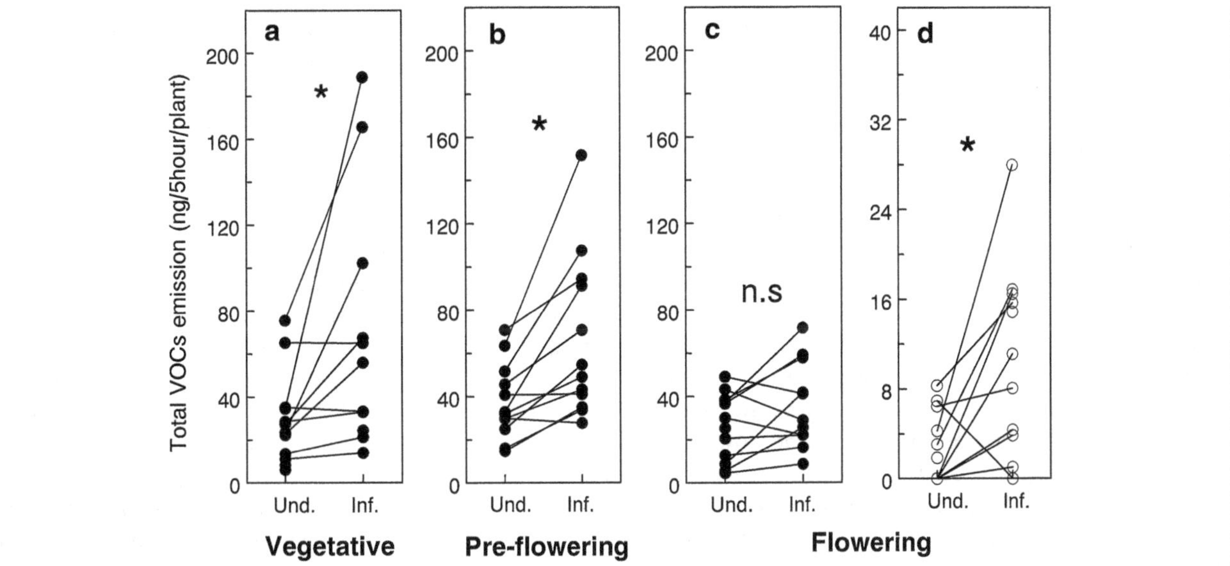


**Figure 6** Total volatile emission of vegetative and flowering plants before and after *Pieris* infestation. **a** Leaf volatiles emitted by vegetative plants (n = 11 pairs), **b** leaf volatiles emitted by pre-flowering plants, **c** leaf volatiles emitted by pre-flowering plants, and **d** floral volatiles emitted by flowering plants. Volatiles were collected for 5 h using super Q filters. *Asterisks* indicate significant increase of volatile emission after infestation (paired t test, $P < 0.05$).

in HIPVs [22, 42–44]. Our findings show clearly that parasitoid attraction is ontogeny-dependant in *B. rapa*: compared to infested vegetative plants, infested pre-flowering plants were more attractive and infested flowering plants less attractive to *C. glomerata* females (Figure 2c, d). Considering that the pre-flowering stage is typically only 7–10 days long (from the production of the stalk to the opening of the first flower, GAD personal observation), this result illustrates the highly dynamic nature of the interactions between plants and members of the third trophic level [45, 46]. Independent of developmental stage, infested plants were always considerably more attractive than undamaged plants (Figure 2) and parasitoids were more attracted to plants damaged by their host than to plants damaged by a non-host herbivore, *S. littoralis* (Figure 3). This supports the idea that volatile cues originating from infested *B. rapa* plants remain herbivore-specific during the transition from the vegetative to the reproductive stage, and represent a reliable signal for parasitoids in search of hosts through plant ontogeny, confirming previous results obtained with the same tritrophic system [47]. Because we kept the herbivores on the plants during the bioassays, we cannot dissociate the effects of plant volatiles from the effects of host-derived cues (e.g. frass odors) on parasitoid attraction, and *Pieris*-derived cues may differ from *Spodoptera*-derived cues, playing a role in the specificity of parasitoid attraction. Herbivores damaged comparable amounts of leaf tissue in all treatments (Additional file 1: Figure S1), which suggests that quantity of damage is not responsible for the differences in parasitoid attraction observed during the different experiments.

Analyses of the volatile blends emitted by *B. rapa* plants before and after *Pieris* infestation showed a significant induction of leaf volatiles after infestation in vegetative and pre-flowering plants, but only a trend toward induction in flowering plants (Figure 6), a pattern that is consistent with recent observations in *B. nigra* [43], suggesting that the decreased attractiveness of infested flowering plants is due to a reduced emission of attractive leaf volatiles. There were qualitative differences between the blends of undamaged and *Pieris*-infested plants at all developmental stages, as illustrated by the results of the PCAs (Additional file 1: Figure S2). Taken together with the fact that parasitoids always preferred *Pieris*-infested plants to undamaged plants, these results show that a quantitative induction of leaf volatiles is not necessary to elicit *C. glomerata* attraction, but that the quantity of volatiles induced still plays a role in the strength of this attraction. Intriguingly, induction of leaf volatiles after infestation was more consistent in pre-flowering plants than in vegetative plants. At the vegetative stage, half of the individual plants tested strongly induced leaf volatiles and the other half of plants showing little or null induction, whereas induction was more consistent in pre-flowering plants (80% of plants tested) (Figure 6a, b). This increased consistency in volatile induction may explain the increased attractiveness of infested pre-flowering plants to *C. glomerata* parasitoids.

Regarding floral volatiles, their emission was increased after *Pieris* infestation (Figure 6d) but, because the number of open flowers at the time of volatile collections was not counted, it is possible that the relative amount of floral volatiles released per flower remained stable or even decreased. Because we used plants that started blooming 2–15 days before the volatile collections, an effect of plant age on the odor blends cannot be excluded. Adding synthetic floral scents to infested vegetative plants made them considerably less attractive to parasitoids (Figure 4), which strongly suggests that floral volatiles do in fact negatively interfere with attractive HIPVs in the headspace and contribute to making infested flowering plants less attractive to *C. glomerata* wasps. In this context, caterpillars feeding on or close to the flowers may avoid detection by parasitoids, a theory that is supported by recent work on *B. nigra* [48]. Further experiments are needed to determine to which extent this interference originates from the presence of certain specific compounds in the blend ("qualitative" interference) or if there is a threshold above which the ratio floral/herbivore-induced volatiles becomes detrimental to parasitoid attraction ("quantitative" interference). In parallel, our hypothesis that florivory is linked to decreased attractiveness was not supported by our results (Figure 5). Au contraire, plants with herbivores restrained to the leaves became less attractive than plants with herbivores left free to feed on both leaves and flowers, giving further support to the idea that induction of volatiles is reduced in the leaves of flowering plants.

Natural selection should favor the expression of plant defenses in tissues that are the most valuable or the most at risk of being attacked by herbivores, which may vary through a plant's lifetime. Reproductive tissues (flowers and seeds) are more valuable than leaf tissue in mature annual plants, as damage inflicted upon them has a direct impact on plant's fitness [10]. Thus, plants should theoretically benefit from expressing high levels of direct and indirect defense in these tissues: work on *Brassica nigra* showed that flowers indeed contain higher levels of glucosinolates, the main class of defensive compounds in crucifers, than leaves [35]. This pattern is consistent with observations from other systems [10]. Our study, however, indicates that putative indirect resistance traits (herbivore-induced volatiles and attraction of natural enemies) are less effective in flowering *B. rapa* plants. Floral VOCs, which are primarily produced to attract pollinators, seem to inhibit the attraction of parasitoids, indicating a trade-off between pollinator and parasitoid attraction. It is also possible that flowering plants may try to repel parasitoids to avoid pollen and nectar robbery. Work done with the same population of *B. rapa* and the same two herbivores showed that herbivore-infested plants become, after several days of infestation, less attractive to pollinators compared to undamaged plants [41], reinforcing the notion that these two types of plant mutualists (parasitoids and pollinators) are attracted and deterred by different plant volatiles, which may create reciprocal interferences in infochemical networks in nature [49].

The decreased production of herbivore-induced leaf volatiles in flowering *B. rapa* plants may be a direct result of resource constraints and reallocation of resources to reproductive organs. For example, the production of a phytohormone associated with direct defense in the leaves of *Nicotiana attenuata* have been shown to decrease as plants enter the reproductive stage [50]. In our study, plants were constrained by the amount of resources present in the soil of their pots, and reached relatively small sizes at maturity (Figure 1). While this situation is not unrealistic and may correspond to nutrient-poor or competition-rich natural environments, repeating similar experiments with plants less resource-limited would be valuable to test this hypothesis. Alternatively, assuming that attracting natural enemies is beneficial for *B. rapa* plants, producing less leaf VOCs at the flowering stage may have an adaptive value, independently of resource constraints. Because the action of *C. glomerata* is not immediate, it may be too late for infested flowering plants to "cry for help" [30], but more effective to invest in rapid pollination and seed production [51]. However, flowering plants could potentially still benefit from the action of parasitoids killing *P. brassicae* larvae quickly [52], or from the action of generalist predators [53]. The reduced induction of leaf volatiles seems to indicate that flowering plants "broadcast" less cues to foraging natural enemies (although these cues appear to remain herbivore-specific), and that the reduced attraction of *C. glomerata* wasps could be generalized to other types of natural enemies. While the ecological relevance of increased attractiveness of *B. rapa* plants in the pre-flowering stage remains speculative, this result illustrates the highly dynamic nature of the interactions between plants and members of the third trophic level [45, 46].

## Conclusion

In summary, our study constitutes a strong example of rapid, short-term temporal shifts in the preferences of a parasitoid driven by plant ontogeny. In *B. rapa*, attractiveness to parasitoids decreases as plants enter the flowering stage, due to a reduced investment in herbivore-induced volatiles and to the interfering effect of floral volatiles on parasitoid attraction. However, despite this reduced investment in volatiles, infested flowering plants remain more attractive to parasitoids than undamaged plants, and the blend emitted retains its specificity to *Pieris* infestation. Thus, infested flowering *B. rapa* plants may still benefit from attracting natural enemies while reducing the costs of herbivore-induced volatiles on other mutualists (i.e. pollinators) [41]. The negative effects of floral volatiles on parasitoid attraction show that infested flowering plants do not seem to constitute an attractive "double reward" (food + host) in *B. rapa*: further experiments with parasitoids at different levels of food satiation may clarify how parasitoids integrate different volatile clues depending on whether they forage for food or for hosts. In the context of insect–plant interactions, research on the influence of plant ontogeny on herbivory has mostly been focused on the effects of direct defenses on herbivores [10, 11, 54], ignoring members of the third trophic level. By showing that the interactions between plants and natural enemies can rapidly change over a plant's lifetime, our study underscores the importance of integrating indirect defense when exploring the strategies plants may rely on to face the challenges posed by herbivores over their lifetime.

## Additional files

**Additional file 1**. Total area damaged by caterpillars (cm$^2$, mean + SE) in infested plant treatments for the different experiments (Figure S1) and principal component analysis (PCA) of VOCs emitted by undamaged and Pieris-infested *Brassica rapa* plants at each plant developmental stage (Figure S2).

### Authors' contributions

GAD, FPS, and TCJT designed the study, GAD and DL collected the data, GAD analyzed the data, GAD wrote the first draft of the manuscript, and all authors contributed to the subsequent versions of the manuscript. All authors read and approved the final manuscript.

### Author details

[1] Institute of Biology, University of Neuchâtel, Rue Emile-argand 11, 2000 Neuchâtel, Switzerland. [2] Institute of Systematic Botany, Zollikerstrasse 107, 8008 Zurich, Switzerland.

### Acknowledgements

We thank Anina Knauer for her help with making the synthetic floral blends, and Xu Hao for taking care of the *Spodoptera littoralis* larvae used in the experiments. Tom de Jong and Nicole van Dam provided *Brassica* seeds. Sandrine Gouinguene and Dani Lucas-Barbosa provided helpful comments on the manuscript. This research was funded by the Swiss National Science Funds (SNF) and by the Eurocores Invavol program (European Science Fundation).

### Competing interests

The authors declare that they have no competing interests.

### References

1. Karban R, Myers JH (1989) Induced plant responses to herbivory. Annu Rev Ecol Syst 20:331–348
2. Heil M (2008) Indirect defence via tritrophic interactions. New Phytol 178(1):41–61
3. Boege K, Marquis RJ (2005) Facing herbivory as you grow up: the ontogeny of resistance in plants. Trends Ecol Evol 20(8):441–448
4. Koricheva J, Barton KE (2012) Temporal changes in plant secondary metabolite production: patterns, causes and consequences. In: The ecology of plant secondary metabolites. Cambridge University Press, Cambridge, pp 34–55
5. Forkner RE, Marquis RJ, Lill JT (2004) Feeny revisited: condensed tannins as anti-herbivore defences in leaf-chewing herbivore communities of Quercus. Ecol Entomol 29(2):174–187
6. De Moraes CM, Mescher MC (2004) Biochemical crypsis in the avoidance of natural enemies by an insect herbivore. Proc Natl Acad Sci USA 101(24):8993–8997
7. Desurmont GA, Hajek AE, Agrawal AA (2014) Seasonal decline in plant defence is associated with relaxed offensive oviposition behaviour in the viburnum leaf beetle *Pyrrhalta viburni*. Ecol Entomol 39(5):589–594
8. Herms DA, Mattson WJ (1992) The dilemma of plants: to grow or defend. Q Rev Biol 67(3):283–335
9. Agrawal AA (2007) Macroevolution of plant defense strategies. Trends Ecol Evol 22(2):103–109
10. McCall AC, Fordyce JA (2010) Can optimal defence theory be used to predict the distribution of plant chemical defences? J Ecol 98(5):985–992
11. Ohnmeiss TE, Baldwin IT (2000) Optimal defense theory predicts the ontogeny of an induced nicotine defense. Ecology 81(7):1765–1783
12. McKey D (1979) The distribution of secondary compounds within plants. Herbivores: their interaction with secondary plant metabolites. Academic Press, New York, pp 55–133
13. McKey D (1974) Adaptive patterns in alkaloid physiology. Am Nat 108(961):305–320
14. Weiner J (2004) Allocation, plasticity and allometry in plants. Perspect Plant Ecol Evol Syst 6(4):207–215
15. Boege K, Dirzo R, Siemens D, Brown P (2007) Ontogenetic switches from plant resistance to tolerance: minimizing costs with age? Ecol Lett 10(3):177–187
16. Kessler A, Halitschke R (2009) Testing the potential for conflicting selection on floral chemical traits by pollinators and herbivores: predictions and case study. Funct Ecol 23(5):901–912
17. Weber MG, Clement WL, Donoghue MJ, Agrawal AA (2012) Phylogenetic and experimental tests of interactions among mutualistic plant defense traits in Viburnum (adoxaceae). Am Nat 180(4):450–463
18. Turlings TC, Tumlinson JH, Lewis W (1990) Exploitation of herbivore-induced plant odors by host-seeking parasitic wasps. Science 250(4985):1251–1253
19. Dicke M, Sabelis MW (1988) How plants obtain predatory mites as bodyguards. Neth J Zool 38(2–4):148–165
20. Villamil N, Márquez-Guzmán J, Boege K (2013) Understanding ontogenetic trajectories of indirect defence: ecological and anatomical constraints in the production of extrafloral nectaries. Ann Bot 112:701–709
21. Cole RA (1980) Volatile components produced during ontogeny of some cultivated crucifers. J Sci Food Agric 31(6):549–557
22. Hare JD (2010) Ontogeny and season constrain the production of herbivore-inducible plant volatiles in the field. J Chem Ecol 36(12):1363–1374
23. Hare JD, Sun JJ (2011) Production of herbivore-induced plant volatiles is constrained seasonally in the field but predation on herbivores is not. J Chem Ecol 37(5):430–442
24. Dannon EA, Tamò M, Van Huis A, Dicke M (2010) Effects of volatiles from *Maruca vitrata* larvae and caterpillar-infested flowers of their host plant *Vigna unguiculata* on the foraging behavior of the parasitoid *Apanteles taragamae*. J Chem Ecol 36(10):1083–1091
25. Lucas-Barbosa D, van Loon JJA, Dicke M (2011) The effects of herbivore-induced plant volatiles on interactions between plants and flower-visiting insects. Phytochemistry 72(13):1647–1654
26. Géneau CE, Wäckers FL, Luka H, Balmer O (2013) Effects of extrafloral and floral nectar of *Centaurea cyanus* on the parasitoid wasp *Microplitis mediator*: olfactory attractiveness and parasitization rates. Biol Control 66: 16–20
27. Wäckers F (2001) A comparison of nectar-and honeydew sugars with respect to their utilization by the hymenopteran parasitoid *Cotesia glomerata*. J Insect Physiol 47(9):1077–1084

28. Géneau CE, Wäckers FL, Luka H, Daniel C, Balmer O (2012) Selective flowers to enhance biological control of cabbage pests by parasitoids. Basic Appl Ecol 13(1):85–93
29. Dicke M (1999) Are herbivore-induced plant volatiles reliable indicators of herbivore identity to foraging carnivorous arthropods? In: Proceedings of the 10th International symposium on insect–plant relationships. Springer, pp 131–142
30. Dicke M, Baldwin IT (2010) The evolutionary context for herbivore-induced plant volatiles: beyond the 'cry for help'. Trends Plant Sci 15(3):167–175
31. Geervliet JBF, Verdel MSW, Snellen H, Schaub J, Dicke M, Vet LEM (2000) Coexistence and niche segregation by field populations of the parasitoids *Cotesia glomerata* and *C. rubecula* in the Netherlands: predicting field performance from laboratory data. Oecologia 124(1):55–63
32. Coleman R, Barker A, Fenner M (1999) Parasitism of the herbivore *Pieris brassicae* L. (Lep., Pieridae) by *Cotesia glomerata* L. (Hym., Braconidae) does not benefit the host plant by reduction of herbivory. J Appl Entomol 123(3):171–177
33. Smallegange RC, Van Loon JJA, Blatt SE, Harvey JA, Dicke M (2008) Parasitoid load affects plant fitness in a tritrophic system. Entomol Exp Appl 128(1):172–183
34. Harvey JA (2000) Dynamic effects of parasitism by an endoparasitoid wasp on the development of two host species: implications for host quality and parasitoid fitness. Ecol Entomol 25(3):267–278
35. Smallegange R, Van Loon J, Blatt S, Harvey J, Agerbirk N, Dicke M (2007) Flower vs. leaf feeding by *Pieris brassicae*: glucosinolate-rich flower tissues are preferred and sustain higher growth rate. J Chem Ecol 33(10):1831–1844
36. El-Sayed A, Suckling D, Byers J, Jang E, Wearing C (2009) Potential of "lure and kill" in long-term pest management and eradication of invasive species. J Econ Entomol 102(3):815–835
37. Xu H, Veyrat N, Degen T, Turlings TC (2014) Exceptional use of sex Pheromones by parasitoids of the genus Cotesia: males are strongly attracted to virgin females, but are no longer attracted to or even repelled by mated females. Insects 5(3):499–512
38. Turlings TC, Davison A, TamÒ C (2004) A six-arm olfactometer permitting simultaneous observation of insect attraction and odour trapping. Physiol Entomol 29(1):45–55
39. D'Alessandro M, Turlings TC (2005) In situ modification of herbivore-induced plant odors: a novel approach to study the attractiveness of volatile organic compounds to parasitic wasps. Chem Senses 30(9):739–753
40. Ton J, D'Alessandro M, Jourdie V, Jakab G, Karlen D, Held M, Mauch-Mani B, Turlings TC (2007) Priming by airborne signals boosts direct and indirect resistance in maize. Plant J 49(1):16–26
41. Schiestl FP, Kirk H, Bigler L, Cozzolino S, Desurmont GA (2014) Herbivory and floral signaling: phenotypic plasticity and tradeoffs between reproduction and indirect defense. New Phytol 203(1):257–266
42. Radhika V, Kost C, Bartram S, Heil M, Boland W (2008) Testing the optimal defence hypothesis for two indirect defences: extrafloral nectar and volatile organic compounds. Planta 228(3):449–457
43. Bruinsma M, Lucas-Barbosa D, ten Broeke CM, van Dam N, van Beek T, Dicke M, van Loon JA (2014) Folivory affects composition of nectar, floral odor and modifies pollinator behavior. J Chem Ecol 40(1):39–49
44. Rostás M, Eggert K (2008) Ontogenetic and spatio-temporal patterns of induced volatiles in *Glycine max* in the light of the optimal defence hypothesis. Chemoecology 18(1):29–38
45. Cisneros JJ, Rosenheim JA (1998) Changes in the foraging behavior, within-plant vertical distribution, and microhabitat selection of a generalist insect predator: an age analysis. Environ Entomol 27(4):949–957
46. Fonseca CR, Benson WW (2003) Ontogenetic succession in Amazonian ant trees. Oikos 102(2):407–412
47. Chabaane Y, Laplanche D, Turlings TC, Desurmont GA (2014) Impact of exotic insect herbivores on native tritrophic interactions: a case study of the African cotton leafworm, *Spodoptera littoralis* and insects associated with the field mustard *Brassica rapa*. J Ecol 103(1):109–117
48. Lucas-Barbosa D, Poelman EH, Aartsma Y, Snoeren TA, van Loon JJ, Dicke M (2014) Caught between parasitoids and predators-survival of a specialist herbivore on leaves and flowers of mustard plants. J Chem Ecol 40(6):621–631
49. Kessler A, Halitschke R (2007) Specificity and complexity: the impact of herbivore-induced plant responses on arthropod community structure. Curr Opin Plant Biol 10(4):409–414
50. Diezel C, Allmann S, Baldwin IT (2011) Mechanisms of optimal defense patterns in *Nicotiana attenuata*: flowering attenuates herbivory-elicited ethylene and jasmonate signalingF. J Integr Plant Biol 53(12):971–983
51. Lucas-Barbosa D, Loon JJ, Gols R, Beek TA, Dicke M (2013) Reproductive escape: annual plant responds to butterfly eggs by accelerating seed production. Funct Ecol 27(1):245–254
52. Harvey JA, Poelman EH, Gols R (2010) Development and host utilization in *Hyposoter ebeninus* (Hymenoptera: Ichneumonidae), a solitary endoparasitoid of *Pieris rapae* and *P. brassicae* caterpillars (Lepidoptera: Pieridae). Biol Control 53(3):312–318
53. Symondson W, Sunderland K, Greenstone M (2002) Can generalist predators be effective biocontrol agents? 1. Annu Rev Entomol 47(1):561–594
54. Van Dam NM, Horn M, Mareš M, Baldwin IT (2001) Ontogeny constrains systemic protease inhibitor response in *Nicotiana attenuata*. J Chem Ecol 27(3):547–568

# 12

# No consistent effect of plant species richness on resistance to simulated climate change for above- or below-ground processes in managed grasslands

Carsten F. Dormann[1*], Lars von Riedmatten[1,2] and Michael Scherer-Lorenzen[3]

**Abstract**

**Background:** Species richness affects processes and functions in many ecosystems. Since management of temperate grasslands is directly affecting species composition and richness, it can indirectly govern how systems respond to fluctuations in environmental conditions. Our aim in this study was to investigate whether species richness in managed grasslands can buffer the effects of drought and warming manipulations and hence increase the resistance to climate change. We established 45 plots in three regions across Germany, each with three different management regimes (pasture, meadow and mown pasture). We manipulated spring warming using open-top chambers and summer drought using rain-out shelters for 4 weeks.

**Results:** Measurements of species richness, above- and below-ground biomass and soil carbon and nitrogen concentrations showed significant but inconsistent differences among regions, managements and manipulations. We detected a three-way interaction between species richness, management and region, indicating that our study design was sensitive enough to detect even intricate effects.

**Conclusions:** We could not detect a pervasive effect of species richness on biomass differences between treatments and controls, indicating that a combination of spring warming and summer drought effects on grassland systems are not consistently moderated by species richness. We attribute this to the relatively high number of species even at low richness levels, which already provides the complementarity required for positive biodiversity–ecosystem functioning relationships. A review of the literature also indicates that climate manipulations largely fail to show richness-buffering, while natural experiments do, suggesting that such manipulations are milder than reality or incur treatment artefacts.

**Keywords:** Climate change manipulation, C-pool, Ecosystem function, N-pool, Productivity, Species richness, Temperate grassland, Vegetation

## Background

The scientific consensus is unambiguous about the role of biodiversity for ecosystem functioning [1]. It is largely based on experiments along species richness gradients from one to tens of species [2]. There is much less experimental attention being paid to the effect of species richness in managed systems, where the range of diversity is different. In managed temperate grasslands, species richness can be as low as five vascular plant species/m$^2$, and as high as 60 species/m$^2$ [3, 4]. Thus, the gradient in grassland species richness generally does not cover the very low end of diversity, which is included in the design of many experimental studies (Cedar Creek: 1–16: [5]; Biodepth: 1–16: [6]; Jena: 1–60: [7]). Since at very low species richness diversity matters most [2], it is not clear whether results from experimental grasslands directly

*Correspondence: carsten.dormann@biom.uni-freiburg.de
[1] Biometry & Environmental System Analysis, University of Freiburg, Tennenbacher Str. 4, 79106 Freiburg, Germany
Full list of author information is available at the end of the article

translate into long-term managed grasslands, that is, whether such controlled experiments have high external validity.

An important effect of plant diversity is the buffering of environmental fluctuations, such as droughts [5]. Under standard management conditions it remains to be investigated whether plant species richness dampens the effect of fluctuations on ecosystem functioning. Central European plant communities are exposed to high inter-annual variability in weather conditions, and different responses of its members will lead to portfolio effects, buffering the effects of environmental variability at the community level [8–12]. The general existence of these mechanisms is beyond dispute, but its relevance for managed systems with at least a moderate number of species (>2) can be questioned. Such systems include most grasslands (used as pastures or meadows) and non-plantation forests.

Plant-species richness in grasslands is determined through a complex interplay of abiotic conditions (e.g. soil type, climate), land use and its history (fertilisation, grazing, management changes) and regional species pool [13]. If a grassland experiences externally induced disturbances, probably all of these factors have some relevance for the system's response [14]. Understanding how much resistance (the degree to which a variable changes, following a perturbation: [15] is transferred to the system by being more species rich allows us to gauge the importance of biodiversity *relative* to the effect of land management and abiotic factors in grassland systems.

Here, we exposed temperate grasslands in three regions in Germany to two pulse per-turbations simulating climate change: advanced spring and summer drought, representing two opposing effects of climate change on vegetation. Earlier spring will extend the vegetation growth period, potentially increasing total productivity, with knock-on effects on plant phenology and below-ground processes. These positives effects may be offset by prolonged drought phases in summer, reducing grassland productivity in the late-summer growth period. Climate-change predictions of precipitation are notoriously uncertain, so our manipulations set out to explore a warm-spring–dry-summer scenario, rather than represent a specific future climate prediction.

Within three land-use types, a realistic but still substantial gradient in plant species richness was present. We measured several above- and below-ground biomass, C- and N-pools to assess ecosystem functioning in two consecutive periods, attempting to test our central hypothesis that plant diversity buffers effects of climatic variability in typical temperate European grasslands. Moreover, we expect different land-use types to respond differently to our climate-change manipulations, thereby revealing how management affects ecosystem resistance. Finally, by assessing ecosystem processes in the vegetation and the soil, we can compare whether biodiversity effects are greater above- or below-ground, linking the presumed buffering mechanism to wider biogeochemical processes.

## Methods

In this study, we imposed a combination of summer drought and an increase in next year's spring temperature in managed grasslands and assessed the effects on vegetation, soil properties and litter decomposition relative to unmanipulated controls. We selected three grassland management types (pasture, meadow, mown pasture), with five replicates each. Across these 15 sites, each land-use type covers a gradient in plant species richness, which we exploited for testing our hypothesis that plant species richness buffers climate change manipulations. The same setup was replicated across three regions within Germany, the locations of the Biodiversity Exploratories, a long-term research platform into the interrelationships of biodiversity, ecosystem functioning and land-use intensity in grasslands and forests. The three locations differ in climate, geology, management of grassland, plant species richness and soil properties (see below). If results are consistent across the three regions, we can claim to have elicited a generalisable response, despite different pathways of how land use affects plant species richness [16, 17].

### Study regions

The experiment took place in managed grasslands of the three Biodiversity Exploratories [18]. Schorfheide-Chorin (SC) in northeast Germany spans an area of 1300 km$^2$. In this young glacial landscape, plots were established on former fens at an elevation of 3–140 m a.s.l. The main soil type is Histosol, rich in organic matter. The annual mean temperature (1981–2010) averages 8–8.5 °C (summer: 18.2 °C; winter: 1.1 °C) and the annual mean precipitation 500–600 mm (driest month: April, 24 mm; wettest month: July, 69 mm). Hainich-Dün (HD) lies in central Germany with an area of 7600 km$^2$ and an elevation of 285–550 m a.s.l. on a calcareous bedrock (Vertisol as main soil type), with an annual mean temperature of 6.5–8 °C (summer: 17.7 °C; winter: 1.5 °C) and an annual mean precipitation of 500–800 mm (driest month: April, 24 mm; wettest month: July, 67 mm). The third exploratory Swabian Alp (SA) in southwest Germany extends over an area of 422 km$^2$ at an elevation of 460–800 m a.s.l. Annual mean temperature is 6–7 °C (summer: 15.8 °C; winter: −0.4 °C) and the annual mean precipitation 700–1000 mm (driest month: February, 47 mm; wettest month: July, 119 mm). The soil type is Cambisol on a calcareous bedrock with karst phenomena. A detailed description of all three exploratories is given by Fischer et al. [18].

We used five replicate plots of similar soil in each of three grassland management regimes: meadow (m, fertilized), mown pasture (mp, fertilized and unfertilized) and pastures (p, unfertilized; grazed by sheep, horse or cattle). Meadows were traditionally restricted to edaphically extreme sites, where livestock trampling could destroy the vegetation. In recent years, overall reduction in cattle grazing has led to emergence of meadows on all soils, particularly in parcels of land that are small and further away from the farm. Depending on soil type, meadows are mown once or twice per year, but up to four times under heavy fertilisation. Pure pastures are nowadays common only as nature management regime, predominantly grazed by sheep. The typical mown pasture receives a late-season cut to prevent spreading of unpalatable herbs, after livestock (mainly cattle) is moved on to other pastures. There is much variation among the three sites in the timing and intensity of grazing both on mown and pure pastures, but within a site management is relatively consistent. By using replicates of land-use types, we aimed to introduce variability in species richness within each management, thereby reducing the correlation between land use and plant species richness inherent in the setup of the Biodiversity Exploratories.

#### Experimental setup of climate change manipulations

Experimental manipulations took place in 2008/2009 and 2009/2010, from the middle of June until the end of July in 2008 and 2009 for the simulated drought treatment and from end of February till end of March in 2009 and 2010 for simulation of increased spring temperatures. Our manipulations thus implicitly include carry-over effects of drought on the spring treatment the following year. Treatments are representing projected qualitative changes in summer drought duration and concurrent changes of the length of the vegetation period. Their length is based on interannual variability of the onset of spring (approximately 4 weeks after begin of snow melt in late February) and length of drought in summer (maximally 6 weeks). We left the rain shelter in place until there were cumulatively 3 weeks of rainfall withheld from the treatment (monitoring the closest German weather service stations for guidance). Thus, all treatments are within the variability witnessed in the region, but aim at representing earlier season and prolonged drought. Constructions were removed when no treatment was applied, but the plot was fenced off for the entire vegetation period, which we do not expect to have substantial short-term impacts. The entire plot was mown in summer after biomass harvest (see below).

Each plot of 5 × 3 m was divided in two subplots, one for the manipulation, the other as control (see Additional file 1: Figure S6). Open-top chambers (OTCs) are commonly used to raise the temperature in climate change experiments with minor effects on gas exchange and ambient precipitation [19]. To increase temperature in early spring, we placed OTCs (2 × 3 m and 1.4 m height) on each treatment subplot. The OTCs were constructed from four PVC tube arches, which supported the 0.2 mm thick greenhouse plastic sheet (UV 5 Coex-foil made of ethylene vinyl acetate copolymers, Folitec Agrarfolien-Vertriebs GmbH, Westerburg, Germany) up to a height of 1.2 m (see [20] Fig. 1D for a similar design; Additional file 1: Figure S7). Soil temperature (10 cm depth) and air temperature (10 cm height) were recorded every half an hour by aluminium-foil-shielded temperature sensors (Thermochron iButton, Maxim Integrated Products Inc., Sunnyvale, CA, USA).

For the drought manipulation the same treatment plots were covered by rain shelters constructed from the same construction as the OTCs, but using the greenhouse foil as a top cover. Air was thus allowed to freely circulate underneath the roof. The air temperature was again recorded by temperature sensors every 30 min at a height of 10 cm. Soil moisture during the drought experiment was measured every 30 min in a depth of 10 cm by a moisture sensor (ECH2O, type EC-5, Decagon Devices Inc., Pullman, USA) and recorded by a data logger (Em5b, Decagon Devices, Pullman, WA, USA). In spring, no soil moisture measurements were possible because sensors were removed from the soil over winter and could not be re-inserted into the frozen ground until later in the season.

Open-top chambers increased spring temperature by 0.5 °C in the air and 0.35 °C in the soil (see Table 1), in a period of absolute temperatures between −5 °C at night and 15–20 °C during the day (Additional file 1: Figure S1). In summer, rain shelters had a warming effect (Table 1) as pronounced as in spring, but at much higher overall temperatures (12–35 °C). Soil moisture during this period was reduced under the shelter by 8–20%vol, with substantial differences between exploratories (Table 1; Additional file 1: Figure S1). In all sites, soil moisture in the control plots averaged to 30–40%vol, but soil moisture was most affected by drought treatment in the Swabian Alp. Both at the SA and HD sites, the drought treatment presumably resulted in soil water potential being more negative than the permanent wilting point (PWP, at pF = 4.2) as estimated from soil texture data with high clay and silt contents. At the SC site with sandy soils, however, soil water potential was still above the PWP (see Additional file 1: Figure S1).

**Table 1 Differences (manipulation—control, given as mean ± standard deviation) in temperature and soil moisture during the different phases of the experiment**

| | Air temperature difference (°C) | | | Soil temperature difference (°C) | | | Soil moisture difference (%vol) | | |
|---|---|---|---|---|---|---|---|---|---|
| | SA | HD | SC | SA | HD | SC | SA | HD | SC |
| Period 1 | | | | | | | | | |
| 2008 summer drought | 1.15 ± 2.179 | 0.32 ± 2.178 | 0.53 ± 2.433 | 0.84 ± 1.617 | −0.032 ± 0.494 | 0.075 ± 0.623 | −20.1 ± 6.53 | −9.3 ± 3.67 | −7.8 ± 2.40 |
| 2009 spring warming | 0.69 ± 1.310 | 0.39 ± 0.607 | 0.36 ± 0.619 | 0.45 ± 0.818 | 0.30 ± 0.585 | 0.20 ± 0.464 | * | * | * |
| Period 2 | | | | | | | | | |
| 2009 summer drought | 0.62 ± 1.333 | 1.30 ± 2.694 | 0.64 ± 1.332 | −0.30 ± 0.601 | −0.50 ± 1.420 | −0.19 ± 0.452 | −20.0 ± 5.24 | −18.8 ± 12.48 | −7.5 ± 2.91 |
| 2010 spring warming | 1.00 ± 1.474 | 0.47 ± 0.713 | 0.60 ± 0.866 | 0.68 ± 0.777 | 0.28 ± 0.235 | 0.39 ± 0.204 | * | * | * |

Exploratories are referred to as *SA* Swabian Alb, *HD* Hainich-Dün, *SC* Schorfheide-Chorin

* Indicates that no sensor were employed because the soil was frozen

### Vegetation recording

In April 2009 and 2010 vegetation recordings were carried out using a 50 × 50 cm frequency frame with a 5 cm grid on each sub-plot. Summer recordings in a pilot study lacked several of the early spring flowers and detected fewer species. These subplots were established centrally in the plot, 1 m from the northern end of the plot's border to reduce edge effects. Subplots were marked by metal pegs in the ground to record in the exact same site in the following year. Plant species were tallied for each grid cell, yielding percentage cover values [21]. Vegetation recordings in 2009 were used to compute Shannon diversity for each plot.

### Plant biomass harvest and analysis

Above-ground biomass was harvested in an area of 50 × 50 cm next to each vegetation-recording subplot before the summer drought treatment 2009, after the summer drought treatment 2008 and 2009 and after the spring warming treatment 2010, by clipping all shoot material above the soil surface and sorting into functional groups (grasses, herbs and legumes). The treatment extended the sampled area on all sides, obviating the necessity of trenching the plot. Treatments showed marked effects with no sign of leakage to outside the treated area.

To determine root growth during the summer drought treatment, we established three in-growth cores [22] per subplot. These 5-cm diameter 10-cm long frames were inserted into the soil using a soil corer and filled with sieved local soil. Plant roots growth into the in-growth core volume was sampled using the soil corer, severing roots around the in-growth core. Root material was obtained by washing in-growth cores over sieves (0.5 cm and 0.2 mm mesh size). This procedure was repeated in summer 2008 and 2009, yielding root production values for the summer drought experiment.

Above- and belowground plant material was dried at 70 °C for 48 h, then weighed to the nearest mg. For C and N analysis, dried plant material was ground in a ball mill and a subsample of 5 mg was processed in an elemental analyser (PerkinElmer 2400 Series II, PerkinElmer, Waltham, MA, USA).

### Soil sampling and analysis

20 g bulk soil was collected by taking five samples of the upper 10 cm randomly from each subplot. 5 g soil were dried at 105 °C for 48 h to determine gravimetric water content. A subsample of 5 mg was analysed for %C and %N after grinding in a ball mill. To measure soil nitrogen content, 5 g soil were extracted with 50 ml of 1 M KCl solution on a shaker for 20 min. After filtering (Black Ribbon filter, Grade 589/1: 12–25 mm, Whatman Ltd., Maidstone, UK) extract was analysed for C and N with vario MAX CNS (Elementar, Hanau, Germany). The percentage of soil organic matter was determined by loss on ignition of 1 g oven-dried soil at 500 °C for 24 h.

### Data analysis

Our main interest was to test the interaction of species richness, our manipulation and land use on the various response variables across the three sites as measured in summer. Differences between the two consecutive phases of the experiment test for consistency of treatment effects, and indirectly allow us to compare effects to inter-annual variability. We regarded cumulative treatment effects as negligible after only two seasons, given the visual similarity of treatment and control *before* the rain shelters were put up. No lasting effect of spring warming was discernible. Spring data analysis of aboveground biomass reflects response to earlier warming, and in our opinion merely demonstrates the effectiveness of this treatment.

Biomass, vegetation and soil C and N pools were expressed on a per $m^2$ basis by extrapolation from the sampled area. We used Linear Mixed Effect Models (lme: [23] with plot identity as a random factor to test the effects of climate change manipulation, land-use management and identity of the three exploratories (as fixed effect) on species richness, plant diversity, functional diversity and on change of plant productivity and C and N pool.

To test our hypothesis that species-rich grasslands will be better able to resist the effects of our warming (spring) and drought (summer) manipulation, we analysed the difference in biomass between treatment and control as log-response ratio (LRR = log(treatment) − log(control)), with approximately normal distribution of residuals. Values above 0 indicate a stronger response of the treatment than the control. In addition to exploratory and land-use type we also included species richness (and interactions) as predictor in the model.

Across and within the exploratories, plant species richness varies with land-use type, but with substantial scatter (e.g. [24]. We thus first analyse the correlation between exploratories, land-use type and plant community composition and richness in our study plots. Results largely refer to spring 2010 and summer 2009, i.e. after two periods of the respective treatment, to represent the largest effects in our study.

Effects on vegetation composition were analysed using Canonical Correspondence Analysis with package *vegan* [25]. All statistical analyses were conducted using R version 3.0.3 [26]. R-code for analysis and figures is available on request from the corresponding author. The data are available freely from the Biodiversity Exploratories data base at https://www.bexis.uni-jena.de (data set number 20186).

## Results

### Species richness, diversity and composition

The number of recorded vascular plant species and Shannon diversity showed a significant interaction between the three study regions and land-use types (species richness *S*: Fig. 1; Table 2; Shannon diversity *H*: Additional file 1: Figure S3). The significant differences of species richness between the land-use types are due to the varying quantity and richness of herbs and, to a lesser degree, legumes, while grass richness was very similar (Table 2). The pastures in the Swabian Alp, in contrast to the pastures of the two other exploratories, are located in semidry grassland and are grazed by sheep, which explains the noticeable higher number of species (32 species compared to only 17 and 11 in Hainich-Dün and Schorfheide-Chorin, respectively; Fig. 1). Legumes were particularly scarce in the Schorfheide-Chorin, both in terms of species numbers and their biomass (see below). Neither total nor functional-group species richness differed

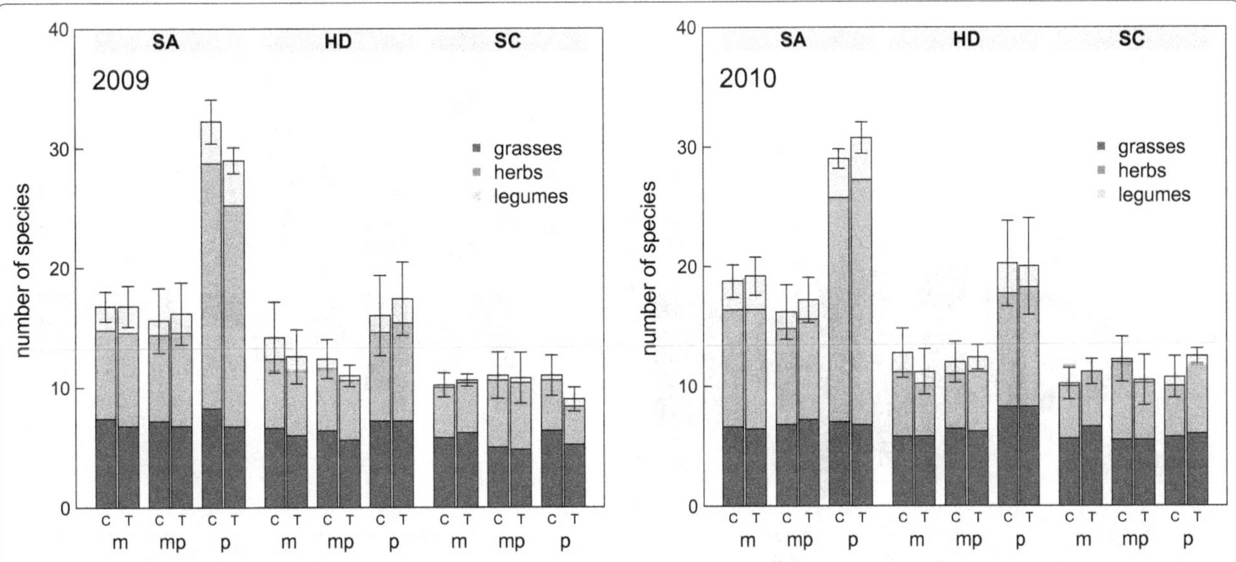


**Fig. 1** Number of vascular plant species, sub-divided by functional groups, found on control (C) and treatment (T) subplots by land-use type (*m* meadow, *p* pasture, *mp* mown pasture) and year (2009 *left*, 2010 *right*). *Bars* represent mean values ± SEM of the total number of species. Exploratories are referred to as SA (Swabian Alp, south-west Germany), HD (Hainich-Dün, central Germany) and SC (Schorfheide-Chorin, north-east Germany)

**Table 2 Mixed-effect model analysis of *species richness* (total as well as of the three functional groups) to exploratories, land use, treatment in the two experimental periods with plot-ID as random effect (see Fig. 1)**

| | numDF | Total | | Grasses | | Herbs | | Legumes | |
|---|---|---|---|---|---|---|---|---|---|
| | | denDF | *F* | denDF | *F* | denDF | *F* | denDF | *F* |
| Period 1 | | | | | | | | | |
| Explo | 2 | 35 | 19.8*** | 39 | 4.79* | 35 | 21.3*** | 39 | 10.6*** |
| LUT | 2 | 35 | 7.81** | 39 | $1.40^{n.s.}$ | 35 | 10.2*** | 39 | 3.46* |
| Treatment | 1 | 43 | $2.00^{n.s.}$ | 43 | 7.08* | 43 | $0.263^{n.s.}$ | 43 | $0.328^{n.s.}$ |
| Explo:LUT | 4 | 35 | 4.78** | | | 35 | 9.33*** | | |
| Period 2 | | | | | | | | | |
| Explo | 2 | 32 | 23.0*** | 36 | $2.04^{n.s.}$ | 32 | 29.5*** | 36 | 13.9*** |
| LUT | 2 | 32 | 13.2** | 36 | $1.60^{n.s.}$ | 32 | 17.5*** | 36 | 4.19* |
| Treatment | 1 | 40 | $0.738^{n.s.}$ | 40 | $2.26^{n.s.}$ | 40 | $0.603^{n.s.}$ | 40 | $0.122^{n.s.}$ |
| Explo:LUT | 4 | 32 | 3.16** | | | 32 | 6.29*** | | |

Numerator degrees of freedom (numDF) are constant across all models, while denominator degrees of freedom (denDF) change with the number of parameters in the model. Significance of *F* value of *P* < 0.001, <0.01, <0.05 and not significant are indicated by ***, **, * and $^{n.s}$, respectively. Models were simplified manually until only significant effects remained (or effects marginal to interactions). Design variables were always included in the model, irrespective of significance. LUT and Explo refer to land-use type and exploratory region, respectively

significantly between treatments (all P > 0.085; Table 2), but by the end of the experiment (2010), Shannon diversity showed a consistent and significantly positive treatment effect ($F_{1,\ 121}$ = 5.50, P < 0.05; Additional file 1: Figure S3). Finally, plant species composition differed substantially among exploratories and land uses, but was affected only very mildly by our treatments (Additional file 1: Figure S2).

### Productivity

Treatment manipulations showed significant effects on above- and below-ground plant productivity measured in summer across all exploratories and land-use types ($F_{1,\ 81}$ = 46.3, P < 0.001; Fig. 2; Table 3). Above-ground biomass was sometimes substantially reduced (e.g. in the Swabian Alp), while root increment showed initially an inconsistent pattern (as indicated by the marginally significant treatment × exploratory-interaction in Table 3; Fig. 2 left; $F_{2,\ 82}$ = 2.82, P < 0.1). At the end of the second season (2009), also below-ground biomass was uniformly reduced due to the summer drought treatment ($F_{1,\ 78}$ = 4.88, P < 0.01; Fig. 2 right; Table 3). In contrast, above-ground biomass remained at control levels in 2009, and none of the three functional

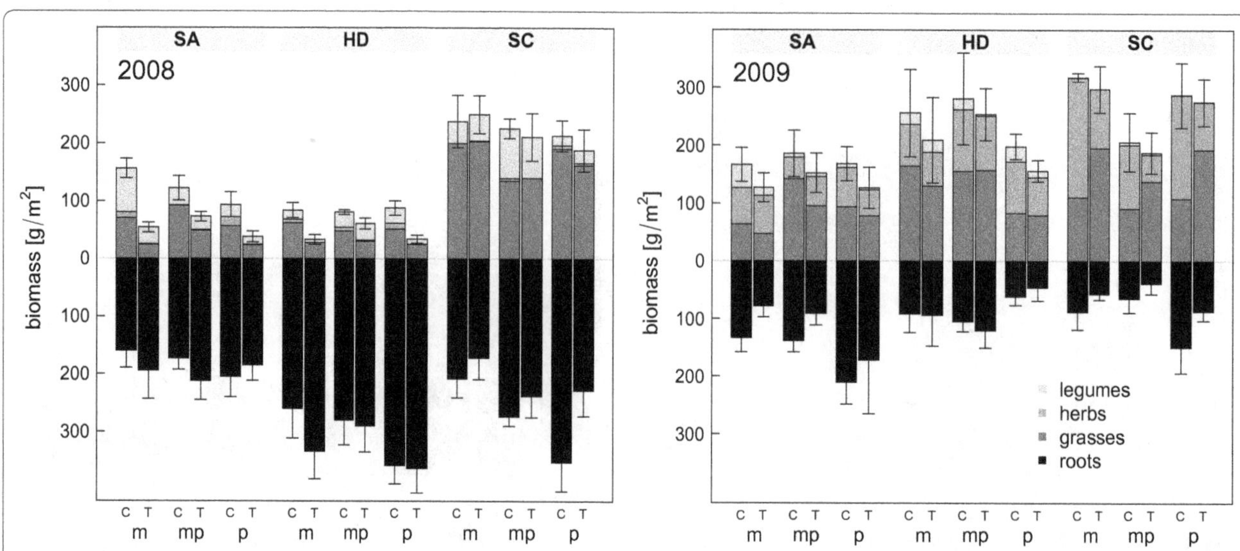


**Fig. 2** Standing above- and below-ground biomass in summer, subdivided by functional groups. *Bars* represent mean values ± SEM of the total above-ground biomass, below-ground biomass respectively. See Fig. 1 for abbreviations

**Table 3 Analysis of *summer biomass* (i.e. harvested around peak biomass) in periods 1 and 2 (see Fig. 2). See Table 2 for explanation of symbols**

| | df | Total | | Grasses | | Herbs[a] | | Legumes[b] | | Roots | |
|---|---|---|---|---|---|---|---|---|---|---|---|
| | | SS | *F* | SS | *F* | SS | *F* | SS | *F* | SS | *F* |
| Period 1 | | | | | | | | | | | |
| Explo | 2 | 28.7 | 90.6*** | 39.8 | 67.6*** | 18.0 | 2.69† | 91.7 | 9.44*** | 4.14 | 15.7*** |
| LUT | 2 | 1.46 | 4.60* | 0.544 | 0.924$^{n.s.}$ | 33.3 | 4.97** | 79.1 | 8.13*** | 1.06 | 4.02* |
| Treatment | 1 | 7.33 | 46.3*** | 7.38 | 25.0*** | 20.2 | 6.02* | 23.6 | 4.85* | 0.0260 | 0.198$^{n.s.}$ |
| Treatment:Explo | 2 | 2.69 | 8.51*** | 2.31 | 3.92* | | | | | 0.742 | 2.82† |
| LUT:Explo | 4 | | | 4.20 | 3.57* | | | 85.2 | 4.38** | | |
| Residuals | | 12.8 | (df = 81) | 22.7 | (df = 77) | 278 | (df = 83) | 384 | (df = 79) | 10.78 | (df = 82) |
| Period 2 | | | | | | | | | | | |
| Explo | 2 | 5.16 | 10.4*** | | 3.98* | 5.24 | 5.58** | 27.85 | 4.29* | 4.93 | 4.91** |
| LUT | 2 | 0.179 | 0.360$^{n.s.}$ | 0.797 | 0.846$^{n.s.}$ | 2.00 | 2.14$^{n.s.}$ | 6.25 | 0.962$^{n.s.}$ | 0.366 | 0.365$^{n.s.}$ |
| Treatment | 1 | 0.575 | 2.31$^{n.s.}$ | 0.020 | 0.0422$^{n.s.}$ | 1.06 | 2.25$^{n.s.}$ | 8.58 | 2.65$^{n.s.}$ | 4.88 | 9.73** |
| Treatment:Explo | 2 | | | 2.90 | 3.08† | | | | | | |
| LUT:Explo | 4 | | | | | 5.51 | 2.93* | | | 8.19 | 4.08** |
| Residuals | | 20.4 | (df = 82) | 37.7 | (df = 80) | 36.6 | (df = 78) | 178 | (df = 55) | 39.12 | (df = 78) |

Biomass data were log-transformed to achieve homogeneity of variances. Models were manually simplified until all model terms were (marginally) significant. Design effects were always kept in the model. LUT and Explo refer to land-use type and exploratory region, respectively

† Indicates a *P* value between 0.05 and 0.1

[a] A constant value of 0.1 was added to all response values in 2008

[b] Response values of 2008 were square-root, rather than log-transformed; for 2009, a constant value of 0.1 was added

groups displayed a significant response to the treatment (Table 3).

Early spring warming led to increased biomass across all exploratories and land-use types (Additional file 1: Figure S4), despite having had reduced biomass at the end of the previous summer. Grasses profited most from elevated temperature manipulations.

Although our design was sensitive enough to detect a significant three-way interaction in summer, and effects of species richness in spring (Table 4), LRR showed no interpretable pattern with respect to species richness. The positive effect of species richness for the response to earlier spring was very small and contributed less than 4% to the variance in the data ($F_{1, 26} = 4.77$, $P < 0.05$). Responses to combined spring and summer manipulations display huge scatter in LRR. Depending on land-use type and exploratory, positive, negative and no correlations with species richness were observed, accompanied by a significant four-way interaction ($F_{4, 25} = 2.90$, $P < 0.05$; Fig. 3; Table 4).

**Table 4 Analysis of *log-response ratios* $\left(\log \frac{\text{treatment biomass}}{\text{control biomass}}\right)$ for treatment effects in spring 2010 and summer 2009 (see Fig. 3). See Table 2 for explanation of symbols**

| | df | Spring[a] | | Summer | |
|---|---|---|---|---|---|
| | | SS | *F* | SS | *F* |
| Explo | 2 | 0.265 | 0.347$^{n.s.}$ | 0.392 | 2.02$^{n.s}$ |
| LUT | 2 | 0.250 | 0.327$^{n.s.}$ | 0.260 | 1.34$^{n.s}$ |
| Species richness[b] | 1 | 1.77 | 4.77* | 0.0177 | 0.182$^{n.s}$ |
| Explo:LUT | 4 | | | 0.342 | 0.883$^{n.s}$ |
| Explo:species richness | 2 | | | 0.0544 | 0.281$^{n.s}$ |
| LUT:species richness | 2 | | | 0.764 | 3.94* |
| Explo:LUT:species richness | 4 | | | 1.125 | 2.90* |
| Residuals | | 9.03 | (df = 26) | 2.42 | (df = 25) |

Models were manually simplified until all model terms were significant. Design effects were always kept in the model. SS were computed sequentially (type I). LUT and Explo refer to land-use type and exploratory region, respectively

[a] Species richness was modelled as log(species richness) in the spring analysis. Using a linear scale yielded similar, but non-significant effects of species richness

[b] Estimate for the spring effect of log(species richness): 0.792 ± 0.363; partial model $R^2_{adj} = 0.036$

### Vegetation C and N pools

Land use had no consistent effect on carbon pools, and our experimental manipulations manifested themselves with similar magnitude in species-rich pastures and less species-rich meadows. After 2 years of our manipulation treatments, carbon pools in the vegetation were overall reduced by about 10%, both above- and below-ground (Fig. 4 left; $F_{1, 74} = 5.43$, $P < 0.05$ and $F_{1, 75} = 0.438$, $P > 0.1$, respectively). Differences between exploratories were substantial, with the Swabian Alp having the lowest carbon pools and Schorfheide-Chorin the highest,

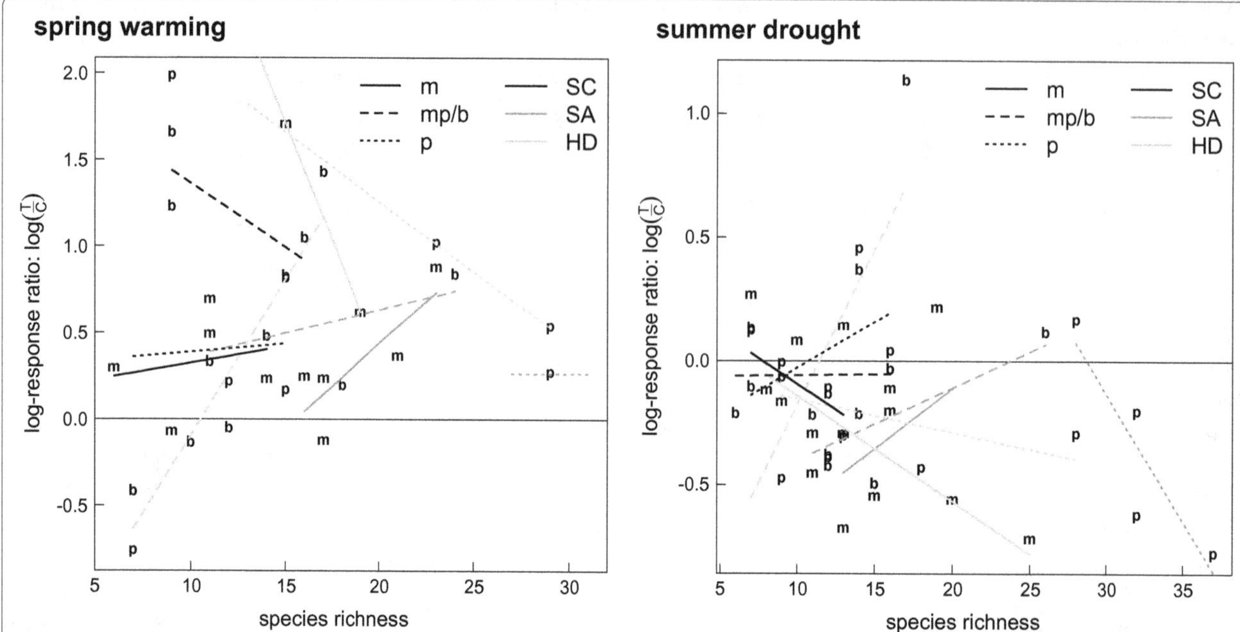


**Fig. 3** Contrast in biomass between treatment and control at the end of a temperature (spring 2010 *left*) and rainfall (summer 2009 *right*) manipulation period, expressed as log-response ratio. *Negative values* indicate a negative treatment effect, 0 indicates no difference between treatment and control. *Lines types* and *letters* refer to land-use types (where "b" stands for "both", i.e. "mown pastures"), *grey shades* represent the three different exploratories. In spring, species richness had a small but significant effect, while neither land-use type, exploratory nor their interaction(s) were significant. In summer, the three-way interaction of species richness, land-use type and exploratory was significant (see Table 3)

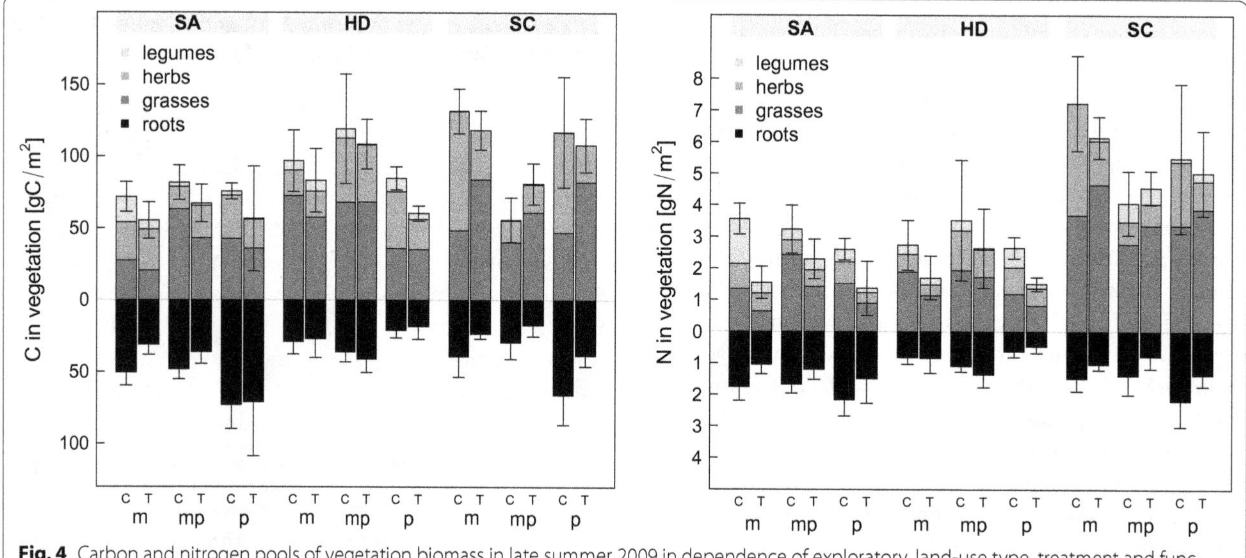


**Fig. 4** Carbon and nitrogen pools of vegetation biomass in late summer 2009 in dependence of exploratory, land-use type, treatment and functional groups. *Bars* represent mean values and ±SEM (N = 5)

although this overall pattern did not hold for mown pastures (significant land use–exploratory interaction: $F_{4, 74} = 2.84$, $P < 0.01$; see Additional file 1: Table S1).

The pattern for nitrogen in vegetation was similar to that of vegetation carbon (Fig. 4 right). The above-ground treatment manipulation effect was even more pronounced here ($F_{1, 74} = 7.43$, $P < 0.01$), as was the difference between Schorfheide-Chorin and the other two exploratories. The significant land use-exploratory interaction ($F_{4, 74} = 4.50$, $P < 0.01$) was here due to mown pastures having higher N-pools in SA and HD, but lower in SC (Fig. 4 left). In line with responses of functional group biomass, the C- and N-pools of legumes was most affected by our manipulations, followed by that of herbs.

### Soil carbon and nitrogen

Despite the effect of the treatment on C and N in plant biomass, we could not detect changes in the pool of soil C and N due to temperature or rainfall manipulation (Fig. 5, $P > 0.48$). As for the vegetation C- and N-pools, there were significant differences between the exploratories and land-use types ($F_{4, 75} = 3.07$, $P < 0.05$ and $F_{4, 73} = 2.83$, $P < 0.05$, respectively), with Schorfheide-Chorin having highest C- and N-levels. The interaction is due to higher C- and N-values for meadows relative to (mown) pastures in SA and HD, but the opposite pattern in Schorfheide-Chorin.

Soil-N correlated significantly with plant species richness, but differently for each exploratory (Exploratory-log(richness) interaction: $F_{2, 73} = 5.81$, $P < 0.01$; Additional file 1: Figure S5 left). Re-scaling the response within sites suggests no correlation (Additional file 1: Figure S5 right).

## Discussion

Our short-term experimental manipulation of temperature and precipitation led to clearly detectable responses in vegetation biomass, both above- and below-ground, for the three functional groups. However, these responses were idiosyncratic across years, regions and land-use types, and no *consistent* correlation with species richness was detectable. Indeed, only for above-ground biomass and soil N did we detect an effect of plant species richness at all, while vegetation and soil C were responded to treatment, land use and location alone (Additional file 1: Table S1). We hence conclude that species richness does not generally increase the ability of these systems to buffer short-term climate fluctuations in our grassland systems, but that its effect depends greatly on the edaphic and management context. The fact that we observed a significant three-way interaction between exploratory, LUT and species richness on log-response-ratios (Fig. 4) shows that our experimental design was sensitive enough to detect such effects. It was the idiosyncratic response across region-land-use type combinations that led us to reject a consistent buffering effect of plant species richness (see also [27]. More specifically, soil type, climate and land use are processes that in our system affect productivity and biogeochemical processes in grasslands more than species richness per se (in contrast to [28].

The design of the Biodiversity Exploratories uses land-use types to realise a gradient in plant species richness [18]. Indeed, land-use type is a significant predictor for species richness of our plots. For this experiment, however, we replicated the same land-use type five times in an attempt to break this strong association. As a consequence, land-use type did not emerge as particularly strong predictor of species richness (Table 1), opening the way for analysing the additional effect of plant species richness (polyserial correlation between land-use type and species richness is only $\rho = 0.325$). We believe that through our experimental design and the apparently sufficiently sensitive measurements we would have been

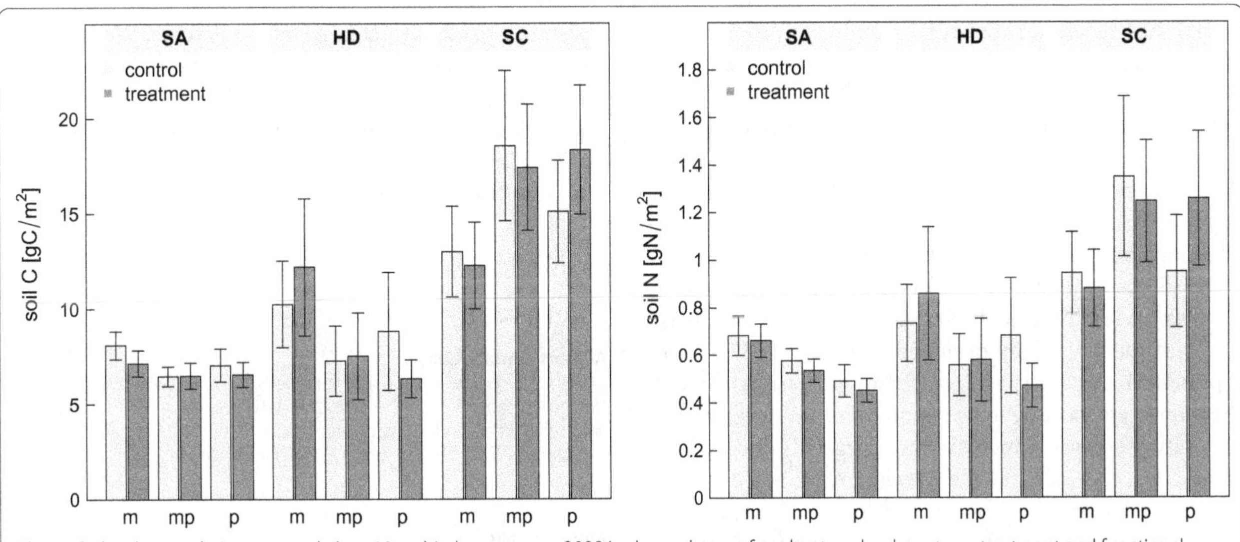


**Fig. 5** Soil carbon and nitrogen pools (top 10 cm) in late summer 2009 in dependence of exploratory, land-use type, treatment and functional groups. *Bars* represent mean values and ±SEM (N = 5)

able to detect a consistent ecologically relevant buffering effect of plant species richness against imposed drought and warming, if it existed.

Across the three regions, species richness gradients differed greatly (e.g. Fig. 4). Extensive sheep grazing led to high species richness in the Swabian Alp, while the rich, organic soils of Schorfheide-Chorin had very low species richness. This is in line with analyses by Socher et al. [16, 17], who report differential effects of management for the three exploratories: the diversity-promoting effect of grazing in the Swabian Alp is inverted into a negative effect in Schorfheide-Chorin. This may be due to grazing disturbance being at a small spatial scale, increasing dominance of tall species in fertile grasslands [29]. We cannot resolve whether it is the regional conditions that led to different correlations between species richness and buffering ability (Fig. 4) or because this finding is due to the fact that the regions cover different ranges along the species-richness gradient. To address this question, the experiment could be repeated on sites specifically selected to yield similar richness ranges in all three regions.

Beier et al. [30] point out that the approach we have chosen ("multi-factor application") inevitably confounds the effects of spring warming and summer drought. This is the price to pay for manipulating a specific scenario (or as they call it, the "inevitable dilemma"). The way spring warming and summer drought act in combination is not obvious. Spring warming led to an earlier growth, which could lead to an overall earlier season of unchanged length. We found no visual evidence for shifted phenology in summer, and also Reyes-Fox et al. [31] showed that warming extends, rather than moves, the season in a temperate grassland. Our experiment is insufficient to tease apart the different effects, and how they may accumulate over time.

The interplay of species richness and drought, in particular, has received substantial attention in the ecological literature. To better place our results in this context, we differentiate studies along two axes: (1) whether drought was manipulated or occurred naturally, and (2) whether diversity gradients were experimentally established or naturally realised in the field. In all of the resulting four combinations one can evaluate the importance of biodiversity for buffering drought effects. Our experiment (drought manipulation in natural diversity setting) showed little ability to buffer drought. This is consistent with most similarly designed studies [32–37], which report no biodiversity effect on drought resistance. Furthermore, also drought manipulations in experimental diversity gradients, which are rare, largely found no buffering effect [38–40], but see Kreyling et al. [41] for some buffering, discussed below). This is in contrast to all findings from measurements taken under naturally occurring drought. Plant species richness reduced the effect of drought under natural diversity settings [5, 42, 43] as well as under experimental biodiversity gradients (reviewed in [44, 45]. Natural diversity settings have the distinct disadvantage of confounding richness and environment. If the process that leads to higher plant diversity is also responsible for the system's resistance to drought, one cannot attribute resistance exclusively to plant diversity (see e.g., [46]). The predominant lack of buffering in drought manipulations even on experimental diversity gradients suggests a different reason, however.

This surprising but rather important finding has apparently not been noticed before, and it demands an explanation that is beyond the scope of this study. An obvious explanation would be that natural droughts are much more severe than manipulations. The EVENT experiment is more radical by simulating a 100-year drought, and indeed finds evidence for buffering of community productivity [39, 41], but not in other responses (reviewed in [47]). More studies need to create such strong manipulations to test whether this is a general pattern. As another potential explanation, we speculate that drought manipulations, typically by rainout roofs, may have treatment artefacts [48] that limit the ability of the vegetation to compensate for loss in productivity of dominant species. For example, even if roof artefacts did not affect community biomass, they altered light quality [49, 50] and may prevent light-demanding drought-tolerant species to gain from competitive release.

In conclusion, our experiment detected an influence of land use, site conditions and species richness and their interactions on the way vegetation and soil C- and N-pools respond to spring warming and summer droughts. This is largely due to manipulations having opposing effects on plant growth, at least in the short term. We failed to find a consistent effect, however, indicating that species richness per se does not contribute substantially to grassland resistance to climate-change manipulations. We speculate that this can be attributed to the relatively high number of species even at low richness levels in our study system, which already provides the complementarity required for positive biodiversity–ecosystem functioning or biodiversity–stability relationships.

**Authors' contributions**

CFD and MSL conceived and designed the experiment. LvR and CFD performed the experiment, LvR and MSL carried out the lab work. CFD and LvR analysed the data and wrote the first draft, all authors contributed to the final manuscript. All authors read and approved the final manuscript.

**Author details**

[1] Biometry & Environmental System Analysis, University of Freiburg, Tennenbacher Str. 4, 79106 Freiburg, Germany. [2] Computational Landscape Ecology,

Helmholtz-Centre for Environmental Research, Permoser Str. 15, 04318 Leipzig, Germany. [3] Geobotany, Faculty of Biology, Schänzlestr. 1, 79104 Freiburg, Germany.

#### Acknowledgements
We thank the managers of the three exploratories, Swen Renner, Sonja Gockel, Andreas Hemp and Martin Gorke and Simone Pfeiffer for their work in maintaining the plot and project infrastructure, and Markus Fischer, the late Elisabeth Kalko, Eduard Linsenmair, Dominik Hessenmöller, Jens Nieschulze, Daniel Prati, Ingo Schöning, François Buscot, Ernst-Detlef Schulze and Wolfgang W. Weisser for their role in setting up the Biodiversity Exploratories project. Fieldwork permits were issued by the responsible state environmental offices of Baden-Württemberg, Thüringen, and Brandenburg (according to § 72 BbgNatSchG). We also like to thank Dörte Radtke for assistance with field sampling and chemical analyses, Astrid Bütof and Helge Bruelheide for encouraging this detailed analysis and Michaela Bellach, Carmen Börschig, Christoph Rothenwöhrer for excellent coordination and mutual assistance during fieldwork. The manuscript has profited greatly from the comments of two anonymous reviewers on a previous version.

#### Competing interests
The authors declare that they have no competing interests.

#### Funding
The work was funded by the DFG Priority Program 1374 "Infrastructure-Biodiversity-Exploratories" (DO 786/4-1).

#### References
1. Cardinale BJ, Duffy JE, Gonzalez A, et al. Biodiversity loss and its impact on humanity. Nature. 2012;486:59–67. doi:10.1038/nature11148.
2. Cardinale BJ, Srivastava DS, Duffy JE, et al. Effects of biodiversity on the functioning of trophic groups and ecosystems. Nature. 2006;443:989–92.
3. Alard D, Poudevigne I. Diversity patterns in grasslands along a landscape gradient in northwestern France. J Veg Sci. 2000;11:287–94.
4. Wilson JB, Peet RK, Dengler J, Pärtel M. Plant species richness: the world records. J Veg Sci. 2012;23:796–802. doi:10.1111/j.1654-1103.2012.01400.x.
5. Tilman D, Downing JA. Biodiversity and stability in grasslands. Nature. 1994;367:363–5.
6. Hector A, Schmid B, Beierkuhnlein C, et al. Plant diversity and productivity experiments in European grasslands. Science. 1999;286:1123–7.
7. Roscher C, Temperton VM, Scherer-Lorenzen M, et al. Overyielding in experimental grassland communities—irrespective of species pool or spatial scale. Ecol Lett. 2005;8:419–29.
8. Hautier Y, Seabloom EW, Borer ET, et al. Eutrophication weakens stabilizing effects of diversity in natural grasslands. Nature. 2014;508:521–5. doi:10.1038/nature13014.
9. Thibaut LM, Connolly SR. Understanding diversity-stability relationships: towards a unified model of portfolio effects. Ecol Lett. 2013;16:140–50. doi:10.1111/ele.12019.
10. Tilman D, Reich PB, Knops JMH. Biodiversity and ecosystem stability in a decade-long grassland experiment. Nature. 2006;441:629–32.
11. Yachi S, Loreau M. Biodiversity and ecosystem productivity in a fluctuating environment: the insurance hypothesis. Proc Natl Acad Sci USA. 1999;96:1463–8.
12. Blüthgen N, Simons NK, Jung K, et al. Land use imperils plant and animal community stability through changes in asynchrony rather than diversity. Nat Commun. 2016;7:10697. doi:10.1038/ncomms10697.
13. Zobel M. The relative role of species pools in determining plant species richness: an alternative explanation of species coexistence? Trends Ecol Evol. 1997;12:266–9.
14. Gross N, Bloor JMG, Louault F, et al. Effects of land-use change on productivity depend on small-scale plant species diversity. Basic Appl Ecol. 2009;10:687–96. doi:10.1016/j.baae.2009.09.001.
15. Pimm SL. The complexity and stability of ecosystems. Nature. 1984;307:321–6.
16. Socher SA, Prati D, Boch S, et al. Direct and productivity-mediated indirect effects of fertilization, mowing and grazing on grassland species richness. J Ecol. 2012;100:1391–9. doi:10.1111/j.1365-2745.2012.02020.x.
17. Socher SA, Prati D, Boch S, et al. Interacting effects of fertilization, mowing and grazing on plant species diversity of 1500 grasslands in Germany differ between regions. Basic Appl Ecol. 2013;14:126–36. doi:10.1016/j.baae.2012.12.003.
18. Fischer M, Bossdorf O, Gockel S, et al. Implementing large-scale and long-term functional biodiversity research: the biodiversity exploratories. Basic Appl Ecol. 2010;11:473–85.
19. Hollister RD, Webber PJ. Biotic validation of small open-top chambers in a tundra ecosystem. Glob Chang Biol. 2000;6:835–42. doi:10.1046/j.1365-2486.2000.00363.x.
20. Marion GM, Henry GHR, Freckman DW, et al. Open-top designs for manipulating field temperature in high-latitude ecosystems. Glob Chang Biol. 1997;3(Suppl.):20–32.
21. Bonham CD. Measurements for terrestrial vegetation. New York: Wiley-Interscience; 1989.
22. Smit AL. Root methods: a handbook. Berlin: Springer; 2000.
23. Pinheiro JC, Bates DM. Mixed-effect models in S and S-plus. New York: Springer; 2000.
24. Blüthgen N, Dormann CF, Prati D, et al. A quantitative index of land-use intensity in grasslands: integrating mowing, grazing and fertilization. Basic Appl Ecol. 2012;13:207–20. doi:10.1016/j.baae.2012.04.001.
25. Oksanen J, Blanchet FG, Kindt R, et al. Vegan: community ecology package. R package version 2.0-10. 2013. http://cran.r-project.org/package=vegan.
26. R Development Core Team. R: a language and environment for statistical computing. Vienna: R Foundation for Statistical Computing; 2014. http://www.R-project.org.
27. Kröel-Dulay G, Ransijn J, Schmidt IK, et al. Increased sensitivity to climate change in disturbed ecosystems. Nat Commun. 2015;6:6682. doi:10.1038/ncomms7682.
28. Hautier Y, Tilman D, Isbell F, et al. Anthropogenic environmental changes affect ecosystem stability via biodiversity. Science. 2015;348:336–40.
29. Gazol A, Tamme R, Price JN, et al. A negative heterogeneity-diversity relationship found in experimental grassland communities. Oecologia. 2013;173:545–55. doi:10.1007/s00442-013-2623-x.
30. Beier C, Beierkuhnlein C, Wohlgemuth T, et al. Precipitation manipulation experiments—challenges and recommendations for the future. Ecol Lett. 2012;15:899–911. doi:10.1111/j.1461-0248.2012.01793.x.
31. Reyes-Fox M, Steltzer H, Trlica MJ, et al. Elevated $CO_2$ further lengthens growing season under warming conditions. Nature. 2014;510:259–62. doi:10.1038/nature13207.
32. Bloor JMG, Bardgett RD. Stability of above-ground and below-ground processes to extreme drought in model grassland ecosystems: interactions with plant species diversity and soil nitrogen availability. Perspect Plant Ecol Evol Syst. 2012;14:193–204. doi:10.1016/j.ppees.2011.12.001.
33. Dreesen FE, De Boeck HJ, Janssens IA, Nijs I. Summer heat and drought extremes trigger unexpected changes in productivity of a temperate annual/biannual plant community. Environ Exp Bot. 2012;79:21–30. doi:10.1016/j.envexpbot.2012.01.005.
34. Grime JP, Fridley JD, Askew AP, et al. Long-term resistance to simulated climate change in an infertile grassland. Proc Natl Acad Sci USA. 2008;105:10028–32. doi:10.1073/pnas.0711567105.
35. Kahmen A, Perner J, Buchmann N. Diversity-dependent productivity in semi-natural grasslands following climate perturbations. Funct Ecol. 2005;19:594–601.
36. van Ruijven J, Berendse F. Diversity enhances community recovery, but not resistance, after drought. J Ecol. 2010;98:81–6.
37. Selsted MB, Linden L, Ibrom A, et al. Soil respiration is stimulated by elevated $CO_2$ and reduced by summer drought: three years of measurements in a multifactor ecosystem manipulation experiment in a temperate heathland (CLIMAITE). Glob Chang Biol. 2012;18:1216–30. doi:10.1111/j.1365-2486.2011.02634.x.
38. De Boeck HJ, Dreesen FE, Janssens IA, Nijs I. Whole-system responses of experimental plant communities to climate extremes imposed in different seasons. New Phytol. 2011;189:806–17. doi:10.1111/j.1469-8137.2010.03515.x.

39. Kreyling J, Beierkuhnlein C, Ellis L, Jentsch A. Invasibility of grassland and heath communities exposed to extreme weather events—additive effects of diversity resistance and fluctuating physical environment. Oikos. 2008;117:1542–54. doi:10.1111/j.2008.0030-1299.16653.x.
40. Vogel A, Scherer-Lorenzen M, Weigelt A. Grassland resistance and resilience after drought depends on management intensity and species richness. PLoS ONE. 2012;7:e36992. doi:10.1371/journal.pone.0036992.
41. Kreyling J, Wenigmann M, Beierkuhnlein C, Jentsch A. Effects of extreme weather events on plant productivity and tissue die-back are modified by community composition. Ecosystems. 2008;11:752–63. doi:10.1007/s10021-008-9157-9.
42. Frank ADA, McNaughton SJ. Stability increases with diversity in plant communities: empirical evidence from the 1988 Yellowstone drought. Oikos. 1991;62:360–2.
43. Hobbs RJ, Mooney HA. Spatial and temporal variability in California annual grassland: results from a long-term study. J Veg Sci. 1995;6:43–56.
44. Isbell F, Calcagno V, Hector A, et al. High plant diversity is needed to maintain ecosystem services. Nature. 2011;477:199–202. doi:10.1038/nature10282.
45. Isbell F, Craven D, Connolly J, et al. Biodiversity increases the resistance of ecosystem productivity to climate extremes. Nature. 2015;526:574–7. doi:10.1038/nature15374.
46. Fry EL, Manning P, Power SA. Ecosystem functions are resistant to extreme changes to rainfall regimes in a mesotrophic grassland. Plant Soil. 2014;381:351–65. doi:10.1007/s11104-014-2137-2.
47. Jentsch A, Kreyling J, Elmer M, et al. Climate extremes initiate ecosystem-regulating functions while maintaining productivity. J Ecol. 2011;99:689–702. doi:10.1111/j.1365-2745.2011.01817.x.
48. Kennedy AD. Simulated climate change: are passive greenhouses a valid microcosm for testing the biological effects of environmental perturbations? Glob Chang Biol. 1995;1:29–42.
49. Kreyling J, Arfin Khan MAS, Sultana F, et al. Drought effects in climate change manipulation experiments: quantifying the influence of ambient weather conditions and rain-out shelter artifacts. Ecosystems. 2016. doi:10.1007/s10021-016-0025-8.
50. Vogel A, Fester T, Eisenhauer N, et al. Separating drought effects from roof artifacts on ecosystem processes in a grassland drought experiment. PLoS ONE. 2013;8:e70997. doi:10.1371/journal.pone.0070997.

# 13

# Surveying the citizen science landscape: an exploration of the design, delivery and impact of citizen science through the lens of the Open Air Laboratories (OPAL) programme

Linda Davies[1*], Roger Fradera[1], Hauke Riesch[2] and Poppy Lakeman-Fraser[1]

**Abstract**

**Background:** This paper provides a short introduction to the topic of citizen science (CS) identifying the shift from the knowledge deficit model to more inclusive, participatory science. It acknowledges the benefits of new technology and the opportunities it brings for mass participation and data manipulation. It focuses on the increase in interest in CS in recent years and draws on experience gained from the Open Air Laboratories (OPAL) programme launched in England in 2007.

**Methods:** The drivers and objectives for OPAL are presented together with background information on the partnership, methods and scales. The approaches used by researchers ranged from direct public participation in mass data collection through field surveys to research with minimal public engagement. The supporting services focused on education, particularly to support participants new to science, a media strategy and data services.

**Results:** Examples from OPAL are used to illustrate the different approaches to the design and delivery of CS that have emerged over recent years and the breadth of opportunities for public participation the current landscape provides. Qualitative and quantitative data from OPAL are used as evidence of the impact of CS.

**Conclusion:** While OPAL was conceived ahead of the more recent formalisation of approaches to the design, delivery and analysis of CS projects and their impact, it nevertheless provides a range of examples against which to assess the various benefits and challenges emerging in this fast developing field.

## Background

The term 'citizen science' is a broad term used to encapsulate a range of different activities, but in its essence, it partners professional scientists with volunteers in shared endeavour to study the physical and biological world. In this paper, we present an introduction to the historical context of citizen science (CS), and provide an overview of one programme cited as such, the Open Air Laboratories (OPAL) network, from concept through delivery and impact. We use OPAL as a framework against which to review the multifarious forms that citizen science activities may take. We compare current thinking on the design, delivery and impact of CS projects with experience gained from the OPAL programme and consider the contribution CS can make to broader scientific endeavour and societal concerns.

*Correspondence: linda.davies@imperial.ac.uk
[1] Centre for Environmental Policy, Imperial College London, South Kensington, London SW7 1NA, UK
Full list of author information is available at the end of the article

## Historical context of citizen science

### *The advent of 'citizen science'*

The contribution by members of the public to the collection, analysis and dissemination of scientific data is not a new occurrence. Volunteers, with no formal qualifications or affiliations, have contributed substantially to scientific discovery. The voluntary efforts of the 'gentleman scientists', such as Benjamin Franklin and Charles Darwin, made significant contributions to the advancement of scientific knowledge across a range of domains while making their living from other or private means [1].

Alongside individual enthusiasts, amateur societies, which have a long and rich history, have also provided mechanisms for public participation in science. Many societies provide forums to bring together professional and amateur members for fieldwork, education, promotion and conservation, while also actively encouraging and supporting involvement from the wider population. These opportunities have spanned a range of disciplines, with particular success in astronomy and environmental studies [2, 3].

While citizen-involved scientific activities continued throughout the twentieth century, there remained a division between the general public and those with high levels of expertise. That level of expertise could be acquired by citizens through the accredited training provided by the professionalised scientific realm or through the expenditure of considerable amounts of time, money and effort in self-directed study. Scientific expertise therefore remained the purview of a minority and those that gained it stood apart from the mass of society [4].

In this paradigm, the public generally had been conceived of as the passive beneficiary of scientific advancement and knowledge, without themselves having a particular voice in either the science itself or its policy applications because, being a lay audience, they lacked the necessary expertise to contribute. This "cognitive deficit" model is a term coined by Wynne [5] as a means of criticising this attitude towards the public (lack of) understanding of science and now is in widespread usage to refer somewhat disparagingly to old-style science communication. It diagnoses a deficiency in public knowledge and understanding of science and proscribes filling this deficit through processes where the public remains the recipient of scientific knowledge (with the process being one directional and educational in nature). Over time this view was challenged by studies that demonstrated the value of local and amateur knowledge to science [6–8] and the important contribution this can make to science policy. In parallel it was increasingly recognised that greater public science literacy does not automatically translate into more deferential support of expert opinion, nor a generally more enthusiastic public towards science [9].

#### *The emergence of the term 'citizen science'*

The term "citizen science" was applied independently at about the same time in the United Kingdom and the United States (mid-1990s). Building on the developments outlined previously, citizen science was promoted by Irwin in the UK who, coming from a background of sociological research, envisioned a new strain of science where the professionals interact with the public to jointly formulate new knowledge and make informed decisions [9]. This tradition advocated a move away from the "deficit" model and instead emphasised that the public should engage with science rather than merely understand it, and also that scientists and experts need to be attentive towards the arguments and contributions the public can make towards science and scientifically informed policy. All this signalling that the communication between public and science should go both ways. As a result and alongside increasing recognition that society could and should play a more active role in the scientific process, new innovative science communication and other public engagement activities, such as science shops [10] and citizen juries [11], foreground democratic and active participation with experts developed. The aim was a critical two-way exchange rather than the mere transfer of knowledge from expert to public.

Independently of Irwin, however, the term "citizen science" was applied in the U.S. by Rick Bonney [12] to refer to a type of public engagement project that he and his colleagues were pioneering at the Cornell Laboratory of Ornithology. They aimed to combine the substantive tradition of amateur participation in ornithology research with an element of science communication and education targetted at those participating. This combination proved to be very successful and became an inspiration for the set-up of many similar projects both within the U.S. and abroad. Contemporary concepts of citizen science to an extent combine the aspirations of both, and citizen science activities arising from the tradition of Bonney can be seen as a possible way in which aspirations for Irwin's citizen scientists can, in part, be realised.

#### *Technological advancements supporting the growth of citizen science*

Alongside changes in perceptions regarding the value to society of a more engaged, scientifically literate citizenship, technological advancements have transformed the public's capability to contribute to scientific activities.

More powerful and internet-connected home computers have greatly increased the capacity of citizens to receive, collect and analyse data [1, 13]. The advent of the internet has improved communications, facilitated the development of new cultural processes, such as the crowdsourcing and sharing of data, and supported the growth of online networks of enthusiastic and interested participants [14]. The increasing sophistication of smartphones has turned every device into a potential mobile sensing station, with capabilities to record, interrogate and transmit global positioning system (GPS) location, time, images, acoustic information and other data

[15–18]. Alongside increasing the capability of citizens to collect data, technology can also greatly improve confidence in those data. Sensors record data with known margins for error, while novel applications of existing technology can support data validation (for example, the submission of high resolution digital photographs for verification by experts) [19, 20].

While many new technologies supporting citizen science are ubiquitous in the developed world, technology can also promote participation in citizen science by citizens in less prosperous parts of the world. Sapelli [21], a mobile platform for data collection and sharing, was designed primarily for non-literate and illiterate users with little or no previous experience with computing technologies, supporting environmental monitoring by indigenous communities, which includes vulnerable groups with little involvement in the management of land on which they live [22].

## The OPAL programme

Open Air Laboratories (OPAL) was designed as an environmental education and research programme delivered through a national network of partners based originally in England (2007–2013) [23] and extended across the United Kingdom (2014-current).

### *Research and outreach drivers*

The main scientific drivers behind OPAL were: (a) the objectives for sustainable development defined at the Rio Summit through the Conventions on Biological Diversity and Climate Change, and Agenda 21 [24]; (b) the UK crisis in taxonomy [25]; and (c) the decline in outdoor learning in the UK [26]. The unprecedented loss of global ecosystems [27] provided further evidence of the urgency of addressing these issues. Following the Rio Summit sustainable development was incorporated into the heart of UK government policy [28]. It was acknowledged then that government alone could not secure a more sustainable future and that everyone had a role to play. Community groups and the voluntary sector inter alia were identified as important participants in this endeavour. As sustainable development became more widely recognised so did the urgency of both the task ahead and the need for greater public awareness and engagement.

In the UK the National Lottery's Big Lottery Fund [29] is recognised as a leading supporter of programmes that improve social well-being and address relevant policy areas. In 2005 they established a major new funding initiative, Changing Spaces, calling for environmental projects that would educate and engage local communities in sustainable development. Emphasis was placed on supporting disadvantaged communities in their local environment but the programme was designed to reach all sectors of society. OPAL was therefore conceived in response to a recognised policy need (sustainable development and the environment) and funded by a national public body.

### *Concept*

In response to this call in 2005, Imperial College London (ICL) proposed a very simple concept: take scientists out of their institutions and into the heart of the community to share knowledge and engage local communities in field-based research.

The three research topics relevant to the identified research and outreach drivers were: pollution (air, water and soil), loss of biodiversity and climate.

The majority of OPAL-England partners were research scientists who were used to meeting regularly to share knowledge and develop collaborative research. They were joined by representatives from local and national government and their agencies and leading environmental organisations, such as the Natural History Museum (NHM), as well as community organisations affected by environmental issues such as the impact of air pollution and loss of biodiversity (parks and conservation managers). These meetings were initially funded through a network grant provided to ICL by the Engineering and Research Council [30] for the Air Pollution Research in London (APRIL) network in 1999 [31]. Davies established the APRIL Natural Environment Group from which the OPAL proposal emerged (APRIL is now managed by the Greater London Authority). The OPAL partnership was therefore largely already established as a collaborative research network familiar with research and policy needs (drivers).

Reflecting the aims and objectives of the Big Lottery Fund, the programme sought to engage a wide audience, particularly people from disadvantaged sectors of society, people not previously engaged with nature, as well as the general public. All partners recognised and supported these aims although for many it was their first experience of working directly with the public.

### *Funding*

ICL was initially awarded £11.8m by Big Lottery Fund to direct and manage the OPAL programme, with additional funds (£1.3m 2010; £1.4m 2012) awarded in subsequent years as the impact of the public participation activities was recognised. In 2014 further funding (£3.0m) was awarded to extend the community engagement work across the UK (OPAL-UK). An overall goal was agreed initially of one million beneficiaries comprising 500,000 through field studies and 500,000 online (with a further 100,000 in-field beneficiaries to be delivered through the OPAL-UK programme). Other targets were agreed and a range of quantitative data was gathered throughout the programme, for example demographic data (i.e. percentage of disadvantaged communities reached and age ranges of participants), media circulation data, web visitors etc., whilst qualitative data were gathered through comment boxes on the website, online and in-field questionnaires, and by social scientists employed to work on the programme.

### *Goals*

There were five key objectives:

1. Supporting a change of lifestyle, a purpose to spend time outdoors, observing and recording the local environment;
2. Developing an exciting and innovative educational programme that can be accessed and enjoyed by all ages and abilities;
3. Inspiring a new generation of environmentalists;
4. Gaining a much greater understanding of the state of the natural environment for research and policy purposes;
5. Building stronger partnerships between the community, voluntary and statutory sectors.

### *Programme design and structure*

The OPAL network is illustrated in Fig. 1. ICL directed and managed the programme guided by an external Advisory Board and supported by a series of regional and (under OPAL-UK) national committees that sought to coordinate activities and, in doing so, maximise programme impacts and support the OPAL objective to promote stronger partnerships between the community, voluntary and statutory sectors.

Under the original programme, OPAL established nine regional teams. Each was based in a university and

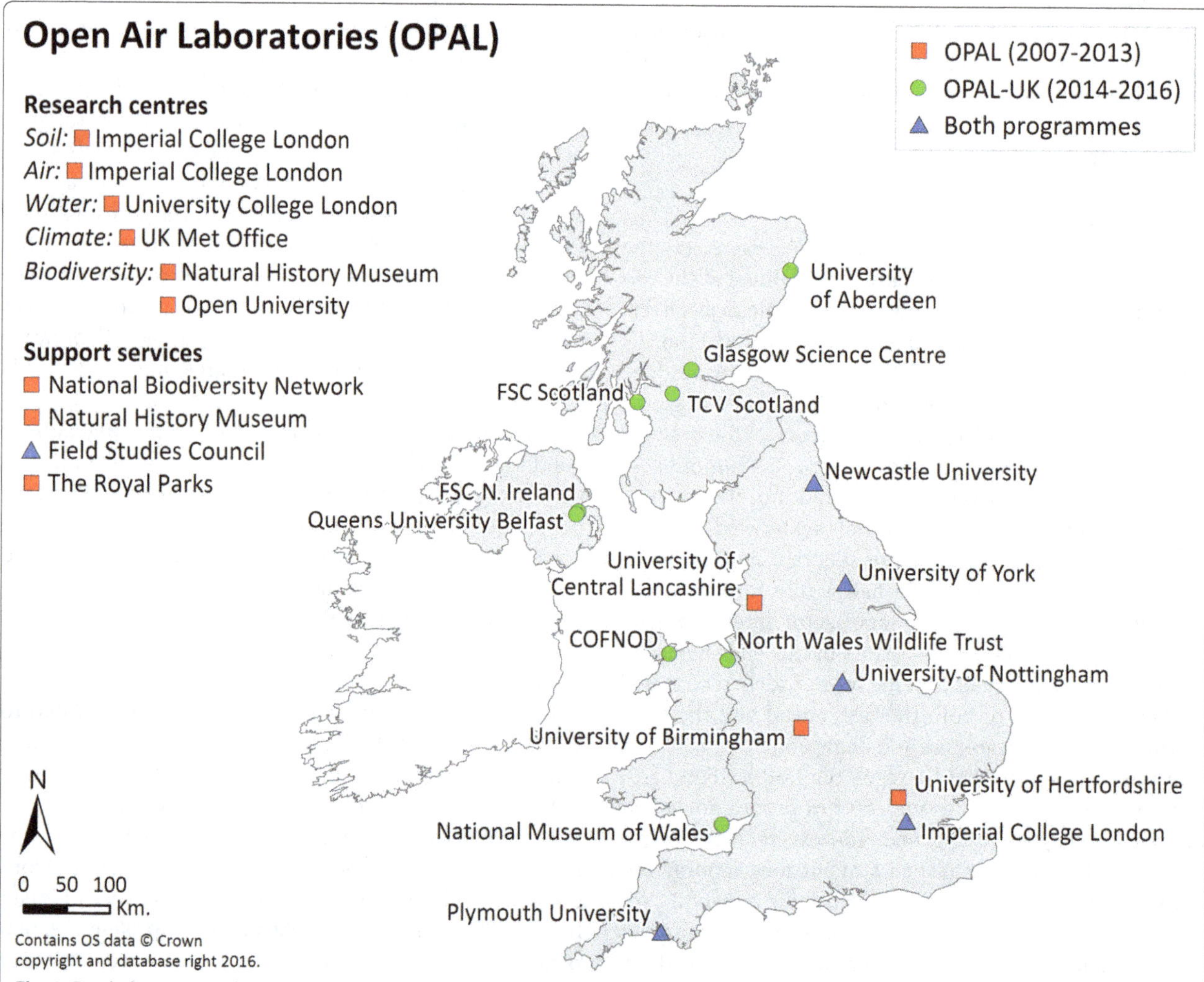


**Fig. 1** Funded partners in the Open Air Laboratories (OPAL) network. Geographic locations of regional partners with engagement staff (Community Scientists) are displayed on the map and partners leading national research centres and providing essential supporting services are listed to the *left*. The period during which partners were active in the OPAL network is indicated

worked directly with local people on research and educational projects of relevance to their region. Community Scientists, a new role created for the programme, worked under the direction of the regional lead scientist and, together with the schools programme [led by the Field Studies Council (FSC)] and public parks programme (led by the Royal Parks), were the main public engagement mechanisms, motivating and engaging local people. Under OPAL-UK, new partners extended public engagement activities to Scotland, Wales and Northern Ireland.

OPAL initially set up research and educational centres (Fig. 1) to provide scientific expertise, carry out research with varying degrees of public engagement (science workshops, public demonstrations, training days, publications in plain English, online progress reports, blogs and attending local and national fairs and events), and deliver research and educational tools. They also led the design and analysis of the OPAL field surveys, OPAL's primary citizen science activities. A large support service underpinned the programme including a national media strategy, web services, data management, and publications.

#### *Engaging participants outdoors*

It was recognised that deprived communities and people from disadvantaged sectors of society were less likely to engage through mainstream media or traditional approaches to public engagement in science so a significant proportion of staff time was spent working to engage these groups. The Index of Multiple Deprivation [32] helped to identify areas to target work and guidance from local authorities and local voluntary sector representatives, including those represented on OPAL regional and national committees, also helped Community Scientists to make contact with minority groups. These and many other innovative approaches were used to build relationships of trust with local communities through repeated face-to-face contact.

#### *Engaging participants through digital tools and media*

In addition to the significant staff resources (the original programme comprised fifteen organisations and over 100 staff employed in either full, part-time or in voluntary capacities) used to achieve OPAL's direct participation objectives, digital tools and traditional media services were used to reach the general public.

The OPAL website [33] provides the main interface for all participants. It houses the OPAL database where all public data are initially submitted, provides instant feedback through interactive visualisations and mapping, as well as presenting research findings in plain English. It also contains all of the educational materials OPAL has developed (free to download), blog posts on community achievements, scientist profiles, and topical news. Further digital projects, such as iSpot and Indicia, were also developed as part of the programme (see below).

A media strategy was designed and led by OPAL partner, the NHM, with their extensive experience of public engagement and all partners, staff and students were encouraged, trained and supported to contribute.

### Classifying citizen science

Citizen science has grown to the extent whereby an understanding of the breadth of projects classified as CS can be helpful to drive the field forward. While elements of volunteer involvement in science have been practiced for centuries, Silvertown [1] notes that the modern use of the term citizen science has only been recognised relatively recently. For example, in January 2009 only 56 articles in the search engine ISI Web of Knowledge were explicitly tagged with the term 'citizen science'; by January 2016 this had risen to over 11,000. Academic publications are not the only indication of the rise of citizen science; the discipline has now reached a maturity where there have been various conferences [34, 35], interest groups [36] and, membership organisations [37–39], seeking to share best practice among practitioners. As the concept of CS has developed a number of classification models have been proposed to understand the diversity of the practice.

At the broadest conceptual scale, Dickinson and Bonney [40] proposed four axes along which environmental CS varies: initiator of projects (academics or public), scale and duration (global/local, short term/long term); types of questions (pattern detection to hypothesis led); and goals (research, education and stewardship).

Reflecting a number of these axes, Prainsack [41], working mainly from the perspective of medical citizen science projects, distinguished them along a number of different dimensions. These include, for example, who has the ability to set the agenda, how the project affects local communities and how open it is with the resulting data and scientific research. Haklay et al. [42] propose a classification framework based on the level of participation from citizens: those requiring the least involvement as (i) 'Crowdsourcing', whereby citizens volunteer computing power or provide and maintain sensors; next (ii) 'Distributed intelligence', whereby the cognitive abilities of participants is utilised to collect or interpret basic data, sometimes with more limited, prior training; next (iii) 'Participatory sensing', where citizens are involved in problem definition and work with scientists to design a data collection methodology; and finally (iv) 'Extreme citizen science', where the relationship between scientist and citizen is collaborative, with opportunities for citizen involvement at all stages of the scientific process, with professional scientists acting "as facilitators, in addition to their role as experts" (p. 12). Wiggins and Crowston [43] identified five types of citizen science projects, including action projects (instigated by the local community to address matters of civic concern), virtual projects (based on internet contributions), investigation projects (driven by scientific aims requiring data collection from the physical world), conservation projects (promoting stewardship of natural resources), and education projects (focusing on education and outreach through formal and informal learning opportunities). For example, some celebrated internet-based and science led projects such as Galaxy Zoo [44] would in this classification fall under both virtual and investigation type.

OPAL, conceived in 2005, can be considered a pioneer in the application of large-scale CS even though it was not explicitly designed to any established framework of criteria for CS. We utilise the aforementioned broad conceptual framework of Dickinson and Bonney [40] (which encapsulates many other more detailed classification systems) and draw on examples from OPAL to investigate the breadth of citizen science in this section.

### *Initiator of project*

Along one of Dickinson and Bonney's four axes—initiator—the Centre for the Advancement of Informal Science Education (CAISE) [45] propose three categories for citizen science projects based on the amount of control that participants have over the different steps of the activity: (i) Contributory projects, where the activity has been designed by professional scientists and to which citizens are invited to contribute data as per the specified methodology; (ii) Collaborative projects, where scientists still lead the project but citizens are invited to refine the design of activities, analyse data, or disseminate findings; and (iii) Co-created projects, where the activities are designed by scientists and citizens working together and "public participants are actively involved in most or all steps of the scientific process" [45].

OPAL is policy driven and the majority of research questions were formulated by academics, their students, or collaborating organisations, and therefore citizens, in the main, acted in a contributory fashion, providing data they collected to answer research questions and using methodologies as defined and developed by professional scientists. OPAL's main mechanism for engaging the public occurs when public participation is intrinsic to research methodology (although not the research questions), namely the national field survey series (the OPAL surveys). The OPAL surveys allow people to work independently at a time, place and pace of their choosing, or directly in the field with OPAL Community Scientists (or other groups trained by OPAL) providing guidance

**Table 1 The OPAL national citizen science surveys**

| Survey name | Launch date | Aim | Approach | Output examples |
|---|---|---|---|---|
| OPAL Soil and Earthworm Survey | 2009 | Which species of earthworm are found in which soil and habitat types | 1. Assessment of site characteristics<br>2. Assessment of soil properties<br>3. Earthworm ID | Hypothesis led and policy links e.g. [71] |
| OPAL Air Survey | 2009 | Bio-indicators assessing local pollution and distribution of lichens and Tar spot on Sycamore | 1. Assessment of site characteristics<br>2. Assessment of tree characteristics<br>3. Identification of indicator lichens/fungus | Hypothesis led e.g. [56] |
| OPAL Water Survey | 2010 | Water quality of ponds | 1. Assessment of site characteristics<br>2. Assessment of water clarity<br>3. pH test<br>4. Identification of indicator invertebrates | Hypothesis led e.g. [59] |
| OPAL Biodiversity Survey | 2010 | Condition of hedges | 1. Assessment of site and hedge characteristics<br>2. Assessment of food resources<br>3. Identification of invertebrates<br>4. Tracking presence of other species | Hypothesis led: e.g. [70] |
| OPAL Climate Survey | 2011 | Human activities and climate | 1. Observations of aircraft contrails<br>2. Measurement of wind speed and direction<br>3. Thermal comfort | Validation e.g. [57] |
| OPAL Bugs Count Survey | 2011 | Impact of a changing environment on urban and rural areas | 1. Assessment of site characteristics<br>2. Assessment of microhabitats<br>3. Identification of invertebrates | Distribution monitoring e.g. [60] |
| OPAL Tree Health Survey | 2013 | Condition of trees and the pests and diseases that affect them | 1. Assessment of site characteristics<br>2. ID of common pests and diseases<br>3. ID of threatening pests and diseases | Policy requirement: e.g. Defra strategy [58] |

and support. OPAL has developed seven surveys to date (and several mini surveys), each focusing on a different environmental topic (Table 1). The surveys often use biomonitoring within their methodologies, an approach long used [46] whereby selected biological organisms can provide information on the state of their environment. OPAL surveys include equipment such as strips for pH measurements and tape measures as well as laminated, illustrated, instruction cards (with policy links and health and safety advice). In terms of their intended audience, the surveys were designed with an educated 13–14 year old in mind or adults new to environmental issues, however younger or less able participants can take part with appropriate support or with materials suitably adapted. Survey data are entered directly by participants to the OPAL database via the OPAL website and analysed by the lead scientist for that topic. When the first OPAL survey was launched (OPAL Soil and Earthworm Survey, 2009) lack of access to a computer proved a problem so a free post address was introduced.

The OPAL surveys were designed to provide a low technology approach to citizen science (and thus reducing barriers to participation, particularly for groups from lower socioeconomic backgrounds, a focus for OPAL's engagement); the opportunity to exploit new technologies and develop digital communities was, however, recognised as an important mechanism for OPAL to deliver its objectives. Some activities were undertaken in response to social and technological developments; for example, the arrival and increased public ownership of smartphones led to developing OPAL survey data submission via mobile phones (first used in 2011 for the OPAL Climate Survey) and a first app (in 2012 for the OPAL Bugs Count Survey). However, an integrated series of digital projects that sought to exploit crowdsourcing capabilities while building a new digital community was built into the OPAL programme by design.

iSpot [47] an online, interactive social network aimed to help the public to correctly identify wildlife and to build and reward the development of taxonomic skills. Participants share photographs of wildlife on the website and a community of amateur experts and professional scientists then provide participants with either verification of their identification or propose new identifications. The online experts providing support were initially OPAL-funded staff members but natural history societies very quickly became interested in the data being submitted by the public and, increasingly, as non-expert users developed their taxonomic skills, they also contributed to verification of records submitted by other users; in so doing iSpot could be considered a CS project that can span both of the CAISE classifications of contributory and collaborative CS. iSpot to January 2016 had >55,000 registered users who supported the identification of >700,000 records (personal communication, Janice Ansine). More than half of the submissions were identified within an hour (and >80 % were named to species level) [48].

While the majority of OPAL's CS activities would fall into contributory or, perhaps, collaborative classifications, there are examples where co-created or entirely citizen-led CS has occurred, often developing organically from OPAL activities. For example, staff members at the OPAL Yorkshire and Humber regional project (University of York), together with a local ranger, were interested in working with local people to monitor the colonisation of flora and fauna onto an ex-coalfield site in Wakefield. This work identified that the pond on the site was infested with invasive crayfish. The local Anglers Association who managed the site were keen to find a way to manage the invasive species and contribute to furthering understanding of this species (as well as others on the site) and so with OPAL staff they applied for a scientific trapping license from the Environment Agency. Another example is La Sainte Union Catholic School, which first used the OPAL Air Survey packs to study local air quality and lichens. The school then contacted the British Lichen Society (BLS) through OPAL and worked with them to develop a project that was awarded a partnership grant by the Royal Society to investigate the relationship between air quality and lichen distribution. Using diffusion tubes they measured levels of nitrogen dioxide as a means of validating the OPAL pollution index based on lichen indicator species employed by the OPAL Air Survey [49].

#### *Project scale and duration*

Revisiting Dickinson and Bonney [40] we look now at spatial and temporal scales and how they can vary between citizen science projects. Some projects may last just one field season (e.g. the Big Bumblebee Discovery [50]) whereas others have continued for decades (e.g. the Christmas Bird Count [51]); some may encourage citizens to examine their local area (e.g. the Hackney Wick Community Map [52] which allowed communities around Hackney Marsh, London to map a site less than 2 $km^2$) while others provide platforms for citizen scientists to work across continents (e.g. iNaturalist [53]).

As part of the OPAL-UK programme, and following testing with local communities for cultural variation and the relevance of indicator species, the seven OPAL surveys were adapted and extended across the UK (including translation of materials into Welsh language) in 2015. However, before this funding was awarded data had already been received from these countries (nearly 800 sites had previously been surveyed). Furthermore OPAL survey data have been received from many other European countries and further afield. Not all survey methodology is transposable to these areas although some data may remain valid (e.g. physical or chemical conditions) and regardless of the research value, participants may receive educational and stewardship benefits. After almost 10 years in operation public participation in OPAL remains high. Efforts to sustain OPAL core activities are ongoing and remain challenging.

### *Types of research*

Just as the research objectives underpinning citizen science activities can vary from hypothesis-led investigations (Conker Tree Science [54]) to pattern recognition exercises (Galaxy Zoo [55]); the approaches and types of scientific questions underpinning each of the OPAL surveys varied considerably.

Table 1 summarises the main type of research questions posed by the OPAL national surveys. These span the range of question types identified by Dickinson and Bonney [40]. For example, the OPAL Air Survey involved elements of hypothesis-led work, investigating whether fungi could be used as a bio-indicator of air pollution [56]. The study partly seeks to understand whether Tar spot fungus appears less frequently on sycamore trees in urban areas than in rural areas where air pollution levels and leaf litter management practices differ. Other studies used publicly collected data for validating computer model predictions: for instance, in the OPAL Climate Survey participants submitted observations of aircraft contrails which were then compared against model predictions of humidity levels at aircraft height [57]. The OPAL Bugs Count Survey placed more emphasis on monitoring species distribution change in urban and rural environments. In addition to scientific questions, environmental policy drivers directly shaped the design of surveys: the OPAL Soil and Earthworm Survey was developed in part to examine whether citizens could contribute data to support soil condition assessments and the OPAL Tree Health Survey supported official government monitoring of tree pests and diseases [58].

To ensure that the quality of data collected was of a usable standard, each OPAL survey was developed through a working group chaired by a scientific lead, supported by other scientists, representatives from natural history societies, government agencies, and other stakeholders. The process involved experts in graphic design, education, communication, web design, social science and public engagement. Drafts were regularly circulated to all OPAL staff for comments and tested with the community before final publication. Mechanisms to minimise error and to help validate records were introduced throughout the programme and ranged from collecting photographic evidence of observations from participants to online quizzes to determine the level of skill of the participant and weighting of data at the analysis stage [59, 60].

The development and delivery of citizen science does not occur in isolation from the social, political and economic conditions surrounding its goals of outreach and research. OPAL contained discrete supporting projects that did not necessarily constitute citizen science of themselves, but were considered to be essential to the processes of enabling citizen science. In response to the acknowledged 'crisis' caused by the then shortage of skilled taxonomists [25], OPAL sought to raise awareness and to increase the profile of natural history societies and conservation groups (voluntary sector) who play a critical role in biological recording and education. The OPAL programme included a dedicated funding scheme, led by the NHM, to help these groups to modernise, recruit new members and raise the profile of their societies with the public. Seventy organisations were awarded grants and a new web interface and database [61] was designed detailing their expertise and their contact details. Many of these organisations provided support to OPAL, particularly to the Community Scientists. At that time no natural history society existed dedicated to the study of earthworms, the biological element within the OPAL Soil and Earthworm Survey, so the Earthworm Society of Great Britain was established in 2009 through an OPAL grant. The society has the aims of (i) conducting research into earthworms; (ii) promoting knowledge and appreciation of earthworms within the non-scientific community; and (iii) educating the non-scientific community in earthworm biology and ecology. The organisation has now established the National Earthworm Recording Scheme and is in the process of developing distribution maps for the 27 species of earthworm; it has also run public events, provided identification training courses and has also developed its own citizen science survey (the Earthworm Compost Survey [62]), thereby continuing to support and foster the conditions for public participation in earthworm ecology research (personal communication, Kerry Calloway).

New software was developed to encourage and facilitate biological recording. The National Biodiversity Network (NBN) [63] manages the national database for biological records in the UK. Through OPAL funding, free, open-source biological recording software known as Indicia was developed [64] and is used by more than 80 societies in the UK and abroad. Indicia required a level of skill beyond that of most OPAL participants and so an easier to use version of Indicia known as Instant Indicia was developed, and also an implementation of Indicia known as iRecord [65], which was designed to allow any member of the public to create their own biological recording

**Table 2 OPAL original programme (2007-2013) impact data**

| Objectives | Target impact | Delivered impact | Source |
|---|---|---|---|
| 1. Spending time outdoors, observing and recording | a. Engagement with 500,000 participants at field events<br>b. Engagement with 500,000 online visitors | a. >850,000 participants at field events (>20 % of regional project engagement with disadvantaged beneficiaries)<br>b. >540,000 visitors to the OPAL website;<br>>1.1 m visitors to iSpot;<br>>520,000 visitors to Indicia sites or iRecord | a. Data provided monthly by OPAL staff<br>b. Web generated data (NHM/ICL; OU; NBN) |
| 2. Creating an educational programme | 240,000 survey packs to be designed, printed and distributed | >275,000 survey packs designed printed and distributed;<br>c50,000 educational resources downloaded | Print run data (FSC) and distribution data (OPAL partners);<br>Web generated data (NHM/ICL) |
| 3. Inspiring a new generation of environmentalists | a. Increasing access to natural history societies (membership at 10 societies increased by 10 %)<br>b. working with schools | a. 32 (46 %) of societies monitored >10 % increase in membership<br>b. working with 3100 schools | a. Data collected by NHM through monitoring associated with OPAL grants programme<br>b. Data provided monthly by OPAL staff |
| 4. Supporting a greater understanding of the state of the environment | No numerical target | >30,000 field surveys submitted (>22,000 further observations of contrail observation sub-activity)<br>>20 research manuscripts citing OPAL methods, using OPAL data or supported by OPAL funding | OPAL website survey entries<br>Publications |
| 5. Building stronger partnerships between voluntary, community and statutory sectors | a. Raising awareness through media engagement > 500,000<br>b. Engaging with community groups (no numerical target) | a. National coverage by >180 radio, TV and print media hits; >100 websites; total circulation figures exceeded 100 million<br>b. Working with >2400 organisations | a. Media cutting service managed by OPAL Communications Project (NHM)<br>b. Data provided monthly by OPAL staff |

account and begin submitting observations of nature. Uptake of iRecord has been extremely positive, with the millionth record submitted in September 2015 [66].

### *Setting and achieving goals*

Defining objectives and monitoring progress against them are important components of any CS programme and are the final classification of Dickinson and Bonney [40]. The OPAL Community Environment Report [67] prepared for the funding body and participants alike summarises OPAL's achievements over the first 4 years of operation covering preliminary research findings, unexpected outcomes, lessons learned and tools and materials designed. All targets agreed with the funder at the outset of the original programme were achieved or exceeded (separate but related outcomes were agreed for the OPAL-UK programme to be delivered by the end of the programme in December 2016). Table 2 provides a summary of the programme's delivery against its targets, updated since publication of the Community Environment Report, with data following completion of the first 6 year programme.

Below, we summarise further examples of impact across the goals for research and outreach. Looking at the interface between these goals, Lakeman-Fraser et al. [68] assimilate the quantitative and qualitative evaluation throughout the original 5 year programme drawing out trade-offs associated with multiple aim projects and identifying key considerations to tackle these challenges when planning and delivering a citizen science project.

*Research* Research outputs span environmental and social science fields [69]. Taking an ecological approach Rose et al. [59], for example, analysed the data collected from the OPAL Water Survey yielding a national assessment of water quality and clarity in England, whereas Gosling et al. [70] investigated the OPAL Biodiversity Survey finding that urban hedges as well as rural hedges can be important habitats for wildlife. A host of other manuscripts have been produced on the scientific outputs of the OPAL programme [56, 59, 60, 71] and the methodological considerations when monitoring data quality [57, 60, 72].

Exploring the societal impacts of citizen science Everett and Geoghegan [73], for example, investigated the motivations and barriers that people face initially engaging with a programme and maintaining enthusiasm for that programme. Other research into this area focused on people from socio-economically deprived backgrounds getting involved for the first time [74], issues that scientists are faced with when getting involved in citizen science [75, 76] and the education and behavioural impact of citizen science involvement [67].

Through OPAL more people have now engaged in activities related to environmental policy and sustainable development objectives (OPAL drivers), particularly with regard to Agenda 21 and the Convention on Biological Diversity (Articles 7,12,13) which promote monitoring, research, education and public awareness [77]. By working together scientists and the public have gathered a wealth of new data about wildlife, its distribution across England and the condition of their habitats. Some of these data are from sites that have previously been difficult for scientists to access such as gardens and inner city areas, allotments and playing fields.

*Outreach* The OPAL mantra is 'Explore Nature' and is all about encouraging people to get outside and learn about the environment on their doorstep. OPAL sought to engage all parts of society through a range of different approaches. For example, media coverage spanned national and local newspapers, television, radio, and online news sources: for example, the OPAL Soil and Earthworm Survey was reported on by, amongst others, the BBC One Show (estimated viewing figures of 4.8 million), BBC Radio 4 (estimated listening figures of 3.3 million), and the Daily Mail (estimated circulation figures of 2.2 million) [78].

Reflecting its funder's mission to focus on "communities and people most in need" [29], particular effort was placed on engaging communities without a large tradition of participation in scientific research, such as those from deprived, low-social capital areas. Traditionally, such groups tend to have fewer cultural resources to fully participate in local environmental decision making, or the social capital to make their voices heard above those of the experts, compounded by a lower access to high level education. Evidence also suggests that groups from lower socio-economic backgrounds tend to live in areas of lower environmental quality (e.g. [79]) and therefore may have greater need to participate in environmental decision making. Citizen science therefore presents a powerful mechanism through which to raise awareness and engage local people in local issues.

Social data taken from the OPAL Community Environment Report [67] indicate that: Half of all participants submitting survey data to the website (8450 from a sample of 16,766 people) state that this was the first time they had carried out a survey; just eight percent (695 from a sample of 9261) said they would not carry out another survey; almost half (43 %) of people interviewed (254 from a sample of 593) said taking part had changed the way they thought about the environment; more than a third of this groups (37 %) said they would change their behaviour towards it; 90 % of participants (13,142 from a sample of 14,621) said they had learnt something new;

83 % of these respondents said they had developed new skills. Approximately 20 % of engagement delivered by OPAL Community Scientists has been with individuals who classify as deprived or in some way hard-to-reach.

A wide range of materials have been developed across all topics for all ages and abilities and are stored on the OPAL website. They are widely used by schools, universities and other educational organisations such as the British Science Association. OPAL has worked with >3100 educational establishments (54 % secondary schools; 43 % primary schools; 2 % universities; and 1 % special schools) and school children contributed survey data relating to c15,000 sites (50 % total submissions). 10 % of the primary schools involved were located in the most deprived 10 % of England (6 % of survey results came from these areas). iSpot continues to be widely used and has been incorporated within the Open University's OpenScience Laboratory initiative that seeks to make practical science available to any student with a connection to the internet [80].

Contributing to a national research programme was a key motivating factor for many participants. OPAL's high quality science programme was said to give confidence to both teachers and students to carry out more fieldwork. Unplanned positive impacts on health and well-being were reported by many group leaders and participants during the programme. The high level of interest from schools was another unexpected outcome with many citing the outdoor learning programme and the opportunity for pupils to contribute to real research and to work with scientists as important factors.

## Conclusions

The OPAL programme provides an encompassing case study that spans a range of approaches within the citizen science landscape. OPAL can be viewed retrospectively as contributing to the democratic ideal of participatory decision-making argued for by Irwin, Wynne and others, through facilitating participatory knowledge production. At the same time, following the ideas of Bonney and his colleagues, OPAL activities also deliver against more explicitly formulated science and education goals.

In the broadest sense, although conceived ahead of the recent upsurge of interest in the classification of citizen science, OPAL does closely follow the key design steps identified by Dickinson and Bonney [40] who proposed the following topics for consideration: choosing a scientific question; forming a project team; developing and refining project materials; recruiting and training participants; accepting, editing and displaying data; analysing and interpreting data, disseminating results, measuring impacts. Whereas in OPAL to date, the research questions and analysis of results have been almost exclusively the province of professional scientists, the national survey series was explicitly designed to engage the widest possible audience in data gathering.

Despite the manifold faces of citizen science, the ever evolving discipline unites academics, educators, community members and policy makers and delivers a raft of benefits for both research and outreach. As we have seen, the approach taken when establishing, designing and delivering citizen science projects can be diverse and deliver a host of different outcomes.This field is evolving rapidly, driven by new technology and experience gained from professional scientists and the public alike as they participate in and contribute to our understanding of CS through projects such as OPAL.

#### Abbreviations

apps: applications; APRIL: Air Pollution Research in London; CS: citizen science; FSC: Field Studies Council; GPS: Global Positioning System; ICL: Imperial College London; NBN: National Biodiversity Network; NHM: Natural History Museum; OPAL: Open Air Laboratories; OU: Open University.

#### Authors' contributions

LD developed the concept and directed the OPAL programme from 2005 to 2013. She prepared the first draft covering these aspects of the paper. RF contributed to all aspects of the paper and drafted the sections on new technologies and data. PLF is responsible for new developments in CS and their relevance to OPAL. HR developed the section on the public understanding of science and the discussion on CS. All authors read and approved the final manuscript.

#### Author details

[1] Centre for Environmental Policy, Imperial College London, South Kensington, London SW7 1NA, UK. [2] Department of Social Sciences, Media and Communications, Brunel University, London, Uxbridge UB8 3PH, UK.

#### Acknowledgements

We would like to thank the Big Lottery Fund for financing this research through the OPAL and OPAL-UK programme grants , Defra for supporting the publication of this paper, Dr. D. Slawson (OPAL-UK), who has provided guidance throughout the development of this manuscript and Laurence Evans for provision of summary statistics and maps. Lastly, we are grateful to the OPAL participants, without whom this citizen science programme would not be possible.

**Open Air Laboratories (OPAL) network, (England, 2007-2013)** Led by Imperial College London, Dr. L. Davies (Director) involving nine regional partners: University of Central Lancashire, Dr. M. Toogood, (North West region); Newcastle University, Dr. A. Borland (North East region); University of York, Professor M. Ashmore (Yorkshire and Humber region); University of Nottingham, Dr. P. Crittenden (East Midlands region); University of Birmingham, Dr. J. Sadler (West Midlands region); University of Hertfordshire, Dr. M. Burton (East of England region); Imperial College London, Professor R. Toumi, (London region); Imperial College London, Dr. S. Power, (South East region); Plymouth University, Dr. M. Donkin (South West region): Research and education centres: Imperial College London, Dr. N. Voulvoulis (Soil); Imperial College London, Dr. S. Power (Air); University College London, Dr. N. Rose (Water); Open University, Professor J. Silvertown (Biodiversity) Natural History Museum, Dr. G. Stevens and Dr. J. Tweddle (Biodiversity) and Meteorological Office, Dr. G. Jenkins (Climate). Essential support services: Natural History Museum, Dr. G. Stevens and Dr. J. Tweddle; National Biodiversity Network, Dr. J. Munford; Royal Parks, Dr. N. Reeve and T. Assarati; Field Studies Council, Dr. R. Farley and Dr. S. Tilling.

**OPAL-UK (United Kingdom, 2014-current)** Imperial College London, Dr. D. Slawson (Director), involving twelve partners across the United Kingdom: The Conservation Volunteers (TCV), K. Riddell and D. Hall (Scotland); University of Aberdeen, Professor R. van der Wal (Scotland); Glasgow City of Science (Glasgow Science Centre), Dr. S. Breslin and Professor T. Howe (Scotland); Field

Studies Council, Dr. S. Tilling, D. Moncrieff and N. Elliot (Scotland and Northern Ireland); Queens University Belfast, Dr. K. Kerr (Northern Ireland); Cofnod, R. Tapping (Wales); North Wales Wildlife Trust, F. Cattanach and N. Hâf Jones (Wales); National Museum of Wales, Dr. R. Bevins and Dr. M. Wilson (Wales); Newcastle University, Dr. A. Borland (England); University of York, Professor M. Ashmore (England); University of Nottingham, Dr. P. Crittenden and Dr. S. Goodacre (England); Plymouth University, Dr. M. Donkin (England).

**Competing interests**

The authors declare that they have no competing interests.

**Declarations**

This article has been published as part of *BMC Ecology* Volume 16 Supplement 1, 2016: Citizen science through the OPAL lens. The full contents of the supplement are available online at http://bmcecol.biomedcentral.com/articles/supplements/volume-16-supplement-1. Publication of this supplement was supported by Defra.

**References**

1. Silvertown J. A new dawn for citizen science. Trends Ecol Evol. 2009;24:467–71.
2. Dickinson JL, Zuckerberg B, Bonter DN. Citizen science as an ecological research tool: challenges and benefits. Annu Rev Ecol Evol Syst. 2010;41:149–72.
3. Fortson L, Masters K, Nichol R, Borne K, Edmondson E, Lintott C, Raddick J, Schawinski K, Wallin J: Galaxy Zoo. Morphological classification and citizen science. In: Adv Mach Learn data Min Astron. 2011(Sandage 1961):1–11.
4. Shapin S. Science and the public. In: Olby RC, Cantor GN, Christie J, Hodge MJ, editors. Companion to Hist Mod Sci. London and New York: Routledge; 1990. p. 990–1007.
5. Wynne B. Knowledges in context. Sci Technol Human Values. 1991;16:111–21.
6. Epstein S. Impure Science: AIDS, Activism, and the politics of knowledge. California: University of California Press; 1996.
7. Irwin A, Wynne B, editors. Misunderstanding Science?: the public reconstruction of science and technology. Cambridge: Cambridge University Press; 1996.
8. Turney J. To Know Science Is to Love It? observations from public understanding of science research. Vol 17. 1973.
9. Irwin A. Citizen Science: A Study of People, Expertise and Sustainable Development. New York: Routledge; 1995.
10. Leydesdorff L, Ward J. Science shops: a kaleidoscope of science-society collaborations in Europe. Public Underst Sci. 2005;14:353–72.
11. Lezaun J, Soneryd L. Consulting citizens: technologies of elicitation and the mobility of publics. Public Underst Sci. 2007;16:279–97.
12. Bonney R. Citizen science: a lab tradition. Living Bird. 1996;15:7–15.
13. Ferdoush S, Li X. Wireless Sensor Network System Design using Raspberry Pi and Arduino for environmental monitoring applications. Procedia Comput Sci. 2014;34:103–10.
14. Burke J, Estrin D, Hansen M, Ramanathan N, Reddy S, Srivastava MB: Participatory sensing. In: Work World-Sensor-Web Mob Device Centric Sens Networks Appl. 2006:117–134.
15. Van der Wal R, Anderson H, Robinson A, Sharma N, Mellish C, Roberts S, Darvill B, Siddharthan A: Mapping species distributions: comparing the spread of UK bumblebees as recorded by the national depository and a citizen science approach. AMBIO Spec Issue- Digit Conserv 2015:in press.
16. Adriaens T, Sutton-croft M, Owen K, Brosens D, Van Valkenburg J, Kilbey D, Groom Q, Ehmig C, Thürkow F, Van Hende P, Schneider K. Trying to engage the crowd in recording invasive alien species in Europe: experiences from two smartphone applications in northwest Europe. Manag Biol Invasions. 2015;6:215–25.
17. August T, Harvey M, Lightfoot P, Kilbey D, Papadopoulos T, Jepson P. Emerging technologies for biological recording. Biol J Linn Soc. 2015;115:731–49.
18. Teacher AGF, Griffiths DJ, Hodgson DJ, Inger R. Smartphones in ecology and evolution: a guide for the app-rehensive. Ecol Evol. 2013;3:5268–78.
19. Graham EA. Using mobile phones to engage citizen scientists in research. Eos (Washington DC). 2011:92.
20. Bell S, Cornford D, Bastin L. The state of automated amateur weather observations. Weather. 2013;68:36–41.
21. Sapelli [http://www.ucl.ac.uk/excites/software/sapelli].
22. Stevens M, Vitos M, Altenbuchner J, Conquest G, Lewis J, Haklay M. Introducing Sapelli: a mobile data collection platform for non-literate users. In: Proc 4th Annu Symp Comput Dev. 2013:17.
23. Davies L, Bell JNB, Bone J, Head M, Hill L, Howard C, Hobbs SJ, Jones DT, Power SA, Rose N, Ryder C, Seed L, Stevens G, Toumi R, Voulvoulis N, White PCL. Open Air Laboratories (OPAL): a community-driven research programme. Environ Pollut. 2011;159:2203–10.
24. United Nations General Assembly. United Nations Framework Convention on Climate Change: Resolution. 1994.
25. Hopkins GW, Freckleton RP. Declines in the numbers of amateur and professional taxonomists: implications for conservation. Anim Conserv. 2002;5:245–9.
26. UK Parliament House of Commons. Select committee on education and skills second report. The Stationery Office by Order of the House. 2005.
27. The Royal Society. Measuring biodiversity for conservation. London. 2003.
28. UK Government. A better quality of life-strategy for sustainable development for the United Kingdom-1999. 2003.
29. BLF missions and values [http://www.biglotteryfund.org.uk/about-big/our-approach/mission-and-values].
30. Engineering and physical research council (epsrc) [http://www.epsrc.ac.uk].
31. Air Pollution Research in London (APRIL) [http://www.april-network.org].
32. English indices of deprivation 2010 [http://data.gov.uk/dataset/index-of-multiple-deprivation].
33. OPAL [http://www.opalexplorenature.org/].
34. 2012 PPSR Conference [http://www.citizenscience.org/community/conference2012/].
35. Citizen Science Association Conference [http://citizenscienceassociation.org/conference/].
36. BES Citizen Science Special Interest Group [http://www.britishecological-society.org/getting-involved/special-interest-groups/citizen-science/].
37. European Citizen Science Association [http://ecsa.biodiv.naturkundemuseum-berlin.de/].
38. Citizen Science Association [http://citizenscienceassociation.org/].
39. Australian Citizen Science Association [http://csna.gaiaresources.com.au/wordpress/].
40. Dickinson J, Bonney R, editors. Citizen Science: public participation in environmental research. Ithaca: Cornell University. 2012.
41. Prainsack B. Understanding participation: the "citizen science" of genetics. In: Prainsack B, Werner- Felmayer G, Schicktanz G. Farnham. Genet as Soc Pract. Farnham: Ashgate. 2014:1–27.
42. Haklay M. Citizen Science and volunteered geographic information—overview and typology of participation. In: Sui DZ, Elwood S, Goodchild MF. Crowdsourcing Geogr Knowl Volunt Geogr Inf Theory Pract. Berlin: Springer; 2013:105–122.
43. Wiggins A, Crowston K. From conservation to crowdsourcing: a typology of citizen science. In Proc Annu Hawaii Int Conf: Syst Sci; 2011. p. 2764–73.
44. Galaxy Zoo [http://www.galaxyzoo.org/].
45. Bonney R, Ballard H, Jordan R, McCallie E, Phillips T, Shirk J, Wilderman CC. Public participation in scientific research: defining the field and assessing its potential for informal science education. A CAISE Inquiry Group Rep. Washington, D.C.: Center for advancement of informal science education (CAISE). 2009.
46. De Temmerman L, Bell JNB, Garrec JP, Klumpp A, Krause GHM, Tonneijck K. Biomonitoring of air pollutants with plants—considerations for the future. In: Klumpp A, Ansel W, Klumpp G, editiors. Urban Air Pollut Bio-Indic Environ Aware; 2004.
47. iSpot [http://www.opalexplorenature.org/ispot].

48. Silvertown J, Harvey M, Greenwood R, Dodd M, Rosewell J, Rebelo T, Ansine J, Mcconway K. Crowdsourcing the identification of organisms: a case-study of iSpot. Zookeys. 2015;146:125–46.
49. Science Live: air quality [http://sse.royalsociety.org/2012/exhibits/air-quality/].
50. Big Bumblebee Discovery [http://jointhepod.org/the-big-bumblebee-discovery].
51. Christmas Bird Count [http://www.audubon.org/conservation/science/christmas-bird-count].
52. Hackney Wick Community Map [http://mappingforchange.org.uk/projects/hackney-wick-community-map/].
53. iNaturalist [https://www.inaturalist.org/].
54. Pocock MJO, Evans DM. The success of the horse-chestnut leaf-miner, Cameraria ohridella, in the UK revealed with hypothesis-led citizen science. PLoS One. 2014;9:1–9.
55. Lintott C, Schawinski K, Bamford S, Slosar A, Land K, Thomas D, Edmondson E, Masters K, Nichol RC, Raddick MJ, Szalay A, Andreescu D, Murray P, Vandenberg J. Galaxy Zoo 1: data release of morphological classifications for nearly 900,000 galaxies. Mon Not R Astron Soc. 2011;410:166–78.
56. Seed L, Wolseley P, Gosling L, Davies L, Power SA. Modelling relationships between lichen bioindicators, air quality and climate on a national scale: results from the UK OPAL air survey. Environ Pollut. 2013;182:437–47.
57. Fowler A, Whyatt JD, Davies G, Ellis R. How reliable are citizen-derived scientific data? Assessing the quality of contrail observations made by the general public. Trans GIS. 2013;17:488–506.
58. Department for Environment Food & Rural Affairs: Chalara Management Plan. 2013(March).
59. Rose NL, Turner SD, Goldsmith B, Gosling L, Davidson T. Quality control in public participation assessments of water quality: The OPAL Water Survey. BMC Ecol 2015, TBC:TBC.
60. Bates AJ, Lakeman Fraser P, Robinson L, Tweddle JC, Sadler JP, West SE, Norman S, Batson M, Davies L. The OPAL bugs count survey: exploring the effects of urbanisation and habitat characteristics using citizen science. Urban Ecosyst 2015, in press.
61. Nature Groups Near You [www.nhm.ac.uk/nature-online/british-natural-history/naturegroups/].
62. Earthworm Compost Survey [http://www.earthwormsoc.org.uk/earthworm-compost-survey1].
63. National Biodiversity Network [http://www.nbn.org.uk/].
64. Indicia [http://www.indicia.org.uk].
65. iRecord [http://www.brc.ac.uk/irecord].
66. Millionth record from Flatford bioblitz [http://www.field-studies-council.org/centres/flatfordmill/news/millionth-record-from-flatford-bioblitz.aspx].
67. Davies L, Gosling L, Bachariou C, Fradera R, Manomaiudom N, Robins S. OPAL Community Environment Report. 2013.
68. Lakeman-Fraser P, Gosling L, Moffat AJ, West SE, Fradera R, Davies L, Ayamba MA, Wal R van der: To have your citizen science cake and eat it? Delivering research and outreach through Open Air Laboratories (OPAL). BMC Ecol 2015, TBC:TBC.
69. OPAL Publications [http://www.imperial.ac.uk/opal/publications/].
70. Gosling L, Sparks TH, Araya Y, Harvey M, Ansine J: Differences between urban and rural hedges in England revealed by a citizen science project. BMC Ecol 2015, TBC:TBC.
71. Bone J, Archer M, Barraclough D, Eggleton P, Flight D, Head M, Jones DT, Scheib C, Voulvoulis N. Public participation in soil surveys: lessons from a pilot study in England. Environ Sci Technol. 2012;46:3687–96.
72. Tregidgo DJ, West SE, Ashmore MR. Can citizen science produce good science? Testing the OPAL Air Survey methodology, using lichens as indicators of nitrogenous pollution. Environ Pollut. 2013;182:448–51.
73. Everett G, Geoghegan H: Motivating participation in citizen science for natural history. BMC Ecol 2015, TBC:TBC.
74. Hobbs SJ, White PCL. Motivations and barriers in relation to community participation in biodiversity recording. J Nat Conserv. 2012;20:364–73.
75. Riesch H, Potter C. Citizen science as seen by scientists: methodological, epistemological and ethical dimensions. Public Underst Sci. 2014;23:107–20.
76. Riesch H, Potter C, Davies L. Combining citizen science and public engagement: the Open AirLaboratories Programme. J Sci Commun. 2013;12.
77. UNEP: Rio Conference Declaration on Environment and Development, United Nations Environment Programme. 1992.
78. Our country needs you to count worms: Volunteers wanted to carry out first earthworm "census" [http://www.dailymail.co.uk/sciencetech/article-1089395/Our-country-needs-count-worms-Volunteers-wanted-carry-earthworm-census.html].
79. Fecht D, Fischer P, Fortunato L, Hoek G, de Hoogh K, Marra M, Kruize H, Vienneau D, Beelen R, Hansell A. Associations between air pollution and socioeconomic characteristics, ethnicity and age profile of neighbourhoods in England and the Netherlands. Environ Pollut. 2015;198:201–10.
80. The Open Science Laboratory [https://learn5.open.ac.uk/course/format/sciencelab/about.php?id=2].

# 14

# Optimal methods for fitting probability distributions to propagule retention time in studies of zoochorous dispersal

Duarte S. Viana[1*], Luis Santamaría[1] and Jordi Figuerola[1,2]

**Abstract**

**Background:** Propagule retention time is a key factor in determining propagule dispersal distance and the shape of "seed shadows". Propagules dispersed by animal vectors are either ingested and retained in the gut until defecation or attached externally to the body until detachment. Retention time is a continuous variable, but it is commonly measured at discrete time points, according to pre-established sampling time-intervals. Although parametric continuous distributions have been widely fitted to these interval-censored data, the performance of different fitting methods has not been evaluated. To investigate the performance of five different fitting methods, we fitted parametric probability distributions to typical discretized retention-time data with known distribution using as data-points either the lower, mid or upper bounds of sampling intervals, as well as the cumulative distribution of observed values (using either maximum likelihood or non-linear least squares for parameter estimation); then compared the estimated and original distributions to assess the accuracy of each method. We also assessed the robustness of these methods to variations in the sampling procedure (sample size and length of sampling time-intervals).

**Results:** Fittings to the cumulative distribution performed better for all types of parametric distributions (lognormal, gamma and Weibull distributions) and were more robust to variations in sample size and sampling time-intervals. These estimated distributions had negligible deviations of up to 0.045 in cumulative probability of retention times (according to the Kolmogorov–Smirnov statistic) in relation to original distributions from which propagule retention time was simulated, supporting the overall accuracy of this fitting method. In contrast, fitting the sampling-interval bounds resulted in greater deviations that ranged from 0.058 to 0.273 in cumulative probability of retention times, which may introduce considerable biases in parameter estimates.

**Conclusions:** We recommend the use of cumulative probability to fit parametric probability distributions to propagule retention time, specifically using maximum likelihood for parameter estimation. Furthermore, the experimental design for an optimal characterization of unimodal propagule retention time should contemplate at least 500 recovered propagules and sampling time-intervals not larger than the time peak of propagule retrieval, except in the tail of the distribution where broader sampling time-intervals may also produce accurate fits.

**Keywords:** Seed dispersal, Dispersal kernel, Probability distribution, Endozoochory, Epizoochory, Gut passage time

## Background

The probability distribution of biological variables is of great importance for modeling and understanding biological phenomena, including the mechanistic basis of ecological processes. Mechanistic models are widely used in seed dispersal ecology (used here as a general term for the dispersal ecology of dormant propagules, including spores, resting eggs and cysts of plants, animals and fungi), as propagule movement is often difficult to track [1–5].

A considerable part of the Earth's biota does not actively move. Instead, they produce dormant propagules that

*Correspondence: dviana@ebd.csic.es
[1] Estación Biológica de Doñana (EBD-CSIC), C/Américo Vespucio, s/n, 41092 Seville, Spain
Full list of author information is available at the end of the article

rely on several types of vectors for their dispersal, such as wind, water and animals [3, 6–10]. Among the various vectors, animals such as birds and mammals disperse a great variety of propagules belonging to different species [8, 10–12]. Propagules are dispersed either externally, entangled in the fur or feathers ("epizoochory" hereafter), or internally, following ingestion and while transiting through the animal's gut ("endozoochory" hereafter).

A key element in the study of animal-mediated dispersal is the estimation of the distance at which propagules are dispersed. Dispersal distance (D) is usually estimated as the product of the vector movement rate (V) and the retention time (R) of ingested or attached propagules ($D = V \times R$). The distribution of dispersal distances, i.e. the dispersal kernel, is a major determinant of the spatial distribution of individuals, populations and species, thus its accurate estimation is of vital importance for studying and modeling metapopulation and metacommunity dynamics [8, 9, 13], as well as the distributions and expansion rates of species [3, 14, 15]. For example, many species distribution models (SDMs), which are used to model how species are distributed along niche gradients, incorporate dispersal kernels to predict range expansions or shifts under different scenarios of environmental change and estimate realistic distributions according to the species' dispersal potential [16, 17].

Together with the vector's movement behaviour, propagule retention time has been found to critically affect several properties of propagule dispersal kernels such as the range and frequency of dispersal events, thus the probability of long distance dispersal [4, 18]. Therefore, the accurate characterization of retention times is of fundamental importance for avoiding the magnification of biases already introduced by assumptions about vector movement when estimating propagule dispersal kernels. This is the reason why numerous empirical studies have investigated the different factors affecting propagule retention time, such as the size and shape of plant propagules [19, 20], the developmental stage of animal propagules [21], or the morphology [4, 22, 23], digestive physiology [24, 25], and activity [26, 27] of animal vectors.

However, obtaining continuous measurements of propagule retention time is, in most cases, extremely difficult owing to the ample time span and variable grain required (from minutes to several days, depending on the animal vector and propagule type), as well as monitoring interferences on the animal vector. In endozoochory studies, the most common strategy to measure retention time is to force-feed captive animals and collect their droppings at given time intervals, often of varying length [e.g., 24, 28]. Similarly, the usual practice in epizoochory studies is to record propagule attachment time at given time intervals, by measuring the number of propagules that remain attached to the fur of captive or semi-captive animals [29, 30], as well as to experimental coats [31], from a sample of propagules placed there by hand at the beginning of the experiment. In both cases, propagule retention time (i.e., defecation or detachment time) is recorded as a frequency at the end of given time intervals, thus as a series of interval-censored data.

Nevertheless, the censored nature of these data is usually not taken into account in studies of animal-mediated propagule dispersal (but see [4]). Although this systematic uncertainty on the precise moment of propagule deposition can severely bias the estimation of dispersal distance, fitting procedures used to characterize the distribution of retention times usually assign the frequency of retrieval to the collection time (i.e., to the upper bound of the time interval). Moreover, in most cases the fitting method is not accurately reported or insufficiently described [e.g. 5, 18, 32–36]. We compared the accuracy and robustness of different methods in fitting continuous probability distributions to propagule retention-time data. Because propagule retention-time data typically show right-skewed distributions with an initial peak (corresponding to the distribution mode) followed by a steep decrease and a long tail, we considered three parametric distributions commonly used to characterize these data: the lognormal, gamma and Weibull distributions. We assessed the performance of five different fitting methods. In the first three methods, we fitted parametric probability distributions to empirical distributions using either the (i) lower, (ii) mid or (iii) upper bounds of the sampling intervals as the data points; and in the other two methods, we fitted cumulative parametric distributions to (upper-bound) data arranged as empirical cumulative distributions, using two different procedures: (iv) maximum likelihood (CD-ML) and (v) non-linear least squares (CD-NLS). To assess the performance of these different methods, we applied them to a simulated dataset (based on empirical distributions; see "Methods" section) and compared the resulting parameter estimates and functions to the original ones. In addition, we assessed the robustness of the five fitting methods to variation in the distribution type (lognormal, gamma and Weibull) or parameter values of the original distribution (from which the simulated dataset was sampled), in the sample size (i.e., number of uptake propagules) and in the length of sampling time-intervals used to generate the simulated dataset.

## Results

### Variation in probability distribution

All three types of probability distributions (lognormal, gamma and Weibull) could accurately characterize the distribution of propagule retention times, as exemplified in Fig. 1. Among the different fitting methods, fittings to cumulative distributions (both CD-ML and CD-NLS)

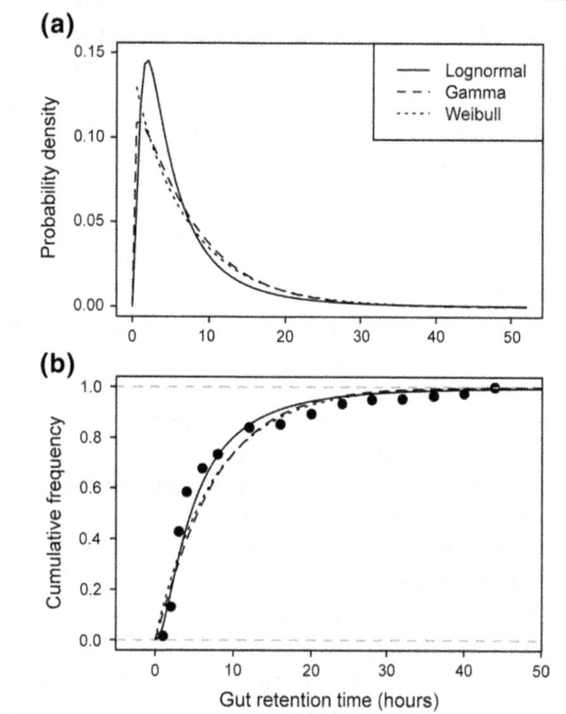


**Fig. 1** Examples of lognormal, gamma and Weibull distributions fitted to gut retention time of propagules ingested by waterfowl **a** and how these parametric distributions fit to empirical data **b**. Data is taken from [4]

provided the most accurate fits for all three types of probability distribution, both in parameter estimates and in the shape of the distribution (KS-statistic; Fig. 2). KS values supported the high accuracy of the estimates obtained with these two methods (KS-statistic <0.05). For fits to interval bounds, those using the upper and lower bounds had worse performances than that using the interval mid-value. The distribution type did affect, however, the accuracy of the different parameter estimates: the location parameter (μ) was less accurately estimated (greater difference between estimated and original parameter values) than the variance parameter (σ) for the lognormal and gamma distributions, while the opposite was true for the Weibull distribution (Fig. 2, upper panels).

### Variation in sample size

Fittings to the cumulative distribution (CD-ML and CD-NLS) were also the most robust against variation in sample size, i.e., variation in the number of retrieved propagules (Fig. 3a, b). Parameters estimated with these two estimation methods were remarkably accurate (Fig. 3a). Despite the low variation in accuracy of CD fits (KS-statistic ranged from 0.04 to 0.08; Fig. 3c), detailed inspection revealed that increasing sample sizes resulted in more accurate CD-ML fits up to N = 500, from which the fitting accuracy levelled off (i.e., reached an asymptotic KS-statistic; Fig. 3c).

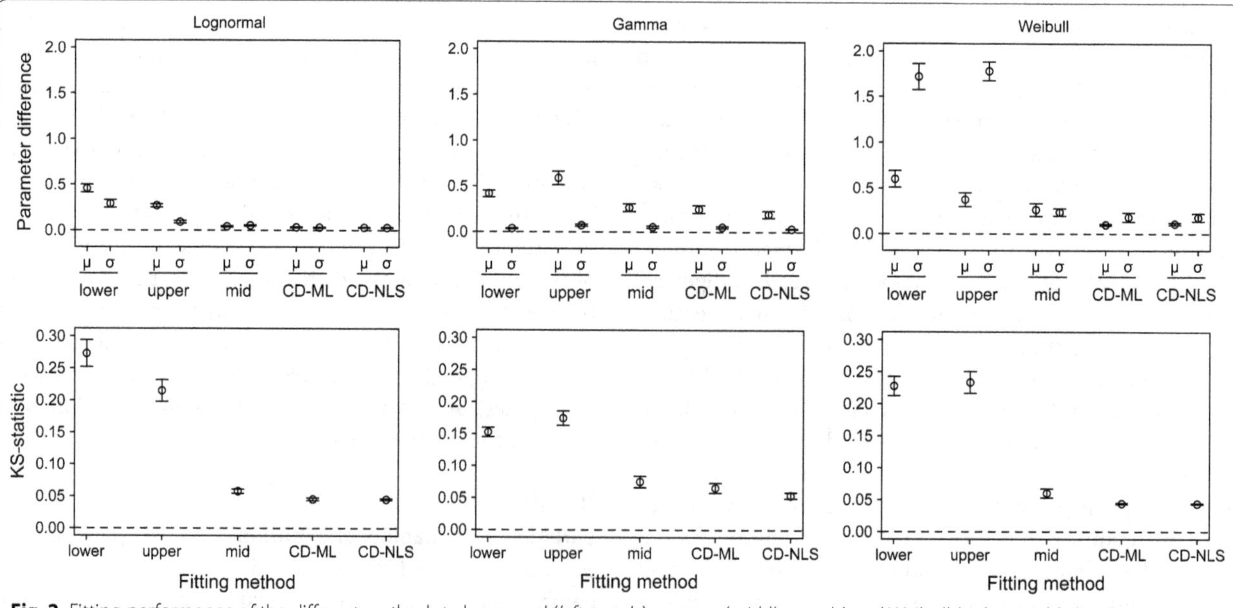


**Fig. 2** Fitting performance of the different methods to lognormal (*left* panels), gamma (*middle* panels) and Weibull (*right* panels) distributions measured by the difference between the original and fitted parameters (mean ± se; *upper* panels) and the KS-statistic (mean ± se; *lower* panels)

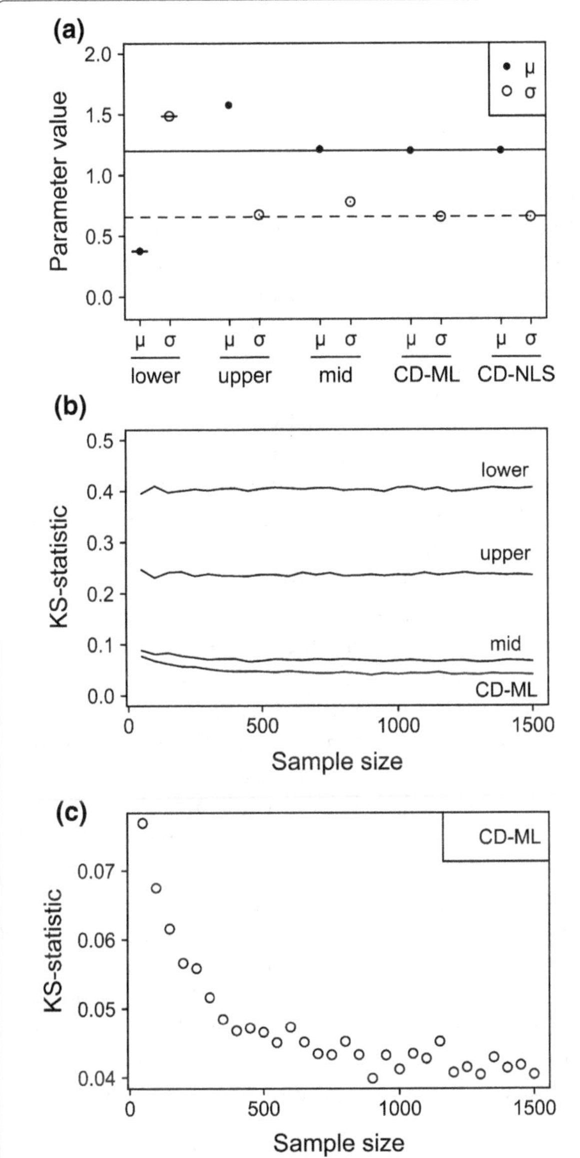


**Fig. 3** Robustness of the different fitting methods to variation in sample size (ranging from 50 to 1500 propagules). Fitting results correspond to the lognormal distribution. **a** Estimated parameter values (mean ± se). The *solid* and *dashed lines* indicate the original values of the location (μ) and variance (σ) parameters, respectively. Where error bars are undistinguishable, it means that standard errors are smaller than the mean-value dots. **b** KS statistic. **c** Fitting performance of the CD-ML method for different sample sizes, estimated by the KS statistic

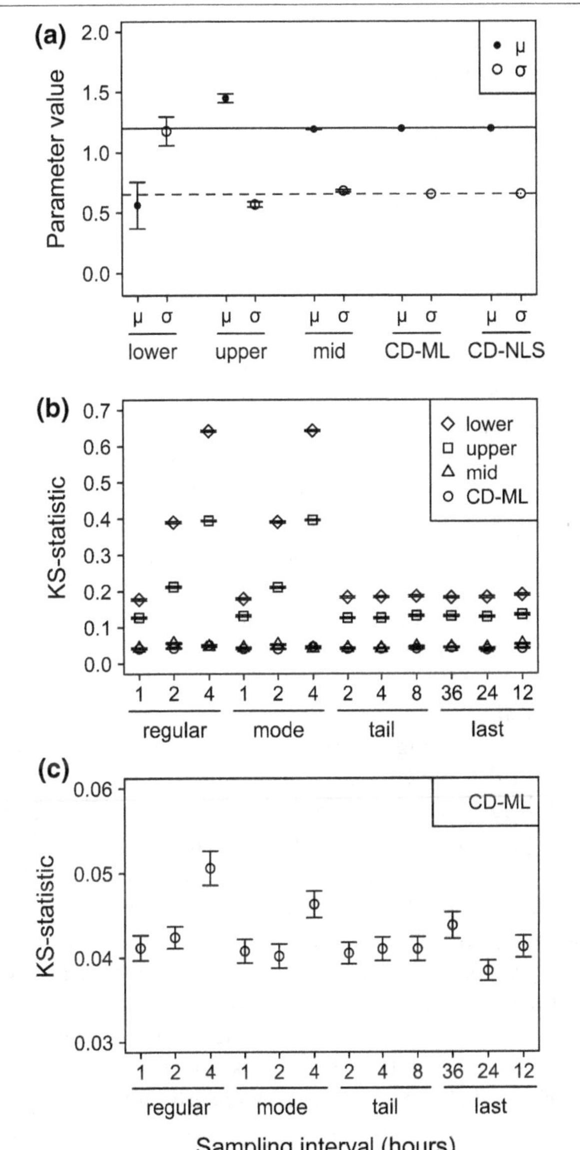


**Fig. 4** Robustness of the different fitting methods to variation in sampling time-interval. Fitting results correspond to the lognormal distribution. **a** Estimated parameter values (mean ± se). The *solid* and *dashed* lines indicate the original values of the location (μ) and variance (σ) parameters, respectively. Where error bars are undistinguishable, it means that standard errors are smaller than the mean-value dots. **b** KS statistic (mean ± se). **c** Fitting performance of the CD-ML method for different lengths in sampling time-intervals, estimated by the KS statistic. Time-intervals varied in a regular way over the whole sampling period (1, 2 or 4 h up to 52 h), around the distribution mode (1, 2 or 4 h during the first 8 h and 4 h afterwards), in the tail of the distribution (1 h up to 8 h and 2, 4 or 8 h afterwards), and the last time-interval (stopping sampling at 36, 24 or 12 h after propagule ingestion up to 52 h, the time of the last sampling)

### Variation in sampling interval

Consistently with the previous results, fittings to cumulative distributions (CD-ML and CD-NLS) provided the most accurate parameter estimates and were the most robust against variations in the length of sampling intervals (Fig. 4a, b). Despite the low variation in performance provided by CD fits, a detailed inspection showed that increasing the length of the initial sampling-time intervals (nearby the distribution's mode) resulted in reduced estimation accuracy, i.e., in increased KS statistics (Fig. 4c).

**Impact of fitting method on estimates of propagule dispersal kernels**

Non-optimal methods used to fit propagule retention time distributions produced severe biases in estimated dispersal kernels (Table 1; difference percentages ranged from 0.1 to 123 %). These biases were strongest for the frequency of long distance dispersal: when using the most common method (fits to upper-bound values), it increased by 123 and 19 % for *Potamogeton pectinatus* and *Scirpus lacustris* seeds, respectively. The magnitude of the bias was also related to the sampling protocol, specifically to the sampling time-interval. Experiments using shorter time-intervals around the distribution mode incurred in smaller biases (0.1–19 % in [37], using *S. lacustris* seeds) than those using more spaced intervals (2.4–123 % in [24], using *P. pectinatus* seeds). Overall, kernel properties related to long distance dispersal (namely dispersal over 100 km and the 99th distance percentile) were the most affected, whereas central tendency measures (i.e., mean and median) were less affected by suboptimal fitting methods.

## Discussion

The use of fitting methods that take into account the interval-censored nature of propagule retention time data proved necessary for a correct estimation of the underlying probability distributions. If we take the example of the lognormal fit to interval upper-bound data, which is the most used fitting method, a difference of 0.27 in the location parameter $\mu$ (i.e., 1.3 h in median retention time) was observed in relation to the original parameter. If the vector flies at an average speed of 60 km/h, a common speed for waterfowl species [38], the difference in median dispersal distance ($= 1.3\ \text{h} \times 60\ \text{km/h}$) would be 78 km, provided that the vector moves linearly until propagule retrieval. Even if actual vector movement distances are incorporated into dispersal distance estimations, considerable biases are also observed, mostly in the estimation of long distance dispersal (Table 1). The magnitude of these biases stress the necessity of using fitting methods that are able to account for the censored nature of propagule retention time data.

The two estimation methods based on cumulative distributions (CD-ML and CD-NLS) produced accurate estimations for lognormal, gamma and Weibull distributions and were remarkably robust against variations in data quality (sample size and sampling-time interval). The CD-NLS method requires more data points than the CD-ML method to be equally robust, as its estimation did not converge for distributions with a low variance (i.e., resulting from either a reduced retention time range or too large sampling time-intervals). We therefore recommend the CD-ML method, which fits the parameters to the censored data by maximum likelihood, as a general approach to characterize the probability distribution of propagule retention time. It can be implemented via the R package *fitdistrplus* [39] and other software packages such as MATLAB and SAS (this is not an exhaustive list).

The accuracy of estimations with the best method (CD-ML) was high enough to ensure very low deviations from the original distribution, as the maximum observed deviation in cumulative probability (compared to the original probability distribution) was only 0.05 (KS-statistic). Although distribution fittings using the interval mid-points also provided satisfactory results, estimated parameters were not as accurate as those using cumulative distributions. In particular, the variance parameter ($\sigma$) was generally overestimated, mostly at low sample sizes. The overestimation of this parameter might result in overestimated dispersal distances and, consequently,

**Table 1 Biases introduced by the choice of methods used to fit propagule retention time, on the dispersal kernels of two plant species (*Potamogeton pectinatus* Pp, and *Scirpus lacustris* Sl) dispersed by the same vector species (mallard *Anas platyrhyncos*)**

| Method | Seed sp. | % LDD | Mean (km) | Median (km) | Q99 (km) |
|---|---|---|---|---|---|
| CD-ML | Pp | 0.15 | 38.3 | 18.1 | 326.3 |
| | Sl | 0.47 | 58.7 | 20.5 | 498.5 |
| Lower | Pp | 0.13 (−12.0) | 48.2 (+25.8) | 18.5 (+2.4) | 435.5 (+33.4) |
| | Sl | 0.40 (−14.0) | 58.6 (−0.3) | 20.3 (−0.9) | 503.7 (+1.0) |
| Upper | Pp | 0.33 (+123.1) | 40.0 (+4.5) | 19.0 (+4.7) | 303.5 (−7.0) |
| | Sl | 0.55 (+19.1) | 61.3 (+4.5) | 20.8 (+1.5) | 506.1 (+1.5) |
| Mid | Pp | 0.12 (−18.9) | 32.4 (−15.3) | 17.5 (−3.2) | 256.0 (−21.6) |
| | Sl | 0.47 (+0.8) | 58.8 (+0.2) | 20.5 (+0.1) | 499.1 (+0.1) |

The comparison is based on four different parameters of the dispersal kernels for which the respective values are given: long-distance dispersal frequency (%LDD; i.e., % of dispersal events with distance >100 km), mean and median distance, and 99th distance percentile (Q99). Values between brackets indicate the magnitude of the bias, i.e., the relative difference (in percentage) between the value obtained using the optimal method (CD-ML) and each of all other fitting methods. Note that the CD-NLS method led to overall similar results to those of CD-ML (but see "Discussion" section) overestimated frequencies of long distance dispersal (as inferred from [4, 18]).

The robustness of the CD-ML method suggests that many hitherto obtained datasets on propagule retention time might be properly used in mechanistic models of propagule dispersal, even if the tail of the distribution is undersampled [e.g., 19]). Our results also suggest that the experimental conditions may and should be designed to optimize the accurate characterization of retention-time probability distributions by taking into account the trade-off between sampling effort and measurement precision. Simulations suggested that (i) at least 500 propagules should be retrieved to obtain more reliable estimates of retention time, and (ii) sampling effort should ensure an accurate characterization of the time peak (i.e., mode) of propagule retrieval by choosing sampling time-intervals of shorter length than the peak retrieval. Although the sampling effort for the distribution tail can be more relaxed, adequate sampling time-intervals should also be used until the end.

## Conclusions

Based on our comparative analysis, we recommend the use of the CD-ML method to fit parametric probability distributions to propagule retention-time data. Because propagule retention time is a key parameter in mechanistic models of passive dispersal [4, 18], an accurate parametric characterization of its probability distribution contributes to more reliable estimations of dispersal kernels and shadows. Given that many plants, invertebrates and microbes (including aquatic taxa) rely on passive dispersal, and that dispersal is a key determinant of biodiversity distribution patterns, the methodology presented in this study will also be useful for modeling population and community dynamics (e.g., meta-population and -community models), as well as species distributions (e.g., species distribution models; SDMs).

## Methods

We assessed the performance of the five different procedures used to fit parametric probability distributions to "empirical" datasets by comparing the distributions obtained from such fits with the original distributions from which the "empirical" datasets were randomly sampled. Original distributions were aimed at representing the propagule retention time of a given vector species; hence, we obtained them from a study in which the gut retention time of seeds fed to several waterfowl species was measured and fitted to three types of probability distribution—lognormal, gamma and Weibull (Fig. 1) [4]. These distributions are suited to characterize the distribution of propagule retention times, and although having a reduced number of parameters, their flexibility allows them to represent a broad variety of curve shapes [4, 33].

The general procedure was as follows. First, we generated the "empirical dataset" by drawing a random sample of the original distribution of propagule retention times and assigning the resulting values to predefined sampling time-intervals (thus simulating empirical sampling in real-world studies). We repeated this procedure 100 times for each original probability distribution to account for random variation. Second, we fitted a probability distribution (of the same type as the original one) to the empirical dataset, using the five procedures outlined above—i.e., fits to either the lower-bound, mid-interval or upper-bound values of the corresponding time interval, or fits to the cumulative distribution of empirical, upper-bound data. In the first three methods, interval bounds (either lower, mid or upper points) were considered to represent a continuous variable and the parameters of the fitted distributions were estimated by maximum likelihood (i.e., calculated according to probability density). For the methods using the empirical cumulative distribution, we estimated the parameters of fitted distributions either by maximum likelihood (method hereafter termed CD-ML), according to the procedure presented in Delignette-Muller and Dutang [40] (see [41, 42] for further details), or by non-linear least squares (method hereafter termed CD-NLS). All fittings were performed in R [43] using package *fitdistrplus* [39] for maximum likelihood estimation (function *fitdist* for the fittings using the time-interval bounds, and function *fitdistcens* for the CD-ML method), and the R base package for the CD-NLS method (function *nls*). We then assessed the fitting performance by comparing estimated and original distribution parameters (difference in value), and by estimating the Kolmogorov–Smirnov (KS) statistic using the package *kolmin* [44] in R [43]. The KS-statistic was obtained by drawing a random sample of the fitted distribution ($N = 500$) and comparing it with the original distribution. It represents the maximum difference in cumulative probability between the reference and focal distributions, thus corresponding to a goodness-of-fit measure ranging from zero (i.e., 100 % accurate) to one.

### Robustness of the different fitting methods

We assessed the robustness of the five fitting methods by simulating natural variation in propagule retention time (i.e., by varying the type and shape of original distributions) and different experimental designs commonly found in the literature, namely variation in sample size and in the length of sampling-time intervals. We applied the procedures described above to the sets of simulated data described below.

### *Variation in distribution type and shape*

We assessed the robustness of the fitting method to variation in the probability distribution by using three distribution types (lognormal, gamma and Weibull) and 30 different sets of parameters (variation in parameter values) to generate the original distributions from which "empirical datasets" were randomly drawn. To obtain a representative set of parameter combinations for each distribution type, we applied a Latin Hypercube Sampling procedure to the range of parameter variation reported in Viana et al. [4], using the R package *lhs* [45].

### *Variation in sample size*

We assessed the robustness of the fitting method to variation in sample size by varying the size of the random samples drawn from each original distribution, i.e., by simulating different numbers of retrieved propagules (ranging from N = 50 to N = 1500). We restricted this analysis to a single type of original distribution, and chose to use the lognormal distribution due to its wide use in studies describing gut retention time in different animal vectors [4, 33]. The lognormal distribution was defined with parameter values corresponding to the mean of the parameters' range reported in Viana et al. [4]. Sampled data for each simulation were assigned to sampling time-intervals corresponding to the lengths that provided the best performance (see Results; below we explain interval length variation).

### *Variation in the length of sampling time-intervals*

We assessed the robustness of the fitting method to different sampling time-intervals by varying, either uniformly or asymmetrically, the length of the time intervals used to sample data from the original distribution. We used both regular intervals over the whole sampling period, using three different lengths (either 1, 2 or 4 h, throughout the whole range of 52 h), and variable interval lengths. The latter varied in length (i) around the distribution mode, defining intervals of 1, 2 or 4 h during the first 8 h followed in all three cases by intervals of 4 h throughout the remaining sampling period, (ii) in the tail of the distribution, defining intervals of 1 h up to 8 h followed by intervals of either 2, 4 or 8 h afterwards, or (iii) only in the last interval, defining intervals of 1 h until 8 h and 4 h until either 12, 24 or 36 h followed by a variable last sampling bout at the end of the sampling procedure (52 h). All these sampling schemes reproduce procedures used in published studies [e.g., 19, 24, 27, 37], though additional variation was introduced in some of them. We used the lognormal distribution as the original distribution and a sample size of 500 propagules.

### *Impact of fitting method on estimates of propagule dispersal kernels*

To assess the bias produced by non-optimal fitting methods on various dispersal kernel properties, we compared the discrepancy in four dispersal kernel parameters (long-distance dispersal frequency, mean and median distance, and the 99th distance percentile) estimated using retention-time distributions fitted according to the five fitting methods described in the previous sections (only lognormal distributions were used). For this purpose, we used two examples [from 24, 37] in which the seed retention times of two aquatic plant species (*P. pectinatus* and *S. lacustris*) in the guts of a single waterfowl species (mallard *Anas platyrhynchos*) were measured using different sampling time-intervals (every 4 h for *P. pectinatus versus* every hour up to 4 h, every 2 h up to 8 h, and every 4 h afterwards for *S. lacustris*). Dispersal kernels were estimated using a realistic distribution of vector movement distances, based on waterfowl banding data (data available in [4]). For each of the four dispersal kernel parameters, we calculated the relative difference (in percentage) between the value obtained using the optimal method (CD-ML; see results) and each of all other fitting methods.

#### Authors' contributions

DSV carried out the statistical analysis and wrote the manuscript. DSV, LS and JF conceived of the study, participated in its design and coordination, and helped to draft the manuscript. All authors read and approved the final manuscript.

#### Author details

[1] Estación Biológica de Doñana (EBD-CSIC), C/Américo Vespucio, s/n, 41092 Seville, Spain. [2] CIBER Epidemiología y Salud Pública (CIBERESP), Seville, Spain.

#### Acknowledgements

This study was supported by project P09-RNM-4744 from Junta de Andalucía. DSV was supported by project RECUPERA 2020, Hito 1.1.1, cofinanced by the European Social Fund (ESF). LS was supported by project GENSABINA, ref. 476/2011, funded by the OAPN (Spanish Ministry of Environment).

#### Competing interests

The authors declare that they have no competing interests.

#### References

1. Cousens RD, Hill J, French K, Bishop ID. Towards better prediction of seed dispersal by animals. Funct Ecol. 2010;24:1163–70.
2. Nathan R, Muller-Landau HC. Spatial patterns of seed dispersal, their determinants and consequences for recruitment. Trends Ecol Evol. 2000;15:278–85.

3. Nathan R, Schurr FM, Spiegel O, Steinitz O, Trakhtenbrot A, Tsoar A. Mechanisms of long-distance seed dispersal. Trends Ecol Evol. 2008;23:638–47.
4. Viana DS, Santamaría L, Michot TC, Figuerola J. Allometric scaling of long-distance seed dispersal by migratory birds. Am Nat. 2013;181:649–62.
5. Will H, Tackenberg O. A mechanistic simulation model of seed dispersal by animals. J Ecol. 2008;96:1011–22.
6. Bilton DT, Freeland JR, Okamura B. Dispersal in freshwater invertebrates. Annu Rev Ecol Syst. 2001;32:159–81.
7. Figuerola J, Green AJ. Dispersal of aquatic organisms by waterbirds: a review of past research and priorities for future studies. Freshwat Biol. 2002;47:483–94.
8. Howe HF, Smallwood J. Ecology of seed dispersal. Annu Rev Ecol Syst. 1982;13:201–28.
9. Levin SA, Muller-Landau HC, Nathan R, Chave J. The ecology and evolution of seed dispersal: a theoretical perspective. Annu Rev Ecol Evol Syst. 2003;34:575–604.
10. Van Leeuwen CHA, van der Velde G, van Groenendael JM, Klaassen M. Gut travellers: internal dispersal of aquatic organisms by waterfowl. J Biogeogr. 2012;39:2031–40.
11. Costa JM, Ramos JA, da Silva LP, Timoteo S, Araújo PM, Felgueiras MS, Rosa A, Matos C, Encarnação P, Tenreiro PQ, et al. Endozoochory largely outweighs epizoochory in migrating passerines. J Avian Biol. 2014;45:59–64.
12. Herrera CM. Seed dispersal by vertebrates. In: Herrera CM, Pellmyr O, editors. Plant-animal interactions: an evolutionary approach. Oxford: Blackwell; 2002. p. 185–208.
13. Jacobson B, Peres-Neto P. Quantifying and disentangling dispersal in metacommunities: how close have we come? How far is there to go? Landscape Ecol. 2010;25:495–507.
14. Clark JS, Fastie C, Hurtt G, Jackson ST, Johnson C, King GA, Lewis M, Lynch J, Pacala S, Prentice C, et al. Reid's paradox of rapid plant migration. Bioscience. 1998;48:13–24.
15. Gillespie RG, Baldwin BG, Waters JM, Fraser CI, Nikula R, Roderick GK. Long-distance dispersal: a framework for hypothesis testing. Trends Ecol Evol. 2012;27:47–56.
16. Engler R, Hordijk W, Guisan A. The MIGCLIM R package—seamless integration of dispersal constraints into projections of species distribution models. Ecography. 2012;35:872–8.
17. Nobis MP, Normand S. KISSMig—a simple model for R to account for limited migration in analyses of species distributions. Ecography. 2014;37:1282–7.
18. Guttal V, Bartumeus F, Hartvigsen G, Nevai AL. Retention time variability as a mechanism for animal mediated long-distance dispersal. PLoS One. 2011;6:e28447.
19. Soons MB, van der Vlugt C, van Lith B, Heil GW, Klaassen M. Small seed size increases the potential for dispersal of wetland plants by ducks. J Ecol. 2008;96:619–27.
20. Sorensen AE. Seed dispersal by adhesion. Annu Rev Ecol Syst. 1986;17:443–63.
21. Charalambidou I, Ketelaars HAM, Santamaría L. Endozoochory by ducks: influence of developmental stage of Bythotrephes diapause eggs on dispersal probability. Divers Distrib. 2003;9:367–74.
22. Charalambidou I, Santamaria L, Langevoord O. Effect of ingestion by five avian dispersers on the retention time, retrieval and germination of *Ruppia maritima* seeds. Funct Ecol. 2003;17:747–53.
23. Karasov WH. Digestion in birds: chemical and physiological determinants and ecological implications. Stud Avian Biol. 1990;13:391–415.
24. Charalambidou I, Santamaria L, Jansen C, Nolet BA. Digestive plasticity in Mallard ducks modulates dispersal probabilities of aquatic plants and crustaceans. Funct Ecol. 2005;19:513–9.
25. Figuerola J, Green AJ. Effects of premigratory fasting on the potential for long distance dispersal of seeds by waterfowl: an experiment with marbled teal. Revue D Ecologie-La Terre Et La Vie. 2005;60:283–7.
26. Kleyheeg E, van Leeuwen CHA, Morison MA, Nolet BA, Soons MB. Bird-mediated seed dispersal: reduced digestive efficiency in active birds modulates the dispersal capacity of plant seeds. Oikos. 2014;124(7):899–907.
27. Van Leeuwen CHA, Tollenaar ML, Klaassen M. Vector activity and propagule size affect dispersal potential by vertebrates. Oecologia. 2012;170:101–9.
28. Westcott DA, Bentrupperbaumer J, Bradford MG, McKeown A. Incorporating patterns of disperser behaviour into models of seed dispersal and its effects on estimated dispersal curves. Oecologia. 2005;146:57–67.
29. Fischer SF, Poschlod P, Beinlich B. Experimental studies on the dispersal of plants and animals on sheep in calcareous grasslands. J Appl Ecol. 1996;33:1206–22.
30. Manzano P, Malo JE. Extreme long-distance seed dispersal via sheep. Front Ecol Environ. 2006;4:244–8.
31. Tackenberg O, Römermann C, Thompson K, Poschlod P. What does diaspore morphology tell us about external animal dispersal? Evidence from standardized experiments measuring seed retention on animal-coats. Basic Appl Ecol. 2006;7:45–58.
32. Kays R, Jansen PA, Knecht EMH, Vohwinkel R, Wikelski M. The effect of feeding time on dispersal of *Virola* seeds by toucans determined from GPS tracking and accelerometers. Acta Oecol. 2011;37:625–31.
33. Rawsthorne J, Roshier DA, Murphy SR. A simple parametric method for reducing sample sizes in gut passage time trials. Ecology. 2009;90:2328–31.
34. Rodríguez-Pérez J, Larrinaga AR, Santamaría L. Effects of frugivore preferences and habitat heterogeneity on seed rain: a multi-scale analysis. PLoS One. 2012;7:e33246.
35. Santamaría L, Rodriguez-Perez J, Larrinaga AR, Pias B. Predicting spatial patterns of plant recruitment using animal-displacement kernels. PLoS One. 2007;2(e1008):1001–9.
36. Uriarte M, Anciães M, da Silva MTB, Rubim P, Johnson E, Bruna EM. Disentangling the drivers of reduced long-distance seed dispersal by birds in an experimentally fragmented landscape. Ecology. 2011;92:924–37.
37. Figuerola J, Charalambidou I, Santamaria L, Green A. Internal dispersal of seeds by waterfowl: effect of seed size on gut passage time and germination patterns. Naturwissenschaften. 2010;97:555–65.
38. Alerstam T, Rosén M, Bäckman J, Ericson PGP, Hellgren O. Flight speeds among bird species: allometric and phylogenetic effects. PLoS Biol. 2007;5:e197.
39. Delignette-Muller ML, Pouillot qR, Denis J-B, Dutang C. Fitdistrplus: help to fit of a parametric distribution to non-censored or censored data. R package version 0.10. http://www.CRANR-projectorg/package=fitdistrplus. 2014.
40. Delignette-Muller ML, Dutang C. Fitdistrplus: an R package for fitting distributions. J Stat Softw. 2015;64(4):1–34.
41. Helsel DR. Nondetects and data analysis. Statistics for censored environmental data. 1st ed. Wiley: Interscience; 2005.
42. Klein JP, Moeschberger ML. Survival analysis: techniques for censored and truncated data. 2nd ed. Springer: Science & Business Media. 2003.
43. R Development Core Team. A language and environment for statistical computing. Vienna: R Foundation for Statistical Computing. ISBN 3-900051-07-0. URL http://www.R-projectorg. 2015.
44. Carvalho L. kolmim: An improved evaluation of Kolmogorov's distribution. R package version 0.2. http://www.CRANR-projectorg/package=kolmim. 2014.
45. Carnell R. lhs: Latin Hypercube Samples. R package version 0.10. http://www.CRANR-projectorg/package=lhs. 2012.

15

# Using multivariate cross correlations, Granger causality and graphical models to quantify spatiotemporal synchronization and causality between pest populations

Petros Damos[1,2,3*]

**Abstract**

**Background:** This work combines multivariate time series analysis and graph theory to detect synchronization and causality among certain ecological variables and to represent significant correlations via network projections. Four different statistical tools (cross-correlations, partial cross-correlations, Granger causality and partial Granger causality) utilized to quantify correlation strength and causality among biological entities. These indices correspond to different ways to estimate the relationships between different variables and to construct ecological networks using the variables as nodes and the indices as edges. Specifically, correlations and Granger causality indices introduce rules that define the associations (links) between the ecological variables (nodes). This approach is used for the first time to analyze time series of moth populations as well as temperature and relative humidity in order to detect spatiotemporal synchronization over an agricultural study area and to illustrate significant correlations and causality interactions via graphical models.

**Results:** The networks resulting from the different approaches are trimmed and show how the network configurations are affected by each construction technique. The Granger statistical rules provide a simple test to determine whether one series (population) is caused by another series (i.e. environmental variable or other population) even when they are not correlated. In most cases, the statistical analysis and the related graphical models, revealed intra-specific links, a fact that may be linked to similarities in pest population life cycles and synchronizations. Graph theoretic landscape projections reveal that significant associations in the populations are not subject to landscape characteristics. Populations may be linked over great distances through physical features such as rivers and not only at adjacent locations in which significant interactions are more likely to appear. In some cases, incidental connections, with no ecological explanation, were also observed; however, this was expected because some of the statistical methods used to define non trivial associations show connections that cannot be interpreted phenomenologically.

**Conclusions:** Incorporating multivariate causal interactions in a probabilistic sense comes closer to reality than doing *per se* binary theoretic constructs because the former conceptually incorporate the dynamics of all kinds of ecological variables within the network. The advantage of Granger rules over correlations is that Granger rules have dynamic features and provide an easy way to examine the dynamic causal relations of multiple time-series variables. The constructed networks may provide an intuitive, advantageous representation of multiple populations' associations that can be realized within an agro-ecosystem. These relationships may be due to life cycle synchronizations,

*Correspondence: petros.damos@ouc.ac.cy; petrosdamos@gmail.com; damos@agro.auth.gr

[1] Department of Environmental Conservation and Management, Faculty of Pure and Applied Sciences, Open University of Cyprus, Main OUC building: 33, Giannou Kranidioti Ave., Latsia, 2220 Nicosia, Cyprus

Full list of author information is available at the end of the article

exposure to a shared climate or even more complicated ecological interactions such as moving behavior, dispersal patterns and host allocation. Moreover, they are useful for drawing inferences regarding pest population dynamics and their spatial management. Extending these models by including more variables should allow the exploration of intra and interspecies relationships in larger ecological systems, and the identification of specific population traits that might constrain their structures in larger areas.

**Keywords:** Population modelling, Graph theory, MVAR models, Correlation and partial, Correlation, Final discovery rate, Synchronization, Causality

## Background

In recent years, there has been growing interest in graphical/causal models for the study of direct and indirect effects of climate on plant phenology and herbivores as well as the lagged effects of trophic or density depended factors on demographic parameters [1, 2]. Graphical models are a merger between probability and graph theory in which nodes represent variables and links non trivial interactions. Such constructs provide an important tool for facilitating communication among scientists, decision makers, and statisticians [3, 4]. Moreover, graphical models may be extremely important as an effective approach for describing multiple correlational associations and synchronization between ecological variables and coping with agricultural systems. In particular, problems in landscape ecology often involve modeling relationships among multiple physical and/ or biological variables that may run on differing spatial scales [5]. However, although these problems are inherently multivariate, researchers commonly rely on univariate methods, such as autocorrelation functions and autoregressive and spatial regression models, to address them [6–9]. Moreover, the time dynamics and causality are often ignored, despite the fact that causality operates and corresponds to a mechanistic perspective of the function of the systems [10].

For insect populations in particular, two principal challenges are understanding the synchronization of insect population life cycles and identifying the causal agents of population progression rates. It is possible that species coexisting together in the same area exhibit synchronous population fluctuations because they are subjected to the same environmental conditions. However, it remains unclear whether there are any population similarities across sites and what specific mechanisms facilitate temporal synchrony or asynchrony in closely related species. Understanding the process of synchronization of dynamics is also a crucial aspect of understanding outbreak dynamics in population ecology, which allows the introduction of management activities and the mitigation of pest expansion [11, 12]. Nevertheless, discovering life cycle synchronization and causality requires statistical tools for separating the endogenous population dynamics from the effects of the time-dependent, and often correlated, forcing variables such as temperature [13–15]. From a statistical perspective, any coupling among ecological variables, also including abiotic drivers, is difficult to quantify and to distinguish from the endogenous correlation structures that are generated by the core ecological system (i.e. pest population).

Recently, networks or graphical models constructed from multivariate time series analysis based on correlations and causality measures have been extended to assess the existence of nonlinear dependences between several variables to offer a means of studying the interactions of complex biological systems [16–19]. Correlation is a normalized version of covariance that measures the linear relationship between serial data and is used to build a correlation network (after thresholding). In ecological studies, correlations tend to detect which populations (or variables in general) may be synchronized [19]. Furthermore, synchrony between populations can be described using cross-correlograms, which are graphs of lag correlations between series vs. lag intervals [20]. In addition, partial correlation measures apply to situations where the relationship between any two variables is influenced by their relationships with other variables. Nevertheless, one disadvantage of the above conventional approaches is that correlation does not mean causation [21, 22]. Correlated occurrences may be due to chance or even due to a common cause but are not necessarily connected by a cause–effect relationship. Thus, the introduction of causality rules may provide a robust means to distinguish whether any two ecological variables interact directly or whether the appearance of a correlation is a result of chance or the variables being forced by a common third variable.

Among the available measures of causality, Granger causality is probably the most popular. Granger causality is a statistical concept first proposed for deciding whether one time-series is useful in forecasting another [23]. Conceptually, Granger causality provides a much more stringent criterion for causation (or information flow) than simply observing high correlation with some lag-lead relationship. Therefore, the rule is particularly designed to address the estimation of causal connectivity to extract the features which characterize the underlying spatiotemporal dynamics rather than just modest correlations [18]. Granger received the 2003 Nobel Memorial Prize in Economic Sciences for applying such methods in stock markets. Since then, the method has been extended to include more variables to detect causality in very complex systems [24] and has been employed in econometrics (i.e., detection and forecasting of stock market interactions) [24, 25] and neuroscience (i.e., identifying directed functional causal interactions from time-series data) [26]. Recently, this method has been implemented in detecting causality in a complex ecosystem to initiate management policies [27] and for differentiating direct causal linkages from indirect causal linkages between multiple ecological state variables [28].

This work combines multivariate time series analysis and graph theory to detect synchronization and causality among certain ecological variables and to represent significant interactions via network projections. The main objective is to introduce sound statistical tools that are useful for the study of time-variant ecological processes and for describing potential interactions through graphical models. Particularly, four different approaches are used to describe multivariate interactions: simple correlations, partial correlations, Granger causality and partial Granger causality are used to describe multivariate interactions. These statistical measures correspond to different ways to compute the relationships between the different variables and to construct the networks using the variables (time-series) as nodes and the significant indices (correlations, Granger rules) as edges. The second objective aimed to apply the method to time series of pest populations in order to detect significant correlations and possible causation between the different variables. Moreover, how the different techniques can be used to build discrete graphical models and related network configurations is illustrated. Finally, efforts are made to compare the different network structures and to interpret some of the ecological processes such as the simultaneous emergence and seasonality of closely related insect species over an agricultural landscape.

The analysis of pest populations' time-series data is of great interest for studying the driving parameters that explain pest population synchronization, and have practical implications for productively introducing time and location specific options for pest management. Moreover, using the proposed techniques to examine the synchrony of multi-species assemblages in ecology, may improve our understanding of how populations interact with long-term changes in their environments. This novel approach is shown to have advantages, not simply because it defines significant correlations among the variables but also because it may potentially capture some meaningful spatial relationships between ecological variables and related topological features. Moreover, conventional approaches do not directly address analyses of multiple ecological time series, while time-lagged causality (i.e., the difference in time units of a series of values and a previous one) is often neglected.

## Description of data

### Moth species

Three moth species were observed during 2003–2011: the peach twig borer *Anarsia lineatella* Zeller (Lepidoptera: Gelechiidae), the oriental fruit moth *Grapholita molesta* Busck (Lepidoptera: Tortricidae; previously known as *Cydia molesta*) and the summer fruit tortrix moth *Adoxophyes orana* (Fisher von Röslerstamm) (Lepidoptera: Tortricidae). The first-generation larvae of *A. lineatella* and *G. molesta* cause similar type of damage as they both attack young shoots, while *A. orana* is a leaf roller. During the farming season, larvae of later generations attack fruits in species-specific ways [29]. Altogether, the above Lepidoptera are widely distributed in Europe, North America and northern Asia and thus are viewed as the most important pests in stone fruit production worldwide [30]. Moreover, these species are polyvoltine and usually have 3–4 generations per year.

### Study area and population monitoring

Observations were carried out in a population sampling network that has been instituted in Northern Greece in particular in the prefecture of Imathia in Veroia (40.32 °N, 022.18 °E). The observation network covers an agricultural landscape that consists of plots in which fruit orchards represent approximately half of the observed field. The moth observation network consisted of 13 observation points. Trap placement sites were selected based on insect-host relationships (all included the main host, peach) and were representative in terms of cultivation conditions and landscape architecture. A cardboard delta trap (pheromone–pheromone traps: Trécé Inc., Salinas, CA, USA) were placed in each patch. Separate traps were used for each moth species, with sticky inserts baited with mixtures of synthetic sex pheromones (i.e., *A. orana*:(Z)-11-tetradecenyl acetate and (Z)-9-tetradecenyl acetate, *A. lineatella:* E)-5-decenyl acetate and (E)-5-decen-1-ol, *G. molesta*: (Z)-8-dodecenyl acetate) [31]. To avoid very strong autocorrelations, moths captured in traps were counted and removed twice a week, to create time series of the moth populations (Fig. 1, Additional file 1). Daily minimum and maximum air temperature data and relative humidity (RH) were obtained by using HOBO data-logging units (Onset Computer Corporation) placed on ALMME® experimental fields and registered during the same period (2003–2011) [32]. All data were collected from field studies that complied with institutional, national and international guidelines.

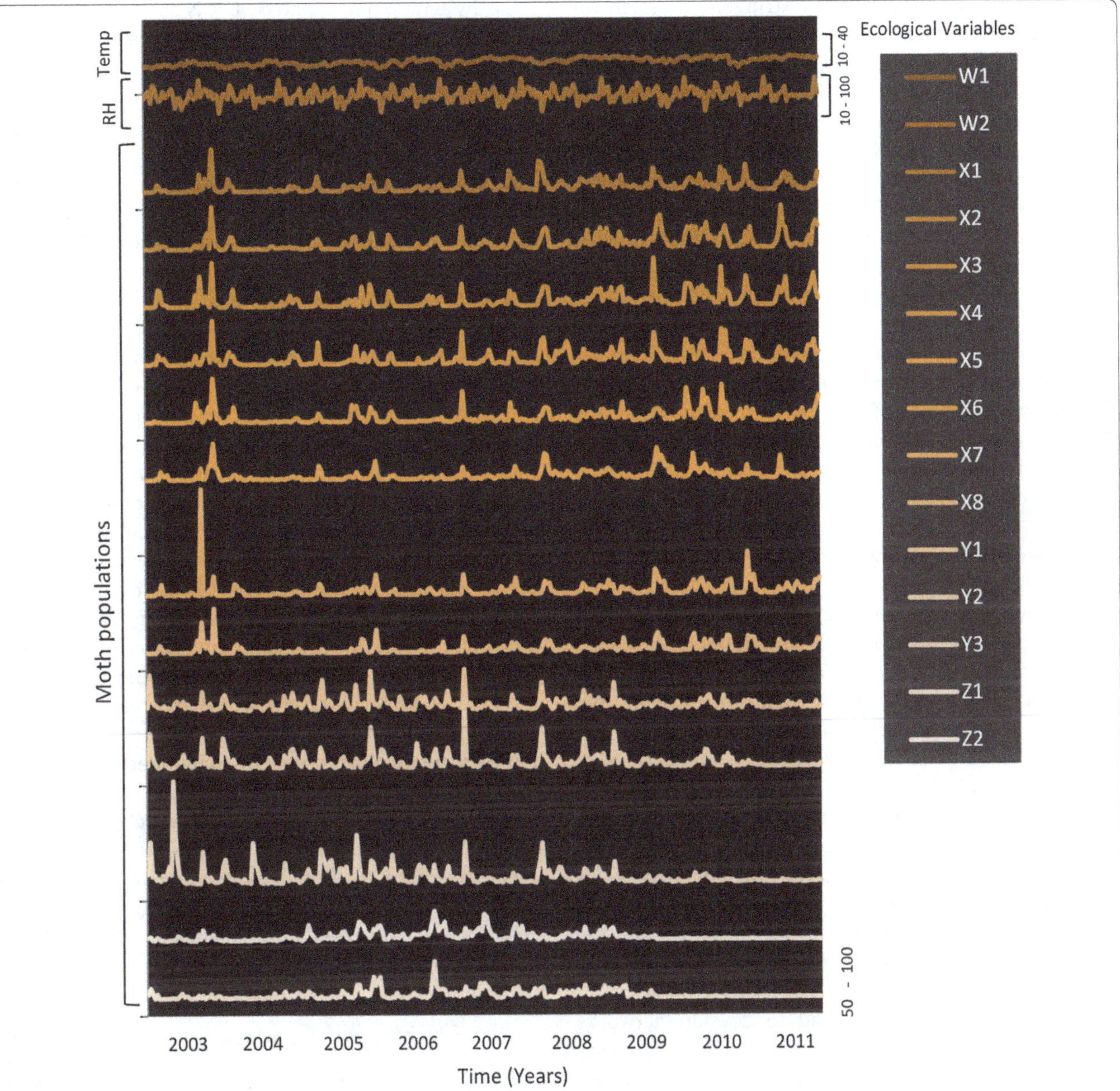


**Fig. 1** Actual ecological time series registered throughout the years 2003–2011 (3-day time intervals). W1: temperature (°C), W2: relative humidity (%), X1–X8: *A. orana*, Y1–Y3: *A. lineatella* and Z1–Z2: *G. molesta* moth populations (individuals)

These data were used to address the problem of transforming ecological time series into the correlation and causal networks given in the following section to propose statistical criteria to establish precise correlations and causal relationships.

## Methods for transforming ecological time series into a causal network

We consider a network, which consists of time-series which are represented as vertices (nodes) and are connected by edges (links) which are estimated through statistical indices. To construct the edges, which represent the link-interactions among the time-series, four different statistical methods were applied: cross correlations, partial cross correlations, Granger causality indices and partial Granger causality indices. Then, statistical significance tests and false discovery rates (FDR) are applicable to the outputs of the four techniques for trimming the networks. The four methods differ in the following ways: cross correlation is a simple measure of similarity of two series (say, X and Y) as a function of the lag of one relative to the other. Partial cross correlations correct the possible delayed effects of an additional variable (say, Z) on the correlation between X and Y. Granger causality provides a much more stringent criterion for causation (or information flow) than simply observing high correlation with some lag-lead relationship. Finally, partial Granger causality addresses the problem of eliminating the confounding influence of exogenous inputs and latent variables. In summary, the measures of correlation that were applied do not consider information from previous time steps (i.e., non-lagged correlation techniques) in contrast to Granger causality, which does considered it. Starting with the most simple and least conservative method, the different approaches are applied to examine how network configurations are affected and which best represents the final ecological relations. It is currently understood, for instance, that many ecological systems exhibit feedback, and it is therefore expected that cross correlations as a symmetric measure may be unsuitable for identifying nontrivial lag-lead relationships.

### Graph theoretic representation of cause-effect ecological network

To continue with the structure of graphs, based on multivariate time series analysis, it is convenient to introduce some graph theoretic matrix notations that mathematically formalize the networks proposed here. By definition, a graph $G$ consists of a set of vertices-nodes ($V(G) = \{v_1, v_2, \ldots, v_n\}$) and a set of edges-links ($E(G) = \{e_1, e_2, \ldots, e_{n'}\}$) in disjoint pair form, $G = (V, E)$. Thus, the graph-network is an ordered pair [V (G), E (G)] [33, 34]. A graph is directed if the edge set is composed of ordered vertex (node) pairs and is undirected if the edge pair set is unordered. Any graph can be represented according to its adjacency matrix. The elements of the matrix indicate whether pairs of vertices are adjacent or not adjacent in the graph.

In the next sections, I intend to develop networks based on four different techniques (i.e., similarity measures) applied to compute the edges $e_{ij}$ between any two nodes $v_i$ and $v_j$. Let $\boldsymbol{E}$ be an *nxn* matrix ($i$ and $j$ are indexes that go from 1 to n, and $e_{ij}$ is a single entry of the matrix $\boldsymbol{E}$), called *similarity-values* matrix, that has as elements the similarity measure values (correlations, partial correlations, Granger causality, partial Granger Causality). Moreover, let $\boldsymbol{E}'$, or *p value* matrix, be a matrix with its elements being the significant probability values of multiple comparison test in respect to the similarity measure (correlations, partial correlations, Granger causality, partial Granger Causality) using either parametric or non-parametric comparisons, respectively. Thereafter, low probability values are considered to have more influence and to consist of the weighted versions of networks, which take into account the varying contributions of each causally significant interaction.

Lastly, an adjacency matrix, $\boldsymbol{E}''$, is considered for each similarity measure, for which the constituents' $e_{ij}$ are the outcomes of the final discovery rate:

$$e_{ij} = \begin{cases} 1 & \textit{if } v_i v_j \in E \\ 0 & \textit{otherwise} \end{cases} \tag{1}$$

$\boldsymbol{E}''$ is a *binary-adjacency* matrix and is used to record the most probable, non-trivial, numbers of edges joining pairs of vertices. Particularly, when the element $e_{ij}$ of the matrix is one there is an edge from vertex $i$ to vertex $j$, and when it is zero there is no edge.

Each matrix (*similarity-values, p-values* and *binary-adjacency*) contains the information about the connectivity structure of the graph and is further used for the extraction of information about the characteristics of the investigated ecological networks. Four types of networks were constructed according to the four different methods (cross correlations, partial correlations, Granger causality and conditional Granger causality) of calculating the elements of the matrix. The number of nodes was the number of the time-series variables (including both; biotic and abiotic variables). The $X$, $Y$ and $Z$ represent populations of *A. orana*, *A. lineatella* and *G. molesta*, respectively, while Temp. and *RH* represent the two abiotic variables: mean temperature and relative humidity.

### Standard measures of correlation for undirected graphs

#### *Cross correlations*

I look at the ecological time series, each of which is observed through successive seasons as a continuous univariate process: weekly counts are a vector that represents population realization available through observation. The Pearson correlation coefficient, $r$, is used as a standard similarity measure [35]:

$$cor(X_i; X_j) = r_{X_i,X_j} = \frac{s_{X_i,X_j}}{s_{X_i} s_{X_j}} \tag{2}$$

where $s_{XiXj}$ stands for the sample covariance of the variables $X_i$ and $X_j$ and where $s_{Xi}$, $s_{Xj}$ are the standard deviations of samples $X_i$ and $X_j$, respectively. Substituting the estimates of the covariance and variance based on a sample gives the Pearson's linear correlation coefficient, with the following estimate:

$$e_{ij(cor)} = r_{X_i,X_j} = \frac{\sum_{t=1}^{n}(X_i - \bar{X}_i(X_j - \bar{X}_j)}{s_{X_i} s_{X_j}} \tag{3}$$

Here, $\bar{X}_i$ and $\bar{X}_j$ are the sample means for the first and second variables, respectively,$s_{X_i}$ and $s_{X_j}$ are the standard deviations for each variable, and $n$ is the series length (here, all successive years in which populations are active are considered) starting from t = 1.

#### Partial correlations

The partial correlation network graphical representation, is defined as the collection of links between those nodes whose partial correlation (as defined below) is not zero. The linkage of these elements may be described in terms of an adjacency matrix that consists of a network with no direction (undirected) [31, 32]. To assess whether non-zero correlations are direct or indirect, causal measures should be considered as very useful because they measure the linear correlation between two variables after removing the effect of other variables and thus also finding spurious correlations and revealing hidden correlations. In particular, *cross correlations* are very suitable for detecting a type of dependence between pairs of variables (e.g., population $X_i$ on population $X_j$, vice versa, or both). However, because we include abiotic variables as well (e.g., $W_1$: temperature and/or $W_2$: relative humidity) it is most probable that both $X_i$ and $X_j$ are dependent on another variable $W_K$ or even $m$ other variables (nodes): $W_K = \{W_{k1}, \ldots, W_{km}\}$, where $K = \{k_1, \ldots, k_m\}$. Thus, we consider the above cases as trivial correlations that most likely do not suggest a link $(i, j)$. To maintain ties among only the ecological variables with direct dependence the following partial correlation measure was introduced:

$$e_{ij(paco)} = \rho_{ij}\big|W_k = \frac{\sigma_{ij}\big|W_k}{\sigma_{ii}\big|W_k\,\sigma_{jj}\big|W_k} \tag{4}$$

where $\sigma_{ij|W_k}$, $\sigma_{ii|W_k}$ and $\sigma_{jj|W_k}$ are components of a partial covariance matrix. The estimate of.$\rho_{ij}|\,W_k$ is the sample partial correlation $r_{ij}|\,W_k$ computationally derived as follows: First the residuals $e_i$ and $e_j$ of the multiple linear regression of $X_i$ on $X_K$ and $X_j$ on $W_K$, respectively, are computed. Next, the correlation coefficient of $e_i$ and $e_j$,.$r_{ij}|\,W_k = r_{ei,ej}$ is computed. Thus, if $X_i$ (i.e., population of species $i$) and $X_j$ (i.e., population of species $j$) are independent but, conditional to $W_k$ (i.e., weather variable) then $\rho_{ij}|\,W_k$ should ideally be zero.

#### Dynamic measures of correlation for directed graphs

In contrast to the rules for known correlations and partial correlation [32], the Granger-Causality approach proposed by Granger [23, 24] was also applied. One important asset, compared to non-lagged cross correlations, is that it provides a stream of interaction (i.e., directed cause—effect associations).

#### Granger causality rules

The Granger causality measure, is based on the general concept of Norbert Wiener [36] that a causal influence should be manifest in improving the predictability of the driven process when the driving operation is followed. A measurable reduction in the unexplained variance of the response process (say, population $Y_i$) as a result of inclusion of the causal (driving) process (say, population $X_i$) in linear autoregressive modeling marks the existence of a causal influence from $X_i$ to $Y_i$ in the time domain $n = 1, 2, \ldots$ [37]. This method requires the estimation of multivariate vector autoregressions (MVAR) as follows:

$$Y_t = \sum_{n=1}^{p} a_n Y_{t-n} + \varepsilon_{r,t} \tag{5}$$

$$Y_t = \sum_{n=1}^{p} a_n Y_{t-n} + \sum_{n=1}^{p} \beta_n X_{t-n} + \varepsilon_{u,t} \tag{6}$$

where $\varepsilon_{r,t}$ *and* $\varepsilon_{u,t}$ are uncorrelated at the same time disturbances-residuals and $p$ is the maximum number of lagged observations included in the autoregressive model. In addition, $\alpha_n$ and $\beta_n$ are coefficients of the model (i.e., the contribution of each lagged variable to the predicted values of $X_i$ and $Y_i$). The GCI is the pairwise linear Granger causality in the time domain and is defined as follows [18]:

$$e_{ij(GC)} = GCI_{X\to Y} = \ln \frac{\sigma^2(\varepsilon_{r,t})}{\sigma^2(\varepsilon_{u,t})} \tag{7}$$

where $\sigma^2(\varepsilon_{r,t})$ is the unexplained variance (prediction error covariance) of $Y_i$ in its own autoregressive model (Eq. 5), whereas $\sigma^2(\varepsilon_{u,t})$ is its unexplained variance when a joint model for both $Y_i$ and $X_i$ is constructed (Eq. 6).

To provide useful heuristics for understanding the empirical ecological time-series, it is necessary to go beyond the simple two-variable vector autoregressive models and to study more variables that incorporate aspects of a more complex system. Thus, if $\boldsymbol{X_k} = [X_1, X_2, \ldots, X_k]^T$ denotes the realizations of $k$ ecological variables and $T$ denotes matrix transposition then the technique can be further extended by using the following multivariate vector linear autoregressive process (MVAR):

$$X_t = \sum_{n=1}^{p} A_n Y_{t-n} + E_t \tag{8}$$

Here $E_t$ is a vector of multivariate zero-mean uncorrelated white noise process, $A_i$ is the $k \times k$ matrices of model coefficients and $p$ is the model order, chosen, in this case, based on the Akaike information criteria (AIC) for MVAR processes. Significant interactions are based on the standard Granger causality index (GCI). It is expected that GCI: $X_i \to Y_i > 0$ when $X_i$ influences $Y_i$, and that GCI: $X_i \to Y_i = 0$ when it does not. In practice, GCI: $X_i \to Y_i$ is compared to a threshold value, and it was identified using parametric and non-parametric methods (i.e., using surrogate data).

**Conditional Granger causality rules**

One disadvantage of the GCI is that indirect partial effects of the other variables are not touched (e.g., examining whether $X_1$ Granger causes $X_2$, by excluding the activities of all other variables $X_3,..., X_i$). This multivariate extension consists of the conditional Granger causality index (CGCI) and is extremely helpful because repeated pairwise analyses among multiple variables can sometimes give misleading results in terms of differentiating between direct and mediated causal influences [38].

As noted previously, the method requires the estimation of multivariate vector autoregressions (MVAR). If we consider $X$ and $Y$ as the driving and response systems, respectively, and conditioning on system $Z$, we have the following:

$$X_t = \sum_{n=1}^{p} a_n Y_{t-n} + \sum_{n=1}^{p} c_n Z_{t-n} + \varepsilon_{r,t} \tag{9}$$

$$X_t = \sum_{n=1}^{p} a_n Y_{t-n} + \sum_{n=1}^{p} \beta_n X_{t-n} + \sum_{n=1}^{p} c_n Z_{t-n} + \varepsilon_{u,t} \tag{10}$$

The CGCI derived if we remove self-dependencies in respect to each variable in the Granger causality index for the two variables $X$ and $Y$ and conditioning on the third variable $Z$ is as follows:

$$e_{ij(CGCI)} = CGCI_{X \to |Z} = \ln \frac{\sigma^2(\widehat{\varepsilon}_{r,t})}{\sigma^2(\widehat{\varepsilon}_{u,t})} \tag{11}$$

where $\sigma^2(=\widehat{\varepsilon}_{r,t})$ and $\sigma^2(\widehat{\varepsilon}_{u,t})$ are variances of error estimators of the above restricted and unrestricted vector autoregressive models [39]. The CGCI was also checked because it is highly useful in bringing out the causal interactions among sets of nodes by eliminating common input artifacts [26].

To remove noise from data and to produce robust estimates of temporal autocorrelations between successive dynamics, data were subjected to prewhitening prior causality analysis to reject as much as possible of the temporal autocorrelated white noise [40]. This must be done if, as is usually the case, an input series is autocorrelated and to avoid having the direct cross-correlation function between the input and response series yield a misleading indication of the relation between the input and response series in an autoregressive moving average (ARIMA) modeling process. This task was performed using the standard MatLab prewhitening filter.

**Significance of similarity measures and network trimming**

The correlation coefficients between the variables form a correlation matrix that represents a weighted network. However, to detect only significant interactions and to trim the networks, it is important to apply a statistical hypothesis test according to a predefined probability that serves as the threshold level. This can be accomplished either by parametric or non-parametric methods, as indicated below.

**Parametric test of correlation**

To reduce the possibility that the observed correlations occurred by chance, a significance test should be performed. Here the null hypothesis $H_0$ that no correlations between any two ecological time series was examined, under the alternative hypothesis of the existence of correlations (two-tailed test). Thus if $\rho_{ij}$ is the true Pearson correlation coefficient, then the hypothesis test for significance is $H_0{:}\rho_{ij} = 0$, with the alternative $H_1{:}\rho_{ij} \neq 0$ and the estimate for $\rho_{ij}{:}r_{ij} = s_{ij}/s_i s_j$. Thus, if we take for granted that each paired data set is normally distributed and stationary, then the paired variables $X_i$ and $X_j$ yield.

$$(X_i, X_j) \cong N([\mu_i, \mu_j], [\sigma_i^2, \sigma_j^2], \rho_{ij}) \tag{12}$$

where $\mu$ and $\sigma^2$ are the standard notations for mean and variance, respectively. The $t$ statistic for a sample $n$ is

$$t = \frac{r_{ij}\sqrt{n-2}}{\sqrt{1-r_{ij}^2}}, \tag{13}$$

where $r_{ij}$ is the correlation coefficient and $H_0$ was rejected if: $|t| > t_{N-2;\, a/2}$ for the $\alpha = 0.05$ significance level.

For each pairwise tests the $p$ values represented the probability of error that was involved in accepting the observed correlations of the ecological time series as valid. If the $p$ values of the test were smaller than the $\alpha$-threshold level (0.05), then the correlations were considered as significantly different from zero and a connection was traced in order to construct the ecological time series network.

**Non-parametric testing of correlation**

Although ecological time series in this work are treated as stationary, bootstrapped randomizations for 100 surrogates were carried out, and non-parametric comparisons were also performed for confirmative reasons. In particular, this method should be applied in cases of small sample size and absence of information concerning deviation from normality and stationarity. Here, we derive the null distribution of $\rho_{ij}$ from resampled pairs consistent with $H_0$: $\rho_{ij} = 0$. Considering the original pair of ecological time-series $(x_i, x_j)$, we generated B randomized pairs $(x_i^{*b}, x_j^{*b})$, $b = 1, ..., B$. Although this random sample permutation destroys the time order, it uses the same distribution as the original time-series. At a next step, the $r_{ij}^{*b}$on each pair $(x_i^{*b}, x_j^{*b})$ and the ensemble $\left\{r_{ij}^{*b}\right\}_{b=1}^{B}$ that forms the empirical null distribution of $r_{ij}$ were computed. $H_0$ was rejected if the sample $r_{ij}$ was not in the distribution of $\left\{r_{ij}^{*b}\right\}_{b=1}^{B}$. The null hypothesis was $H_0$:$\rho_{ij}|K = 0$ and the alternative $H_1$:$\rho_{ij}|K \neq 0$. The analysis was carried out using the MatLab procedure; in all cases randomizations for 100 surrogates were performed.

### Significance test for GCI and CGCI

If the variable $X$ does not Granger cause $Y$ then the contribution of $X$-lags in the unrestricted model (Eqs. 5 and 9) should be insignificant, and the model parameters should therefore be insignificant. The null hypothesis is $H_0$: $b_i = 0$, for all $i = 1, ..., p$ and the alternative is $H_1$: $b_i \neq 0$, for any of $i = 1, ..., p$. A rejection of the null hypothesis implies that there is Granger causality and this can be evaluated using the $F$-test (Snedocor-Fisher) [41]:

$$F = \frac{(SSE^r - SSE^u)/p}{SSE^u/ndf}. \tag{14}$$

Here, SEE is the sum of square errors, $ndf = (n-p)-2p$ is the degrees of freedom, $n-p$ is the number of equations, and $2p$ is the number of coefficients in the unrestricted model (Eqs. 6 and 10).

### False discovery rates and true correlations

The false discovery rate (FDR) multiple testing procedure was applied to correct the false significant correlations of multiple comparisons, which can arise from incorrect rejections of false positives. The FDR was applied for conceptualizing the rate of *type I* errors in null hypothesis testing when conducting the multiple comparisons of the series and for further trimming the networks. The FDR was suggested by Benjamin and Hochberg [42] to address the problem with performing multiple simultaneous hypothesis tests. The FDR is a powerful concept by which one can retain the statistical power that would be lost to simultaneous comparisons made with Bonferroni-type procedures.

In particular, as the number of hypotheses increases, so does the probability of wrongly rejecting a null hypothesis because of random chance [42]. Therefore, to correct for multiple testing, as the same test applies for any $i$ and $j$ in $\{1, ..., n\}$, one can use the correction of the false discovery rate (FDR) [43]. According to the procedure, first the $p$-values of $m = n(n-1)/2$ tests are set in ascending order $p(1) \leq p(2)... \leq p(m)$. Next, the rejection of the null hypothesis of zero correlation, at the significance level $\alpha$, is decided for all variable pairs for which the $p$-value of the corresponding test is less than $p(k)$, where $p(k)$ is the largest $p$-value for which $p(k) \leq k{\cdot}\alpha/m$ holds [44].

## Results

### Weighted networks using cross and partial cross correlation

Successive captures of moth traps throughout the observation period and weather data are presented in Fig. 1. Figure 2a depicts the *similarity-values* matrices $\boldsymbol{E}$ (i.e., significant measure values, whereas non-significant values are set to zero) and the related weighted networks, showing each with the statistical similarity measures that were applied to construct them, i.e., the cross correlation networks (CRCO) and the partial correlation networks (PACO). Figure 2b depict the significant *p-values* matrix ($\boldsymbol{E}'$) (left) and the resulting weighted networks for the CRCO and PACO similarity measures (right). These graphical depictions provide a first model of a significant interaction flow based on their correlation. The structure of these weighted networks was established using the cut-off threshold $\alpha = 0.05$. Figure 2c presents the *binary-adjacency* matrices ($\boldsymbol{E}''$) and the related network configurations obtained after the final discovery rate analysis (FDR). These are binary values (0 or 1), and the associated network is therefore not weighted. However, vertices having higher degrees and clustering coefficients are represented as larger nodes and with darker colors, respectively.

Because the CRCO and PACO-FDR networks represent only the significant correlations among the variables, some very interesting information can be identified. In particular, in the PACO-FDR networks, significant correlations are generally observed among populations of the same species.

In addition, the weather variables, temperature and relative humidity, are equally interlinked. For example, we discover that the nodes Y11, Y12 and Y13, which represent populations of *A. lineatella*, form a triangle. The nodes Z14-15, which correspond to populations of *G. molesta*, are connected, whilst populations of *A. orana*, nodes X3, X4, X5, X6, X7, X8, X9 and X10, define a subgraph. Eventually, the two weather variables are also connected, a pattern that again is clearer when observing the PACO-FDR networks because the components of the weighted networks are fully linked.

### Granger related binary causal networks

Figure 3 shows the adjacency matrices (left) and the respective directed causal networks (right) constructed based on the Granger causality (GCI) and the conditional Granger causality index (CGCI).

In particular, Fig. 3a depicts the *similarity-values* matrices ($\boldsymbol{E}$) with the significant measure values of the

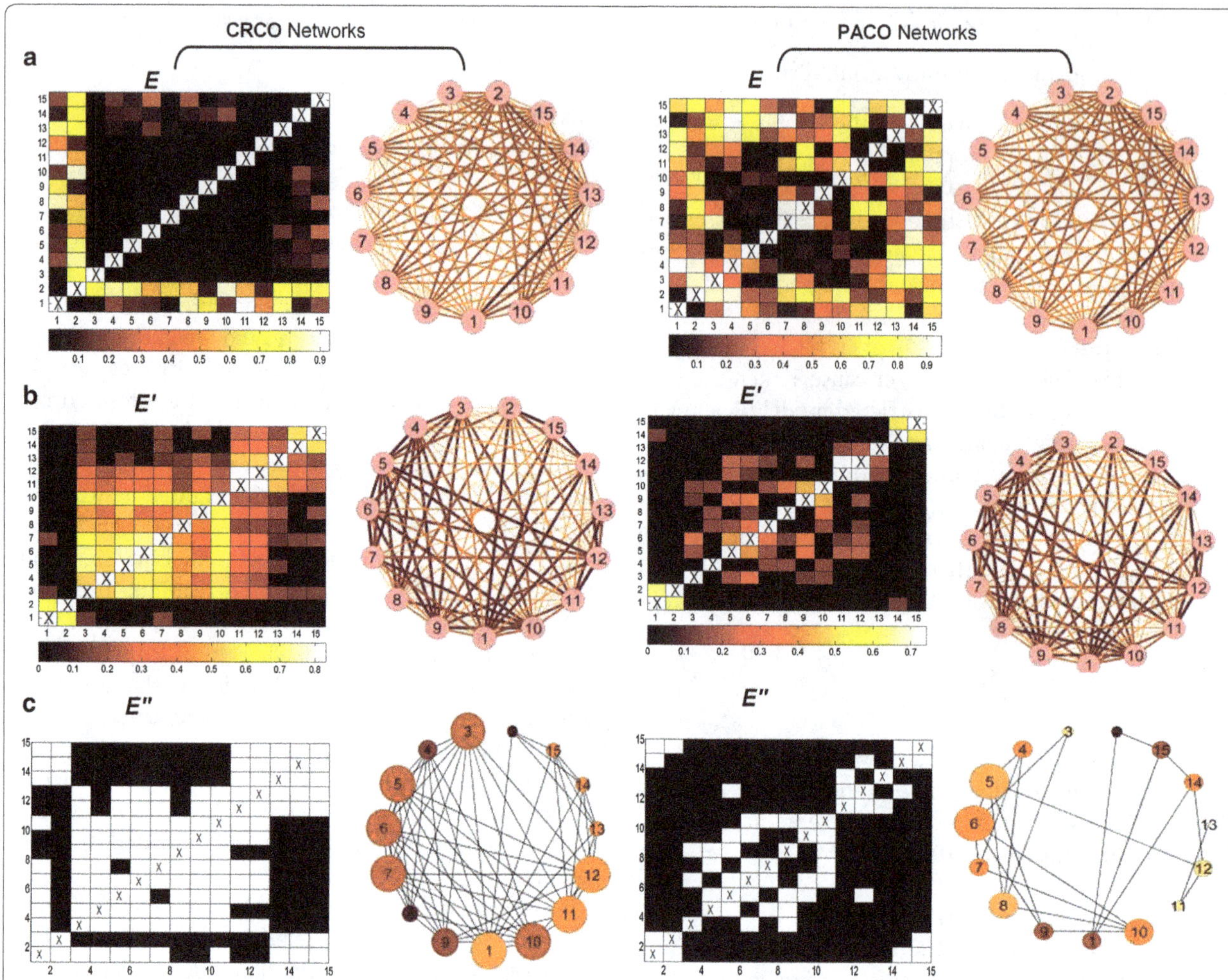

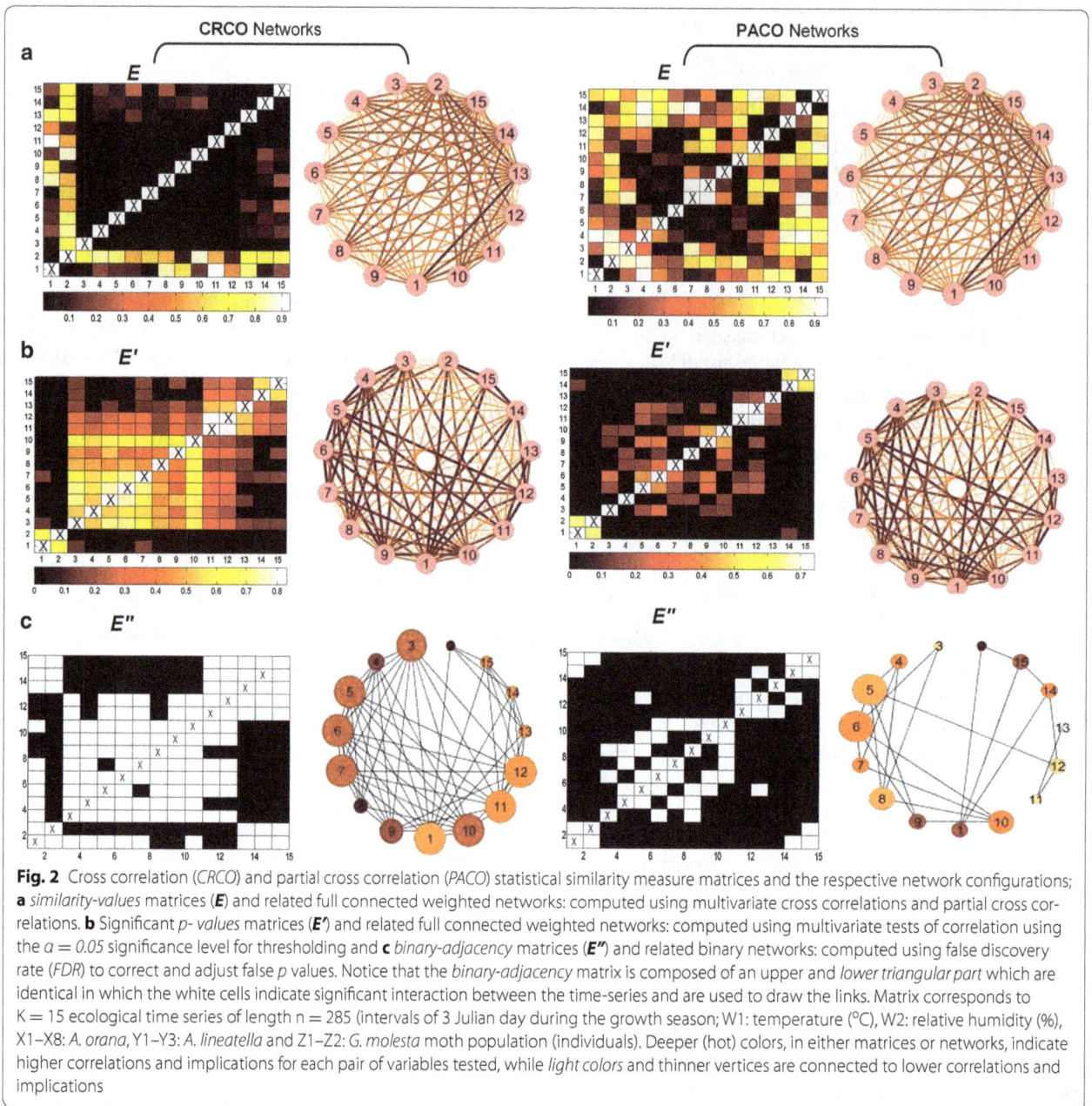

**Fig. 2** Cross correlation (*CRCO*) and partial cross correlation (*PACO*) statistical similarity measure matrices and the respective network configurations; **a** *similarity-values* matrices (***E***) and related full connected weighted networks: computed using multivariate cross correlations and partial cross correlations. **b** Significant *p- values* matrices (***E'***) and related full connected weighted networks: computed using multivariate tests of correlation using the $a = 0.05$ significance level for thresholding and **c** *binary-adjacency* matrices (***E"***) and related binary networks: computed using false discovery rate (*FDR*) to correct and adjust false *p* values. Notice that the *binary-adjacency* matrix is composed of an upper and *lower triangular part* which are identical in which the white cells indicate significant interaction between the time-series and are used to draw the links. Matrix corresponds to K = 15 ecological time series of length n = 285 (intervals of 3 Julian day during the growth season; W1: temperature (°C), W2: relative humidity (%), X1–X8: *A. orana*, Y1–Y3: *A. lineatella* and Z1–Z2: *G. molesta* moth population (individuals). Deeper (hot) colors, in either matrices or networks, indicate higher correlations and implications for each pair of variables tested, while *light colors* and thinner vertices are connected to lower correlations and implications

GCI and the CGCI, while Fig. 3b displays the *p value* matrices (***E'***) after the pairwise multivariate analysis and the parametric hypothesis testing. The GCI and CGCI causal networks are directed, and their matrices are therefore not symmetrical, in contrast with Fig. 2. This is associated with the fact that Granger measures produce non-symmetric adjacency matrices; thus, the associated networks are able to designate the direction of causalities. Moreover, the GCI and CGCI causal networks display a less dense form than the CRCO and PACO networks and remove the links between species and the Granger rules, thereby showing which of the correlated variables are cointegrated and share common stochastic drift. This represents an advantage of the Granger causality measures compared to simple non-laged cross correlations.

Furthermore, and as noted above, to avoid the multiple testing problems, p values were estimated using the false discovery rate as shown in the *binary-adjacency matrices* (***E''***). In Fig. 3c, larger nodes represent variables having higher degrees, and deeper colors represent higher clustering coefficients. Thus, variables 5 and 10, which correspond to populations of *A. orana*, have

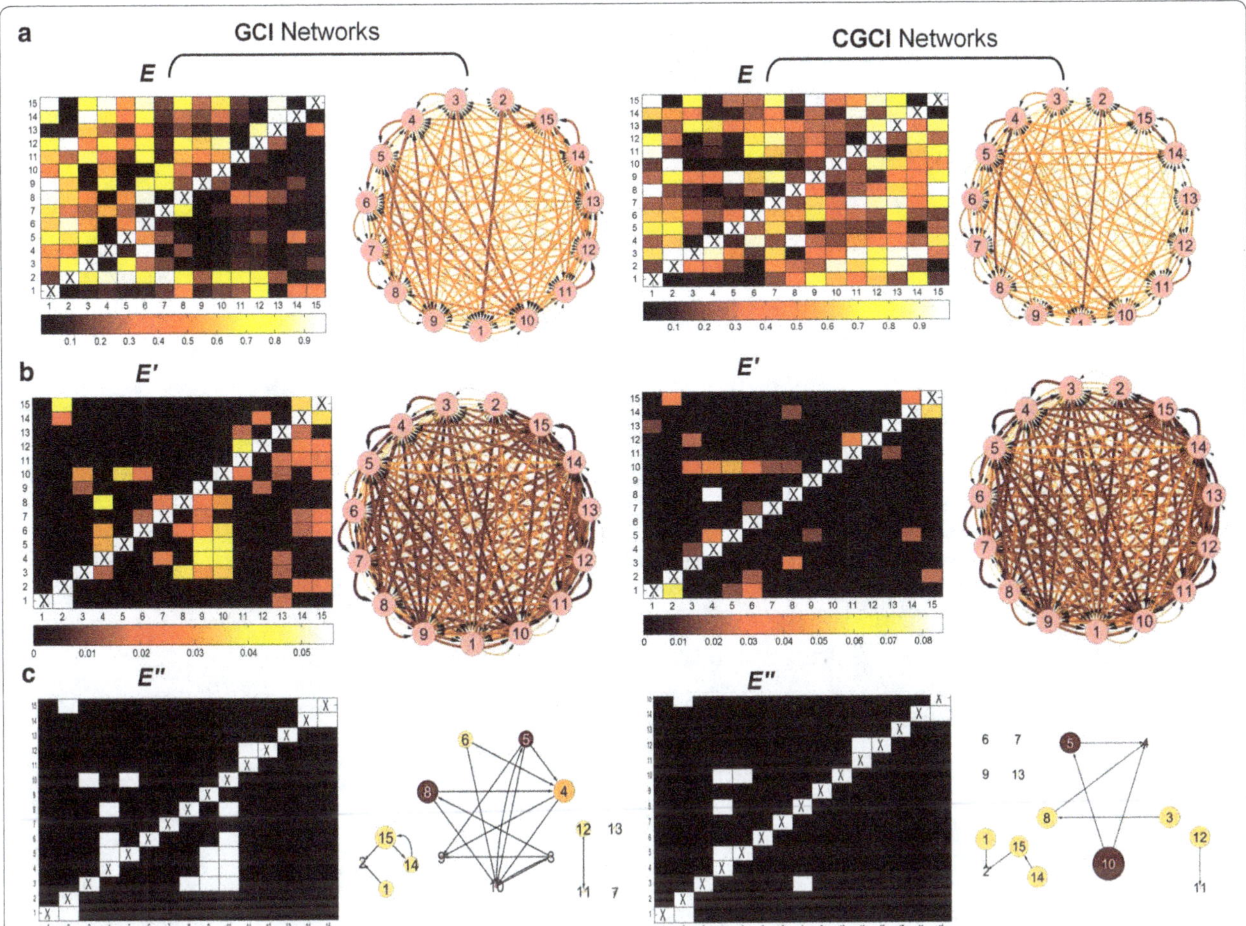

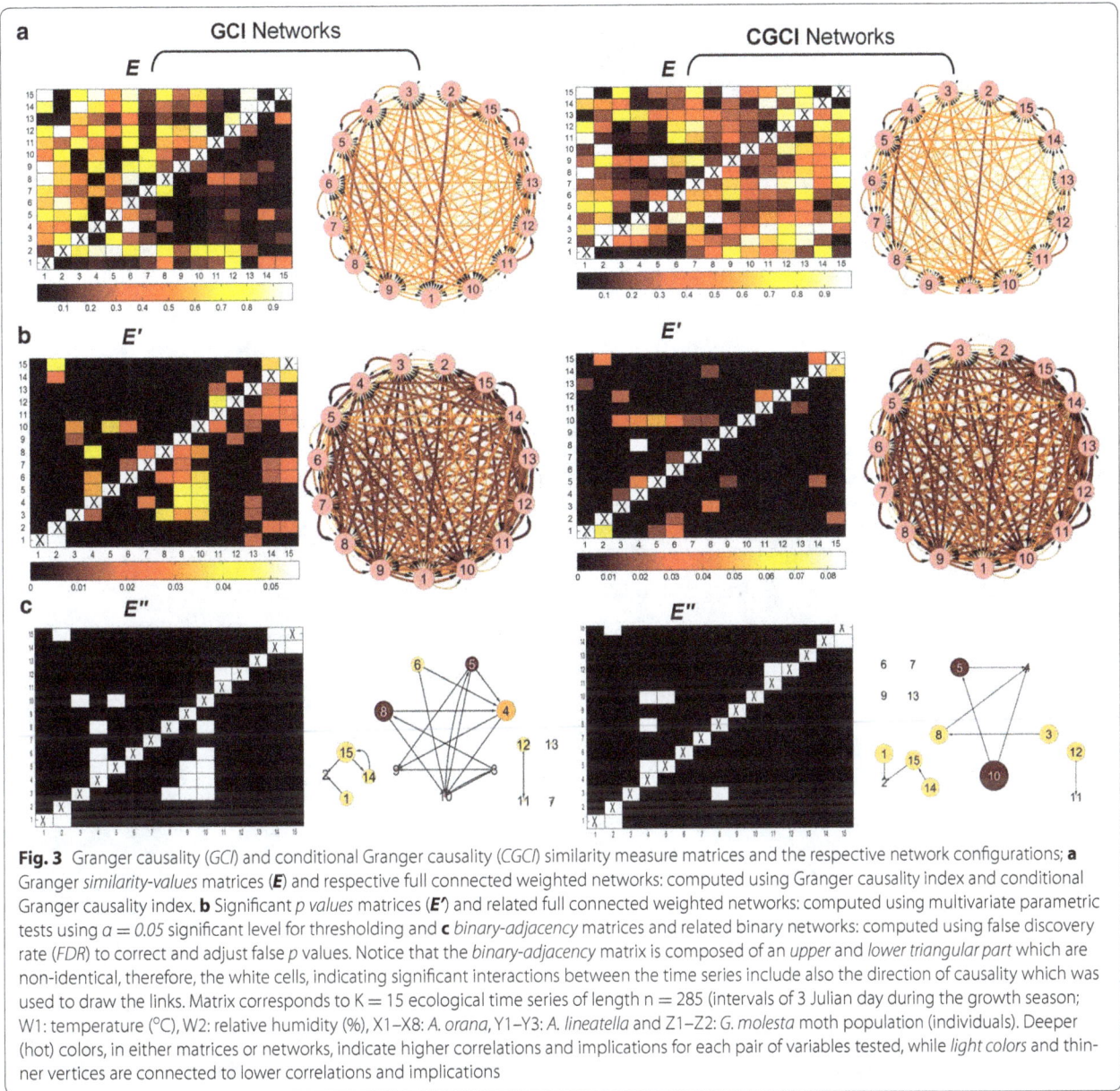

**Fig. 3** Granger causality (*GCI*) and conditional Granger causality (*CGCI*) similarity measure matrices and the respective network configurations; **a** Granger *similarity-values* matrices (***E***) and respective full connected weighted networks: computed using Granger causality index and conditional Granger causality index. **b** Significant *p values* matrices (***E'***) and related full connected weighted networks: computed using multivariate parametric tests using *α = 0.05* significant level for thresholding and **c** *binary-adjacency* matrices and related binary networks: computed using false discovery rate (*FDR*) to correct and adjust false *p* values. Notice that the *binary-adjacency* matrix is composed of an *upper* and *lower triangular part* which are non-identical, therefore, the white cells, indicating significant interactions between the time series include also the direction of causality which was used to draw the links. Matrix corresponds to K = 15 ecological time series of length n = 285 (intervals of 3 Julian day during the growth season; W1: temperature (°C), W2: relative humidity (%), X1–X8: *A. orana*, Y1–Y3: *A. lineatella* and Z1–Z2: *G. molesta* moth population (individuals). Deeper (hot) colors, in either matrices or networks, indicate higher correlations and implications for each pair of variables tested, while *light colors* and thinner vertices are connected to lower correlations and implications

higher out-degrees and clustering coefficients. From a graph-theoretic standpoint, these nodes can be viewed as hubs. A hub contains multiple links and is of exceptional interest for any network configuration. Moreover, based on the CGCI-FDR networks, at least two subgraphs are observable. The first consists of populations of *A. orana*, while the second includes variables 14 and 15, which both belong to the species *G. molesta*. Moreover, the two abiotic variables, nodes 1 and 2, are related.

### Force-directed causal network configurations

Force-directed network layouts are constructed based on forces assigned to the set of edges and nodes to obtain interpretable community structures in multipartite networks. Here, I have used the default Barnes–Hut approximation algorithm in Cytoscape [45].

Figures 4 and 5 depict the force-directed network layouts that correspond to the matching networks given in Figs. 2c and 3c. However, the GCI and CGCI networks of Fig. 5 have been forced to show only non-incidental connections. Based on these configurations, the interaction patterns of the biological variables are more easily indicated. All partial configurations clearly show the presence of sub graphs that consist of populations belonging to the same species, and they enhance the fact that Granger rules provide a robust method for removing trivial links and trimming the networks.

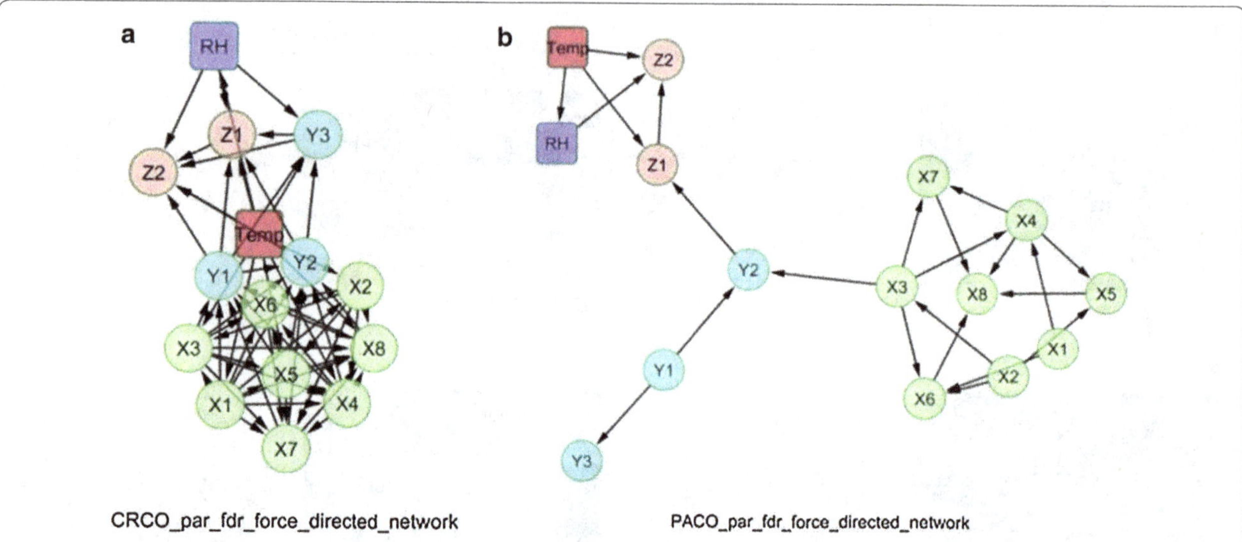

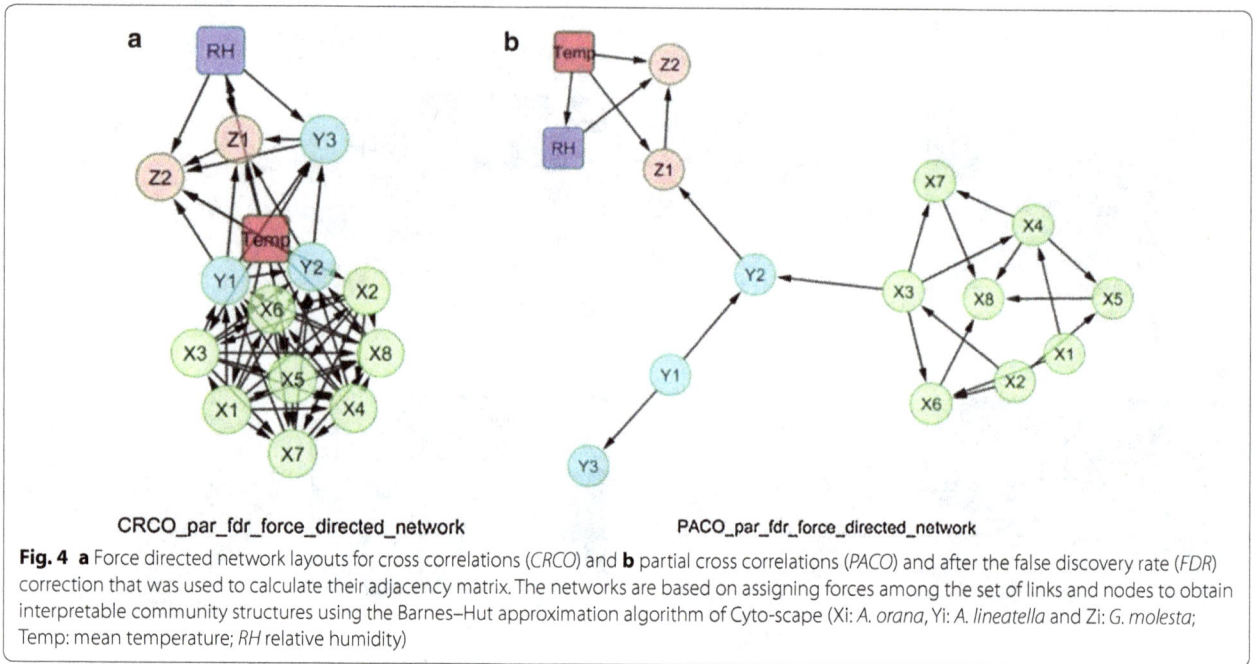

**Fig. 4** **a** Force directed network layouts for cross correlations (*CRCO*) and **b** partial cross correlations (*PACO*) and after the false discovery rate (*FDR*) correction that was used to calculate their adjacency matrix. The networks are based on assigning forces among the set of links and nodes to obtain interpretable community structures using the Barnes–Hut approximation algorithm of Cyto-scape (Xi: *A. orana*, Yi: *A. lineatella* and Zi: *G. molesta*; Temp: mean temperature; *RH* relative humidity)

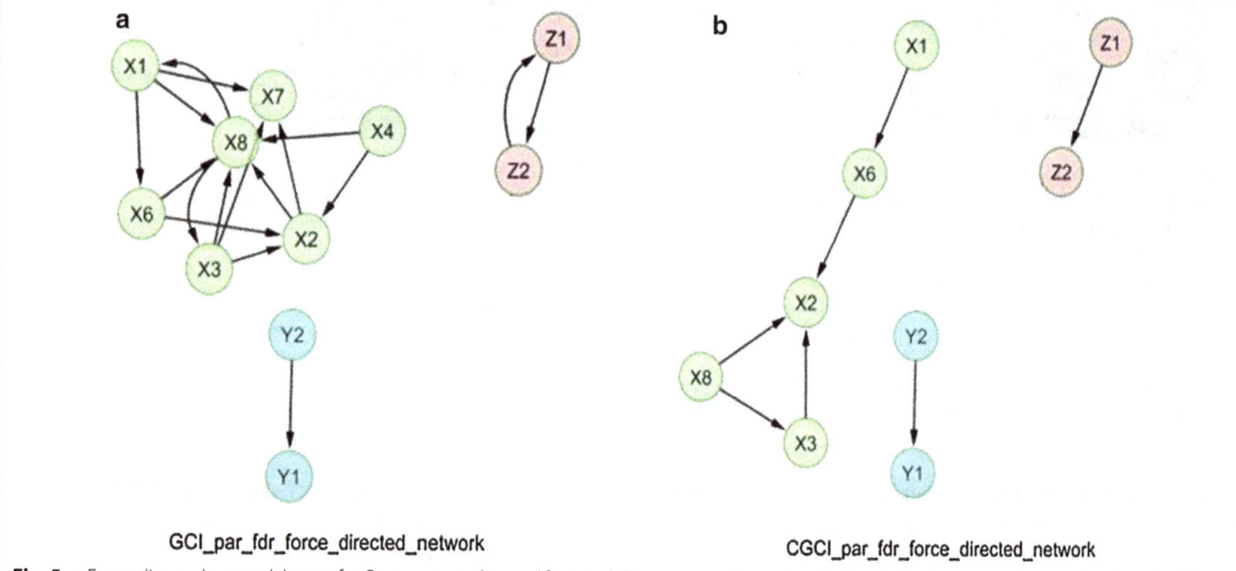

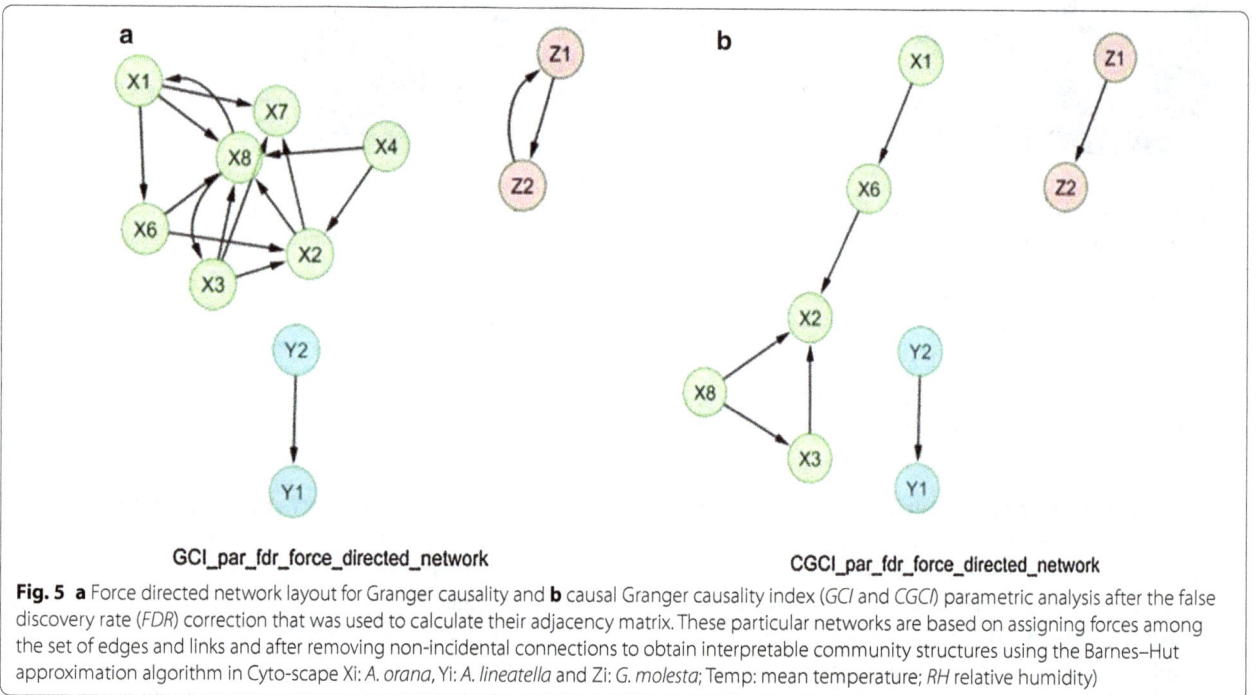

**Fig. 5** **a** Force directed network layout for Granger causality and **b** causal Granger causality index (*GCI* and *CGCI*) parametric analysis after the false discovery rate (*FDR*) correction that was used to calculate their adjacency matrix. These particular networks are based on assigning forces among the set of edges and links and after removing non-incidental connections to obtain interpretable community structures using the Barnes–Hut approximation algorithm in Cyto-scape Xi: *A. orana*, Yi: *A. lineatella* and Zi: *G. molesta*; Temp: mean temperature; *RH* relative humidity)

The weather variables have no significant driving role in most cases and especially in the case of the partial correlation analysis. This attribute probably suggests that the effect of environmental variables is diminished after the current network analysis, although these variables (especially temperature) generally have a substantial impact on insect population ecology [30]. However, in some cases, incidental connections have also appeared. This oxymoronic result was not unexpected, considering that the methods that were applied to introduce links are strictly statistical in nature and not based on biology.

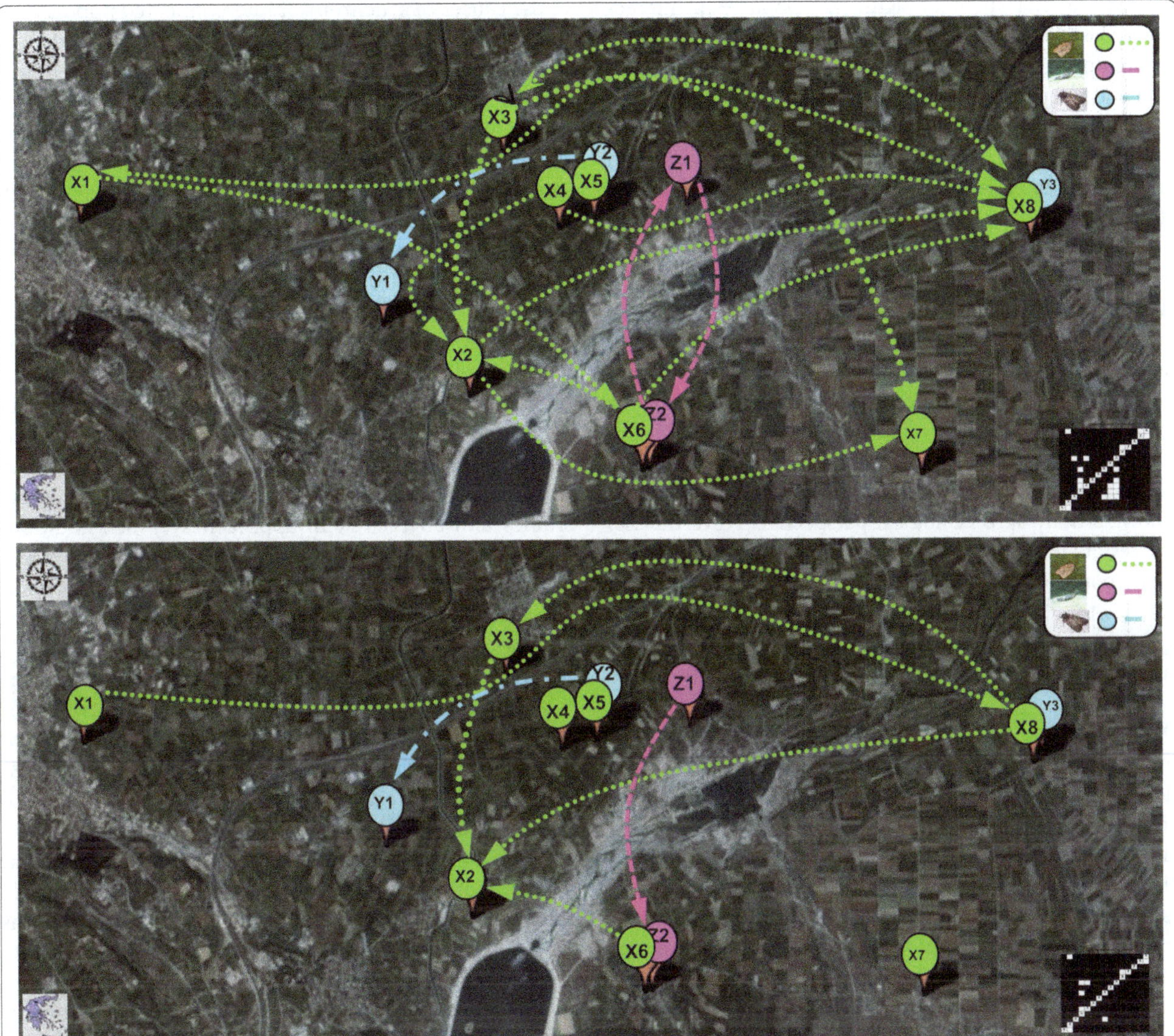


**Fig. 6** Google maps layer of the observation region and related Landscape projections of causal population networks, CGI-FDR network (*upper picture*) and CGCI-FDR network (*lower picture*), with respect to sampling points and region-specific topology (scale: 1:5000; area: A~8 × 16 km$^2$; Xi: *A. orana*, Yi: *A. lineatella* and Zi: *G. molesta*; weather variables are not included)

## Graph-theoretic and landscape-related network projections

Because the GCI and CGCI networks are directed, they are the only networks that have been overlaid on the observation region to identify the location of the causal forcing variables. Figure 6 depicts the topological projections of the GCI-FDR and CGCI-FDR causality networks over the agricultural landscape. Each node represents a site where moth populations were observed, while the arrows represent the Granger causal relations. A higher number of interactions is observed in the GCI network (upper picture) than in the CGCI-FDR causality network (lower picture). This happens because prior to the projection, indirect preferential effects of any other variable were removed based on the false discovery rate procedure. The observation of significant interspecies associations was dependent on the type of analysis. Moreover, based on the topological projections of Fig. 6, it is apparent that some nodes, which represent landscape-related population activity, point to some particular locations and not to all sites where observation was performed. For instance, in the CGCI-FDR network, it is apparent that most populations of *A. orana* and *A. lineatella* of the West side point to those of the east side. Furthermore, Fig. 6 more clearly indicates the landscape-related locations where populations act as hubs and may function as 'hot spots' of significant population interactions. Based on the landscape properties, the experimental area can be split into two major areas, northwest and southeast, separated by a river (Aliakmon). Based on the current analysis in these areas, the intra-species interactions seem to act across the physical river border and not just inside each sub-region. The GCI network shows 10 maximum relations across the river and 4 within sub-regions relations, while the CGCI network shows 5 relations across the river and only 2 within sub-regions. However, this may have resulted from several factors, including the inherent properties of a species population in relation to environmental conditions [46, 47]. Warmer temperatures and increased precipitation, for instance, can reduce the rates of population growth [48]. Additionally, because there is no similar analysis in the literature, some of the answers are difficult to translate from a strictly agro-ecological perspective. Nevertheless, the results provide partial support to the hypothesis that moth population emergence in one location is synchronized with that of populations located nearby, that emerge a few days later, and this dynamic can be described via the landscape-projected cause-effect graphical models.

## Discussion

In this work, the question of evaluating significant relationships between ecological time-series has been addressed based on multivariate time-series analysis and graph-theoretical approaches. In particular, spatial associations among a set of ecological entities have been studied, including two abiotic variables and thirteen biotic (moth population) variables. The most recent statistical measures have been adopted to assess the significances of correlations and to construct causal networks, which may provide information on landscape-related species synchronization and causation.

The detection of spatial synchrony is of great ecological interest since it provides answers on which factors affect the observed patterns of spatial population synchrony. For example, although a positive environmental correlation with populations across sites has been observed in both, the similarity-values networks and probability-values networks its effects degenerate in the binary-adjacency networks. These results indicate not only that environmental correlation and population spatial dynamics should be looked at in combination, but also that we may derive to different conclusion depending on the rigor of statistical criteria used to define non trivial relations. Additionally, because correlation does not imply causation way may obtain different interpretation when using causality measures over simple correlations. Therefore, causality networks not only show, which population may be spatial synchronized, but provide some evidence of the causes of synchrony of interconnected populations (i.e. growth, dispersal or noise).

The current network construction approach differs to that of most ecological network analyses, because it offers both heuristics and practical methods to define the strength of associations among the ecological variables of interest. The core idea is to integrate knowledge of each population time series in the description and interpretation of all others. These data are then used to determine the order of the multivariate autoregressive model (MVAR) used for the correlation similarity and Granger causality measures. By removing the correlations inherent to the time-series, as well as the partial series of multivariate interactions, we derived the final network configurations. We conclude that although the constructed networks do not come in any ecological category in a narrow sense, traditional population community studies and all ecosystem approaches (which additionally emphasis on energy influxes, including biomass and nutrient cycles [49, 50]), they may provide a new qualitative method for building and studying complex ecological network organizations.

Furthermore, in contrast to other studies based on correlations *per se*, the current approach differs in two ways. First, it defines an a posteriori correlation between ecological variables to detect synchronization, using statistical hypothesis testing; second, it provides a means to discover the existence of causal dependences between the different variables. One additional advantage is that all four methods of constructing networks can be implemented in the absence of any restrictive presuppositions regarding the underlying dynamical structure of the data set. Therefore, the use of the current multivariate techniques may contribute to detecting population synchronizations as well as the relative importance of biotic and abiotic processes to the distribution of population abundance and its dynamics.

From an ecological standpoint although synchronized fluctuation in abundance among spatially segregated populations and ecological variables is common in nature, the identification of the relative contribution of each factor to patterns of synchrony remains a challenge [51, 52]. Furthermore, when the processes that contribute to synchrony are studied in isolation, the synchrony patterns can be ascribed to the underlying cause [53]. However, under field conditions, ecological systems are more complicated, with intrinsic and extrinsic causal processes appearing together [54] and therefore, the proposed method represents a robust approach for detecting multivariate spatial synchronization and causality between ecological time-series.

Moreover, based on the current case study results, we can extract some very interesting ecological information.

For example, these results show that although the environmental variables are to a high degree correlated with the populations, they are not acting as the main driving causal forces. Therefore, the results support the hypothesis that synchrony and causality can also be induced by other factors (i.g., local dispersal among populations, competition, host allocation and many more). Furthermore, these results show that spatial synchrony and causation are more likely between populations of the same species.

However, it is also important to note that the transformation of connectivity values from a continuous to a binary scale generally entails difficulties and has some drawbacks. In particular, while the binary scale clearly enhances contrast, it also hides potentially vital information as connectivity values move below or above threshold levels. Still, the general appearance of the weighted population graphs is not qualitatively very different from the binary ones. In reality, very similar connectivity patterns are replicated in both graph categories (i.e., the circular as well as the force-directed layout).

Comparing the basic methods that have been applied to construct the networks (Correlations and Granger rules of connectivity) it is important to note that each method may provide a different range of interpretation. Although both methods may account for synchronization, correlation shows which variables are synchronized but not which drives the others as in the case of the Granger rules. Moreover, the advantage of the Granger statistical rules is the incorporation of a simple test to determine whether one series (population) is being caused by another series (i.e., environmental variable or nearby population) even when they are not correlated. Therefore, the Granger method is particularly suitable for estimating the directed connectivity of time-series data in order to extract features that characterize the underlying spatiotemporal dynamics. For example, the landscape-related causality networks are used for studying the relationship between simultaneously recorded populations and provide insight into which population-locations act as driving variables (i.e., hubs) and may used to predict an increase or decrease in linked population locations.

Moreover, partial correlation and partial causality measures are able to remove the correlation between any two variables that is present just because they are correlated with a third variable [55, 56]. For instance, the fact that two different species are each correlated to temperature does not mean that they are also related to each other. Finally, the FDR networks also provide the most robust network construction in terms of statistical power. The FDR approach is a new statistical procedure, and this technique is even more valuable for cases in which multiple calculations are performed on large-scale data sets. Moreover, the key advantage of FDR is that it takes into account a priori fast control of the mean fraction of the false rejections made over the total number of rejections performed and thus avoids bias. Actually, the FDR procedure is quick and easy to compute and can be trivially adapted to work with correlated data as well [57]. Consequently, if we mean to choose among the methods applied to build the networks, those using partial correlation measures are more reliable because they take out the effect of the random variables which may create links just by chance. Moreover, this study brings out the significance of the false discovery rate to capture only the real non-trivial links to generate accurate and informative networks. The usefulness of the FDR approach is all-important as the number of time series increases.

Despite the promising outcomes of the current analysis, there are also some restrictions. First, the case subject field was studied based on a relatively modest percentage of the population and biotic components; thus, the interpretation of causal influence is determined solely by the measured variables. We cannot exclude, for instance, potential modification of the results if additional variables are included. For example, as the network size is changed (i.e., by adding more populations variables and species), it may be possible that new edges will appear even if the underlying network topology remains identical [58, 59]. Thus, some caution must still be exercised when considering overall patterns in the moth population network structures, because they rely upon a relatively small number of variables observed in that particular area of research. Therefore, we should consider that the current network construction method should be regarded more as a method for the investigation of population synchronizations and causality between discrete variables rather than as a way to understand actual interactions between species. Only assumptions can be made regarding whether the same relationships also hold in other areas despite several other factors that affect actual population dynamics and are that are not considered in the present study; therefore, the ecological inference remains somewhat rudimentary.

From a stricter ecological standpoint, therefore, it is rather difficult to interpret why moth populations are mostly correlated along an east–west axis through physical barriers rather than into nearby regions within the same habitat. One hypothesis is that because the 'driving population variables' benefit from local micro environmental conditions, the moths emerge a few days earlier and cause a lagged emergency in the adjacent population. There are no similar studies that would allow a direct comparison, but for poikilothermic organisms and insects in particular, moderate temperatures and rich humidity favor reproduction and development

[15]. In closely related species such as the codling moth *Cydia pomonella*, geostatistical spatial analysis revealed that high populations are captured at sites most suitable for the pest and the host [60]. Moreover, similar spatial analysis performed in South Italy for *A. lineatella* and *G. molesta* showed that the main 'hot spot' for both lepidopterous pests was in a stone fruit orchard in the northern zone of the study area; other infested areas were in peach orchards and, in the case of *A. lineatella*, also in plum orchards. However, in contrast to the current study, that analysis showed a river acting more as a barrier than as an ecological corridor [61]. Still, it is noteworthy to say that the current analysis differs from the affront mentioned geostatistical approaches in terms of their objectives, implementation and interpretation of the outcomes, and especially because they are using probabilistic models based on the spatial domain without reference to time. The Granger rules provide attention to extrapolations for making forecasts, and although geostatistics may be interested in extrapolation, the methods are optimized for interpolation. Hence, because there are no related subjects using the current approach, comparison of the current results on an ecological-behavioral basis is rather difficult. Therefore, the study additionally emphasizes that the new methodological approach can be used to develop ecological networks and does not provide strict ecological answers regarding the function of the study system.

In summation, the constructed networks may provide an intuitive, advantageous representation of the relationships among multiple pest population that may be present within the agro-ecosystem. For example, the present work not only validates the hypothesis that populations belonging to the same species are correlated, by showing synchronized population dynamics, but also that some of them may have prominent roles in population dynamics as driving variables. For pest control, the incorporation of spatial information into an Integrated Pest Management program (IPM) is valuable for site specific pest management or for precise timing of targeting specific populations. The detection of spatial synchronization between the different pest populations may be useful for minimizing direct control tactics.

## Conclusions

The proposed multivariate modeling approaches provide a means to detect significant correlations and causality in ecological time-series. Regardless of the fact that correlations by themselves are crude, they may provide a novel basis for the study of non-trivial spatiotemporal relations in ecological time-series. The advantage of the Granger rules over correlations is that the Granger rules have dynamics features and thus provide an easy way to analyze the dynamic causal relations of a heterogeneous system of time-series variables.

In particular, causal measures overcome certain limitations in studying ecological relations because (i) they introduce a mathematical context and properties that serve as a core to construct network relations, (ii) they incorporate causal relations with a probabilistic and no subjective nature and (iii) they are closer to reality, because they not only incorporate the dynamics of all kinds of ecological variables but can be projected over the same landscape layer, allowing the development of inferences (e.g., populations evolve simultaneously with the surrounding environment).

Moreover, the Granger rules provide a means to detect synchronization and to detect which of the series may act as forcing variables. It is shown that populations of the same species are synchronized to a high degree, while populations of some locations exert a causal effect on others and may be used to predict the dynamics of the latter. The landscape-projected graphical models also describe the spatial patterns of causation and may be useful for site-specific management. The constructed networks may have practical utility because they suggest where to monitor populations to predict, for example, increases or declines of populations at other locations at the end of an edge in the graphs.

By extending these models, specifically by including more variables, it should be possible to explore interspecies and interspecies relationships with larger ecological systems and identify specific population traits that might constrain their structures in larger areas.

#### Abbreviations

FDR: false discovery rate; CROCO: cross correlations; PACO: partial cross correlations; MVAR: multivariate vector autoregression process; CGI: granger causality index; CGCI: conditional Granger causality index; ARMA: autoregressive moving average model; IPM: integrated pest management.

#### Author details

[1] Department of Environmental Conservation and Management, Faculty of Pure and Applied Sciences, Open University of Cyprus, Main OUC building: 33, Giannou Kranidioti Ave., Latsia, 2220 Nicosia, Cyprus. [2] WebScience, Mathematics Department, Faculty of Sciences, Aristotle University of Thessaloniki, University Campus, 59100 Thessaloniki, Greece. [3] Laboratory of Applied Zoology and Parasitology, Department of Crop Production (Field Crops and Ecology, Horticulture and Viticulture and Plant Protection), Faculty of Agriculture, Forestry and Natural Environment, University Campus, 59100 Thessaloniki, Greece.

**Acknowledgements**
The author acknowledges the help provided by the Agronomists of the public confederation ALMME® in collecting part of the data that were used to illustrate some representative region specific ecological networks and thanks D. Kugiumtzis for enabling this work. I would also like to thank the two anonymous reviewers, as well as the editor Cang Hui, for their valuable suggestions, which have significantly improved the contents of the MS and American Journal Experts for proof-reading.

**Competing interests**
The author declares no competing interests.

**Funding**
No funding was obtained for this study.

**References**

1. Mysterud A, Yoccoz NG, Langvatn R, Pettorelli N, Stenseth NCh. Hierarchical path analysis of deer responses to direct and indirect effects of climate in northern forest. Philos Trans R Soc Lond B Biol Sci. 2008;363:2359–68.
2. Gimenez O, Anker-Nilssen T, Grosbois V. Exploring causal pathways in demographic parameter variation: path analysis of mark-recapture data. Methods Ecol Evol. 2012;3:427–32.
3. Shipley B. Cause and correlation in biology: a user's guide to path analysis, structural equations and causal inferences. Cambridge: Cambridge University Press; 2000.
4. Clark JS, Gelfand AE. A future for models and data in environmental science. Trends Ecol Evol. 2006;21:375–80.
5. Irvine KM, Gitelman AI. Graphical spatial models: a new interpreting spatial pattern. Environ Ecol Stat. 2011;18:447–69.
6. Turchin P, Taylor AD. Complex dynamics in ecological time series. Ecology. 1992;73:289–305.
7. Buonaccorsi JP, Elkinton JS, Evans SR, Liebhold AM. Measuring and testing for spatial synchrony. Ecology. 2001;82(6):1668–79.
8. Haddow AD, Bixler D, Odoi A. The spatial epidemiology and clinical features of reported cases of La Crosse Virus infection in West Virginia from 2003 to 2007. BMC Infect Dis. 2011;11:29.
9. Damos P. A stepwise algorithm to detect significant time lags in ecological time series in terms of autocorrelation functions and ARMA model optimization of pest population seasonal outbreaks. Stoch Environ Res Risk Assess. 2015. doi:10.1007/s00477-015-1150-1.
10. Aalen O, Røysland K, Gran JM, Ledergerber B. Causality, mediation and time: a dynamic viepoint. J R Stat Soc Ser A. 2012;175:831–61.
11. Bjørnstad ON, Liebhold AM, Johnson DM. Transient synchronization following invasion: evisiting Moran's model and a case study. Popul Ecol. 2008;50:371–89.
12. Bjørnstad ON, Ims RA, Lambin X. Spatial population dynamics: analysing patterns and processes of population synchrony. Trends Ecol Evol. 1999;14:427–31.
13. Moran PAP. The statistical analysis of the Canadian lynx cycle. II. Synchronisation and meteorology. Aust J Zool. 1953;1:291–8.
14. Ranta E, Kaitala V, Lindstrom J, Linden H. Synchrony in population dynamics. Proc R Soc London Ser B. 1995;262:113–8.
15. Damos P, Savopoulou-Soultani M. Temperature driven models for insect development and vital thermal requirements. Psyche. 2012;2012:123405. doi:10.1155/2012/123405.
16. Ives AR, Dennis B, Cottingham KL, Carpenter S. Estimating community stability and ecological interactions from time series data. Ecol Monogr. 2003;73:301–30.
17. Detto M, Boher G, Nietz JG, Maurer KD, Vogel CS, Ghough CM, Curtis PS. Multivariate conditional Granger causality analysis for lagged response of soil respiration in a temperate forest. Entropy. 2013. ISSN 1099-4300.
18. Zhang WJ. Constructing ecological interaction networks by correlation analysis: hinds from community sampling. Netw Biol. 2011;1:81–98.
19. Land R, Engen S, Saeth BE. Stochastic population dynamics in ecology and conservation. Chennai: Oxford University Press; 2003. p. 89–90.
20. Raimond S, Liebhold AM, Srazanac S, Butler L. Population synchrony with and among *Lepidoptera* species in relation to weather, phylogeny, and larval phenology. Ecol Entomol. 2004;29:96–105.
21. Buchanan M. Cause and correlation. Nat Phys. 2012;8:852.
22. Sugihara G, et al. Detecting causality in complex ecosystems. Science. 2012. doi:10.1126/science.1227079.
23. Granger CWJ. Investigating causal relations by econometric models and cross-spectral methods. Econometrica. 1969;37:424–38.
24. Granger CWJ. Seasonality: causation, interpretation, and implications. In: Zellner A editor. Seasonal analysis of economic time series. 1979. p. 33–56.
25. Hiemsta C, Jones JD. Testing for linear and non-linear Granger causality in the stock price volume relation. J Finance. 1994;49(5):1639–64.
26. Ding MY, Chen Y, Bresslier S. Granger causality: basic theory and application to neuroscience. In: Schelter S, Winterhalded M, Timmer J, editors. Handbook of time series analysis. Wienheim: Wiley; 2006. p. 438–60.
27. Sugihara G, May R, Ye H, Hsieh C, Deyle E, Fogarty M, Munch S. Detecting causality in complex ecosystems. Science. 2012. doi:10.1126/science.1227079.
28. Wootton JT, Emmerson M. Measurement of interaction strength in nature. Ecol Evol Syst. 2006;36:419–44.
29. Damos P, Savopoulou-Soultani M. Temperature dependent bionomics and modeling of *Anarsia lineatella* (Lepidoptera: Gelechiidae) in the laboratory. J Econ Entomol. 2008;101:1557–67.
30. Damos P, Savopoulou-Soultani M. Development and statistical evaluation of models in forecasting major lepidopterous peach pest complex for integrated pest management programs. Crop Prot. 2010;29:1190–9.
31. Fuende A, Bing N, Hoeschele I. Discovery of meaningful associations in genomic data using partial correlation coefficients. Bioinformatics. 2004;20:3565–74. doi:10.1093/bioinformatics/bth445.
32. Kolaczyk ED. Statistical analysis of network data. Berlin: Springer; 2009.
33. Bondy JA, Murty USR. Graph theory with applications. North-Holland: Elsevier Science Publishing Ltd; 1982.
34. Bonnington CP, Little CHC. Foundations of topological graph theory. New York: Springer; 1995.
35. Wei WWS. Time series analysis. Univariate and multivariate methods. 2nd ed. New York: Peasron Education Inc; 1994.
36. Wiener N. The theory of prediction. In: Beckman EF, editor. Modern mathematics for engineers. New York: MacGraw-Hill; 1956.
37. Krumin M, Shohan S. Multivariate autoregressive modeling and Granger causality analysis of multiple spike trains. Comput Intell Neurosci. 2010. article ID 752428, 9 pages.
38. Wen X, Rangarjan G, Ding M. Multivariate Granger causality: an estimation framework based on factorization of the spectral density matric. Phil Trans R Soc A. 2013;371:20110610.
39. Papana A, Kyrtsou C, Kugiumtzis D, Diks C. Simulation study of direct causality measures in multivariate time series. Entropy. 2013;15:2635. doi:10.3390/e15072635.
40. Warner RM. Spectral analysis of time-series data. New York: Guliford Press; 1998. p. 116.
41. Piot-Lepit I, M'Barek R. Methods to analyse agricultural commodity price. New York: Springer; 2011. p. 159.
42. Benjamini Y, Hochberg Y. Controlling the false discovery rate: a practical powerful approach to multiple testing. J R Stat Soc B (Methodological). 1995;57:289–300.
43. Pike N. Using false discovery rates for multiple comparisons in ecology and evolution. Methods Ecol Evol. 2011;2:278–82.
44. Siggiridou E, Kugiumtzis D, Kimiskidis VK. Correlation networks for identifying changes in brain connectivity during epileptiform discharges

and transcranial magnetic simulations. Sensors. 2014;14:12585–97. doi:10.3390/s140712585.
45. Cystoscape. Network data integration, analysis, and visualization in a box, version 3.3. http://www.cytoscape.org.
46. Damos P, Soulopoulou P. Do insect populations die at constant rates as they become older? Contrasting demographic failure kinetics with respect to temperature according to the Weibull model. PloS One. 2015;10(8):e0127328. doi:10.1371/journal.pone.0127328 **(eCollection 2015)**.
47. Damos P. Mixing times towards demographic equilibrium in insect populations with temperature variable age structures. Theor Popul Biol. 2015;. doi:10.1016/j.tpb.2015.04.005.
48. Aheer GM, Alin A, Akram M. Effect of whether factors on populations of *Helicoverpa armigera* moths at cotton-based agro-ecological sites. Entomol Res. 2009;39:36–42.
49. Odum EP. Fundamentals of ecology. Philadelphia: Saunders; 1953.
50. Lindeman RL. The trophic-dynamic aspect of ecology. Ecology. 1942;23:399–418.
51. Liebhold A, Koenig WD, Bjørnstad ON. Spatial synchrony in population dynamics. Annu Rev Ecol Evol Syst. 2004;35:467–90.
52. Vasseur DA, Fox JW. Phase-locking and environmental fluctuations generate synchrony in a predator-prey community. Nature. 2009;460:1007–10.
53. Post E, Forchhammer MC. Synchronization of animal population dynamics by large-scale climate. Nature. 2002;420:168–71.
54. Gouhier TC, Guichrd F, Menge BA. Ecological processes can synchronize marine population dynamics over continental scales. PNAS. 2009;107:8281–6.
55. Gue S, Seth AK, Kendrick KM, Zhou C, Feng J. Partial Granger causality-eliminating exogenous inputs and latent variables. J Neourosci Methods. 2008;172:79–93.
56. Opgen-Rhein R, Strimmer K. From correlation to causation networks: a simple approximate learning algorithm and its application to high-dimensional plant gene expression data. BMC Syst Biol. 2007;1:37. doi:10.1186/1752-0509-1037.
57. Zehetmayer S, Posch M. False discovery rate control in two stage designs. BMC Bioinform. 2012;13:81. doi:10.1186/1471-2105-13-81.
58. Dunne JA, Williams RJ, Martinez ND. Food-web structure and network theory: the role of connectance and size. Proc Natl Acad Sci USA. 2002;99:12917–22.
59. Amaral LAN, Scala A, Barthelemy M, Stanley HE. Classes of small-world networks. PNAS. 2000;97:11149–52.
60. Tramaterra P, Gentile P, Sciaretta A. Spatial analysis of pheromone trap catches of codling moth *Cydia pomonella* L. (Lepidoptera: Tortricidae), in two heterogeneous agroecosystems, using geostatistical techniques. Phytoparasitica. 2004;32:325–41.
61. Sciarretta A, Trematerra P. Geostatistical characterization of the spatial distribution of *Grapholita molesta* and *Anarsia lineatella* males in an agricultural landscape. J Appl Entomol. 2006;130:73–83.

# 16

# Initiating and continuing participation in citizen science for natural history

Glyn Everett[1*†] and Hilary Geoghegan[2†]

**Abstract**

**Background:** Natural history has a long tradition in the UK, dating back to before Charles Darwin. Developing from a principally amateur pursuit, natural history continues to attract both amateur and professional involvement. Within the context of citizen science and public engagement, we examine the motivations behind citizen participation in the national survey activities of the Open Air Laboratories (OPAL) programme, looking at: people's experiences of the surveys as 'project-based leisure'; their motivations for taking part and barriers to continued participation; where they feature on our continuum of engagement; and whether participation in an OPAL survey facilitated their movement between categories along this continuum. The paper focuses on a less-expected but very significant outcome regarding the participation of already-engaged amateur naturalists in citizen science.

**Results:** Our main findings relate to: first, how committed amateur naturalists (already-engaged) have also enjoyed contributing to OPAL and the need to respect and work with their interest to encourage broader and deeper involvement; and second, how new (previously-unengaged) and relatively new participants (casually-engaged) have gained confidence, renewed their interests, refocussed their activities and/or gained validation from participation in OPAL. Overall, we argue that engagement with and enthusiasm for the scientific process is a motivation shared by citizens who, prior to participating in the OPAL surveys, were previously-unengaged, casually-engaged or already-engaged in natural history activities.

**Conclusions:** Citizen science has largely been written about by professional scientists for professional scientists interested in developing a project of their own. This study offers a qualitative example of how citizen science can be meaningful to participants beyond what might appear to be a public engagement data collection exercise.

## Background

Citizen science is defined here as the participation of non-professional scientists in observation and recording for professional science projects [1]. Citizen scientists have been heralded as one solution to a crisis of monitoring and shortage of data in the field [2–6]. Historically, networks of natural historians have made essential contributions to the acquisition of taxonomic data [7]. Notwithstanding other monitoring activities, the Audubon Christmas Bird Count is widely regarded as the first 'citizen science' exercise in the field of natural history, starting in 1900 and continuing through to the present day [8, 9].

Since the mid-1930s, volunteer naturalists– rather than professional taxonomists—have formed an 'army of new recorders' [10] recruited by initiatives such as the British Trust for Ornithology's Nest Record Scheme and the Royal Society for the Protection of Birds' Big Garden Birdwatch. With millions of people contributing to such schemes on an annual basis [2], a recent report regarding the state of UK taxonomy stated that: 'The voluntary sector, with its core of expert amateur naturalists, is an important repository of taxonomic expertise. The volunteers monitor changes in their local fauna and flora, provide records for biological recording schemes, and generate data for Biodiversity Action Plans' [7].

Today there is a concern (in the UK and the US at least) that we are seeing a 'decline in numbers of both amateur

*Correspondence: glyn.everett@uwe.ac.uk
†Glyn Everett and Hilary Geoghegan contributed equally to this work
[1] Faculty of Environment and Technology, University of the West of England, Frenchay Campus, Coldharbour Lane, Bristol BS16 1QY, UK
Full list of author information is available at the end of the article

and professional taxonomists' [11] and that volunteer efforts in the area of biodiversity recording have been subject to a general decline in numbers. It has been suggested, in a study conducted for the House of Lords in the UK and elsewhere, that the relative strength of the amateur naturalist community as a 'workforce' of taxonomy [11] is fading and that the ability to recruit and train new generations of naturalists is a struggle [12, 13]. Indeed, much has been written about the decline, death or 'impending extinction' of natural history as both an academic subject and amateur enthusiasm [14–18]. For Anna Lawrence [19], 'specialist amateurs are on the decline while more generalist volunteers and environmental enthusiasts are on the rise'.

Notwithstanding professionals working in this area, it appears that our fascination with natural history has shifted from one of keen amateurism to a casual leisure interest with fewer people actively recording and contributing data. This is a concern for many, who argue that there is a 'dearth of basic knowledge' just as our need for knowledge is increasing due to the loss of biodiversity [20, 21]; many biologists today refer to the past 500 years as that of a sixth mass (and first grand anthropogenic) extinction [22–25]. Central to any understanding of and response to changes in flora and fauna is the participation of an adequately trained group of taxonomists, whether amateur or professional, to develop and maintain our understanding of the state of biodiversity.

## A continuum of engagement

In the new context of citizen science and public engagement with science, we know very little about who participates in natural history and what motivates their continued volunteering, whether as an attractive but unpaid leisure activity or an accredited profession. A small number of authors have recently produced interesting work around motivations. For example, Dana Rotman et al. [26] argue that 'volunteers participate in scientific activities out of interest, curiosity and commitment to conservation and related educational efforts'. Extending this further, Daniel Batson et al. [27] identify egotism, collectivism, altruism and principlism (upholding moral principles) as central underlying motivational factors for involvement with citizen science; whilst Jordan Raddick et al. [28] have studied motivations for involvement with GalaxyZoo, finding that contributing, learning, discovering, teaching others and perceiving the beauty and vastness of space were significant motivational factors for participants.

In this paper, we build upon these recent studies by drawing together recent work on the sociology of science and leisure studies in order to develop a continuum of engagement in citizen science for natural history, from the *previously-unengaged* participant who has never undertaken any citizen science work through the more *casually-engaged* participant who has been involved to a lesser degree in natural history or science in the past, to the strength and commitment of involvement frequently displayed by the *already-engaged* participant who in this instance may be described as a traditional amateur naturalist. We acknowledge the contribution of amateur naturalists to citizen science, and consider how participation can work to move people along this continuum in surprising and productive ways. We do so by examining the motivations behind citizen participation in the activities of the Open Air Laboratories (OPAL) programme, an England-wide, biodiversity monitoring and engagement project which began in 2007. Before we move on to our case study, we briefly outline the intellectual context for our research and findings.

## Citizen science and natural history

Although citizen science initiatives have exploded in number over the past 10–15 years, the practice has remained relatively under-represented in the peer-reviewed academic literature (cf. [9]: using Google Scholar, 2000–2009 produces 3420 results containing the phrase 'citizen science', whilst 2010–2014 produces 8750). Much of this work on citizen science has largely been written by professional scientists for professional scientists, in order to improve and argue for best practice in public involvement with projects, and allay fears surrounding data quality and reliability (see [5] for a review of citizen science environmental monitoring, cf. [29–32] for OPAL-related papers in this regard). A body of work is now emerging from within the social sciences on the more qualitative dimensions of what it means to participate in citizen science, shining a more critical light on how volunteering is understood not merely as an opportunity to increase data collection and manpower, but as a fundamental way in which people can work with and know the natural world [3, 33–36].

Recent work by sociologists of science and others has argued against the dichotomy of professional science's interest in data versus humanistic concerns around motivation and participation [37, 38]. Indeed, this work and our paper seek to bridge the gaps between personal, embodied and emotional experiences of citizen science, wider political agendas, pressing environmental concerns and the demands for improved and increased scientific data and knowledge of the world. In order to make sense of the engagement continuum proposed above, which begins to account for the ways in which participants might remain or be transformed from previously-unengaged into casually- and perhaps already-engaged participants, we can usefully consider work around volunteering and leisure.

### Leisure studies

Leisure studies scholars identify volunteering as both unpaid work and attractive leisure. This offers a way of making sense of our continuum, specifically from the 'serious leisure' perspective, whereby leisure is categorised as either serious, casual or project-based. Leisure is understood by Robert Stebbins [39], as ranging from:

- *Serious leisure*: systematic pursuit of an amateur, hobbyist or volunteer activity sufficiently substantial, interesting and fulfilling for the participant to find a (leisure) career there, acquiring a combination of its specialist skills, knowledge and experience.
- *Casual leisure*: immediately, intrinsically rewarding, relatively short-lived pleasurable activity, requiring little or no special training to enjoy it.
- *Project-based leisure*: short-term, reasonably complicated, one-shot or occasional, though infrequent, creative undertaking carried out in free time or time free of disagreeable obligation.

We argue that citizen science activities, such as OPAL, form a major part of project-based leisure, whereby people are asked to participate in a scientific project that responds to either a pressing scientific question (such as the Soil and Earthworm Survey mapping worm populations) or urgent environmental challenge (such as the Tree Health Survey asking the public to report on tree health and harmful pests and diseases). However, our results reveal that OPAL is not only a form of project-based leisure; it also recruits individuals who may undertake forms of serious and casual leisure in the field of natural history and other associated topics. The empirical material here thus enables us to ask and understand: (i) how individuals encounter and experience the survey as a form of project-based leisure; (ii) what motivates them to take part and whether people volunteer as part of leisure, work or a sense of collective responsibility, and (iii) where volunteers feature on our continuum of engagement and in turn whether their participation facilitates their movement between the categories of previously-unengaged, casually-engaged and already-engaged. Furthermore, the inclusion of leisure studies perspectives ensures that the wide-ranging trials, tribulations, and commitments associated with citizen science are no longer overlooked in the desire to gather data for professional science projects.

In the race to herald citizen science as the panacea to many of science's data problems, the figure of the amateur naturalist—as a serious leisure participant—cannot and should not be overlooked [40]. We begin by introducing OPAL, following this with a discussion of several instances of amateur involvement in OPAL. We then conclude the paper by arguing that this study offers a qualitative example of how citizen science can be meaningful to individuals beyond any public engagement and data collection exercise.

## Methods

As Davies et al. [41] outline in the first paper in this supplement, OPAL is one of the largest citizen science for natural history programmes ever attempted in England (cf. [1, 40, 42–44]). Unlike other biodiversity-focussed initiatives such as those of the BBC (Springwatch, Autumnwatch) and the RSPB's Big Garden Birdwatch, OPAL differs in both its provision of materials asking people to follow an accessible yet formalised scientific methodology, and the diversity of fields covered. Further, OPAL's team of regional community scientists act as key agents on the ground in the communication of science and engagement with the public. In this paper, we draw on qualitative research into OPAL activities, specifically focussing on those of OPAL North West (OPAL NW).

OPAL NW was one of nine OPAL regions in England operating during the programme's first phase in 2007–2013. The NW team had the responsibility of distributing surveys and coordinating activities in the North West, as well as carrying out social research in the North West and West Midlands exploring how the thinking and behaviour of OPAL participants changed over time. The social research involved recorded focus groups, recorded in-person interviews in the two regions and telephone interviews with respondents from across the country, as wll as an online survey. All interactions took place around the principal 'OPAL national citizen science' surveys, and the link to the online survey was made available after people entered their data for these. The online survey was used to gain quick feedback from a maximum number of people close to the time of their doing a survey; it also allowed contacts to be gathered for later telephone interviews. Focus groups were used in addition to interviews to deepen understanding by drawing out reflections that might not have come out in a one-to-one interaction.

Five focus groups were held with 50 participants in total in the North West and West Midlands, and over 100 interviews were conducted. Six hundred online surveys were completed nationally, using mostly closed-response, agree-disagree questions with several free-text boxes where respondents could express briefly how they felt about activities. Fifty events or survey activities were also attended to enhance understanding and gain interview contacts. The research presented here is not intended to be representative of all OPAL participants; rather, it represents the views of a broad collection of participants in the North West and West Midlands that reveal the multiple ways in which people have engaged in the OPAL surveys.

The data was transcribed and then analysed in SPSS and NVivo as it became available using a Grounded Theory approach [45]; specifically, data-codes of significance are allowed to emerge from repeated readings of the transcriptions, rather than being imposed upon the data. In the following Results section, focus group data is marked as such and all named interviewees (using pseudonyms) are from either face-to-face or telephone interviews.

## Results

### The previously-unengaged participants

Feedback from OPAL participants reveals that the programme succeeded in engaging many people who previously had had no involvement with natural history. More than half of over 500 online survey respondents aged over 18 reported that OPAL was the first time they had participated in any such activity. The comments below from one online survey question illustrate some of the things people enjoyed about the activities and some reflections upon the motivations for their participation:

> *Q: What did you most enjoy about the OPAL survey activity?*
>
> *'Seeing my garden through different eyes', 'Learning about the natural world', 'I enjoyed seeing what was in the lake, being out in the fresh air, and doing the water sampling', 'Being able to identify what we found and feeling that by taking part we would be contributing to something useful', 'Participating was very interesting and I learned a few things. As a retired person it was nice to feel that I was part of a team of volunteers contributing to an important study', 'Learning something new and investigating familiar surroundings and seeing it in a different light', 'The chance to learn something new and to do something useful at the same time.'*

These rich quotes relating to satisfaction with being outside, learning, observing new things and contributing data and time to a scientific project are representative of the general thrust of feedback and strongly supportive of Rotman et al., Batson et al. and Raddick et al.'s [26–28] findings. However the more in-depth data gathered from focus groups and interviews pointed at times to different elements in the overall picture. Interestingly, although three different methods of qualitative engagement were pursued in this research, no significant differences appeared between what people told us in focus groups, face-to-face and telephone interviews. The online survey did not elicit in-depth reflections, rather 'vox-pop' quotes, but this would be to be expected in such a more restricted interaction.

As outlined earlier, the social dimensions and motivations surrounding participation in citizen science remain still relatively unexplored. For this reason, the following section will consider one of the key challenges that emerged, namely a lack of time. For many OPAL participants, the experience of doing a survey was, as the quotes above suggest, so satisfying that they wanted to go on to do more. However as with all voluntary activity, it is exactly that: voluntary. Participants donate their time, energy and skill and are free to withdraw it at any time [46]. As the following examples attest, while the head and heart might be willing, often other pressures took priority such as family, leisure and work:

> *'I mean, my life is incredibly busy at the moment. I think it's the sort of thing I'd like to do when I'm retired' [Bernice, 35-44]*
>
> *'I would like to do more but I don't have the time to commit, so I think I would say at this point no.' [Janet, 25-44]*
>
> *'I think my life is pretty full at the moment. I don't feel that taking on anything else, I don't think I would be able to do it justice' [Patricia, 45-54]*

Perceived lack of time is clearly a major factor influencing participation in projects where there is a commitment to being outdoors doing fieldwork. Even participants keenly aware of the environmental concerns underlying certain surveys often did not feel they could allow themselves to participate:

> *'My day-job stops me doing more. If I had a job in environment and conservation I'd do more. I do as much as I can, I have very little free time. And my wife, although she works in gardening, planting trees and so on, she's working all hours God sends as well, so I really don't think we've got any time.' [Dave, 35-44]*
>
> *'They're all interesting. For me, if I was going to get involved in anything like that, it's the time aspect ... they're all something I'd like to be involved in, but the practicalities of it, with the other commitments in my life.' [Allotment-holders Focus Group]*

These respondents struggle to justify contributing the spare time they *do* have to the OPAL surveys, juggling other pressures. However, the one-off, project-based nature of OPAL means the activities facilitate participation for time-pressed individuals.

**The casually-engaged participants**

As mentioned, a key part of OPAL's remit has been to engage the previously unengaged in natural history. A less expected but very significant outcome of OPAL's work has been a further engagement of the casually-engaged amateur naturalist community. A key mechanism for enthusing the previously unengaged has been to draw on the success and passion of existing natural history societies and networks. In so doing, OPAL has come to the attention of many already casually-engaged individuals—developing, broadening and deepening their interests:

> *'I've been involved with stuff to do with wildlife for a long time, but it's been good, for really opening my eyes to what's local to me ... getting involved with OPAL encouraged me to want to brush up my knowledge ... it's enabled me to get back to doing something I loved doing a while ago, and I've kind of drifted – it certainly has got me more involved in things.' [Cecilia, 35-44]*

> *'I think OPAL goes into more depth which is good, and feels more 'sciencey' [sic] – new word. It's got me interested in going a bit further with researching, rather than just plopping about in a field or puddle, nice as these activities are. For me personally, as a failed science/biology student at school, it's been a nice experience.' [Diana, 35-44]*

These interviewees highlight how OPAL has offered them significant experiences observing and monitoring nature, which has in turn given rise to increased confidence, renewed interest, refocused activity and validation. The power of citizen science with respect to empowerment cannot be underestimated. For many participants, increased confidence came from the purpose and satisfaction derived from contributing to a much larger dataset for a scientific project, valuing their records as 'real science':

> *'I do care about the local environment, and I felt that I was going to be doing something useful ... It's something where I thought I could contribute to something bigger ... which could create a database of, if lots of people got involved, the whole country.' [Barbara, 35-44]*

> *'It's given me a bit more confidence to do that sort of thing than I had before, because I feel I'm contributing ... it's a confidence booster really, because it helps me understand that I'm not as decrepit as I think I am sometimes.' [Abigail, 65+]*

Citizen science projects like OPAL clearly have a role to play in re-engaging those who have lost touch, or confidence in their abilities. The following respondent, for example, re-engaged with natural history through OPAL following the life event of having children:

> *'I am very interested in the OPAL programme because of the opportunities it offers for education, re-acquainting myself with lost skills and giving a sense that one is making a difference by contributing to a wider research base.' [Neil, 45-54]*

The surveys further worked to engage those who had previously spent time outdoors for reasons other than natural history, key to arguments for the potential value in piggy-backing on the pre-existing interests and activities of the casually-engaged:

> *'I was fascinated by [the OPAL Soil and Earthworm survey], because as an angler I knew there were lob worms and I knew there were brandlings, and the rest were just variations on a theme.' [Paul, 55-64]*

> *'Before attending the OPAL activities and workshops, I went outside to enjoy the countryside, which usually involved following a ramblers trail ... Post-OPAL interaction, I am now an active paid member of The Yorkshire Naturalists Union, Bumblebee Conservation Trust, Bat Conservation Trust ... that's only a selection of the activities!' [Louis, 18-24]*

It is clear from what has been said that participation in the OPAL surveys has empowered some previously-unengaged or casually-engaged individuals; in the next section we will highlight how OPAL has had comparable effects upon the already-engaged.

**The already-engaged participants**

Participation in OPAL surveys has enabled the casually-engaged to broaden and deepen their interest and enthusiasm for natural history. For many already-engaged participants, the surveys offer a means of reframing their natural history activities for a different purpose and taking them out of their comfort zone to consider new areas they are unfamiliar with:

> *'I would always have been doing natural history type things. I probably wouldn't have done the pond-dipping, to be fair, without OPAL encouraging me – and having the nice little pack of stuff certainly encouraged me to go out and do the survey.' [Martin, 55-64]*

The 'little pack of stuff' is important to highlight further: as mentioned earlier, the OPAL survey packs, developed by the Field Studies Council, are regarded as relatively unique for incorporating a field notebook, field guide and other useful kit (such as a magnifying glass, compass, pencil and tape measure):

*'Well that's what seduced me with OPAL really ... the materials were so beautiful, I thought: 'Oh, I'd really like to study this, so I get a better knowledge of what I'm looking at.' [Brenda 55-64]*

Even for some already-engaged participants, the OPAL surveys (literally or figuratively) expanded their toolkits:

*'I've always been interested in doing surveys ... OPAL is just another string to my bow really, where I can seek advice or gain experience doing surveys. OPAL to me is another useful tool.' [Martin, 35-44]*

We have already highlighted how participation in citizen science can offer a way of renewing a pre-existing interest for the casually engaged. For the already engaged, OPAL surveys can go a step further:

*'It's suddenly opening the box – it's bottomless isn't it? And I think that's the beauty of it really, I'll never learn as much as my enthusiasm wants me to learn ... I've taken on too much now and I think my enthusiasm has outstripped my ability!' [Adrian, 55-64].*

Enthusiasm is infectious [47]. Participation in one OPAL survey begets increased participation in other surveys and so a widening of interests:

*'I'd most definitely like to know more – and organisations like OPAL have certainly helped me along that path ... it's an eye-opener, things I love learning ... I've got nothing but admiration and praise for OPAL. I just wish we could reach all the people.' [Steve, 55-64]*

Participation is a social activity, whether between people and people, or between people and the natural world. For many respondents, OPAL worked as a means of opening up and building social networks:

*'What OPAL's done for me is, whereas before I was a solitary naturalist, it's introduced me to a lot more people who feel the same, who have got the same interests, so in that respect I think it's absolutely brilliant.' [Colin, 55-64]*

*'[OPAL's] helped me to see where I want to go with my career, it's pushed me towards volunteering things ... because of OPAL I met the nature person from the Council, and I'm doing a project with him now, [OPAL's] kind of connected us.' [April, 18-24]*

Already-engaged individuals are likely to have developed some of the core skillsets required to undertake biodiversity monitoring activities and species identification. These participants will therefore be more likely to undertake the surveys with the required determination and patience to produce good quality results, as well as to recognise the importance of submitting these results.

Some of the respondents featured in this section form part of what Stebbins [39] describes as 'serious leisure' participants who are making a leisure career out of their interest, what might be termed a vocation. Their years of established experience in observation and recording and their associated networks remain invaluable to the continuing success of citizen science initiatives such as OPAL. This enthusiasm and experience can be key to encouraging previously-unengaged and more casually-engaged people to carry out surveys and increase their knowledge and abilities. OPAL has invested significantly in establishing good relationships with natural history societies, and these societies have in turn provided training and support for the more casually-engaged, as demonstrated by Leanne, who ran a small community group for her village:

*'I did the surveys for their educational aspects. They were great, professionally presented, everything in there, that made a big difference. But they were also good just for getting people involved, opening their eyes so they could see what was around them ... With one group, we worked through the lichen survey and then they wanted to know more, so they got more materials and kept practising their ID skills. They have since done a lichen survey of the whole site!' [Leanne, 45-54]*

These already-engaged participants will bring years of established experience in observation and recording to the areas they now turn their eye to, as well as their networks of contacts who may also become interested. For new societies established alongside the OPAL programme such as the Earthworm Society of Great Britain, this will likely prove invaluable.

## Conclusions

OPAL's aim of increasing participation in natural history is regarded by the environmental community, both amateur and professional, as sorely needed [26]. Long-term programmes of engagement such as OPAL are required in order to generate and retain significant attention and commitment to citizen science. Our research has demonstrated the potential for productive feedback to encourage advancement along our continuum between previously-unengaged, casually-engaged and already-engaged citizen science participants, producing opportunities for knowledge- and skill-sharing and thereby widening and deepening, as well as increasing, participation.

Our research echoes the academic literature on motivation identified earlier in this paper [26–28], revealing that there is no one-size-fits-all solution to increasing motivation for and participation in citizen science.

However, our study identified the importance of projects like OPAL that combine public engagement and scientific endeavour in order to accommodate differing levels and rates of participation. Paying close attention to the new, relatively-new and established natural history participants identified here, OPAL and projects like it should continue to develop a range of approaches for different age-groups and demographics, designing and targeting their activities accordingly (see Davies et al. for examples of the approaches OPAL has engaged with thus far [41]).

Many of the issues highlighted in this paper are beyond the control of OPAL and its community scientists, survey-designers and project partners. OPAL is of course making strong contributions to encouraging shifts in thinking for people to find the time to engage in monitoring activities, creating the spaces and conditions for participation through project-based leisure that tackles important environmental questions [43], for example the health of the nation's trees. However, as this paper has argued, interest, motivation and a sense of collective responsibility can never be guaranteed (Ibid.). The full potential of citizen science is yet to be realised, however the example of OPAL reveals the power of participation in citizen science to move volunteers between the categories of previously-unengaged, casually-engaged and already-engaged. The success of this continuum of engagement should not be underestimated as the rewards for participation range from a personal sense of achievement to the contribution to 'real' scientific research.

### Abbreviations

ID: identification; OPAL: open Air Laboratories; RSPB: Royal Society for the Protection of Birds; SPSS: statistical package for the social sciences.

### Authors' contributions

GE undertook the interviews and focus groups quoted and their qualitative analysis, and drafted the manuscript. HG reviewed and developed the manuscript and added theoretical perspective and structuring. Both authors co-developed, read and approved the final manuscript.

### Author details

[1] Faculty of Environment and Technology, University of the West of England, Frenchay Campus, Coldharbour Lane, Bristol BS16 1QY, UK. [2] Department of Geography and Environmental Science, University of Reading, Whiteknights, Reading RG6 6DW, UK.

### Competing interests

The authors declare that they have no competing interests.

### Declarations

This article has been published as part of *BMC Ecology* Volume 16 Supplement 1, 2016: Citizen science through the OPAL lens. The full contents of the supplement are available online at http://bmcecol.biomedcentral.com/articles/supplements/volume-16-supplement-1. Publication of this supplement was supported by Defra.

### References

1. Bonney R, Ballard H, Jordan R, McCallie E, Phillips T. Public participation in scientific research: defining the field and assessing its potential for informal science education. A CAISE inquiry group report. Washington: Center for Advancement of Informal Science Education; 2009.
2. Roy HE, Pocock M, Preston CD, Savage J, Tweddle J, Robinson LD. Understanding Citizen Science and Environmental Monitoring. London: NERC Centre for Ecology & Hydrology and Natural History Museum; 2012. p. 1–179.
3. Cooper CB, Dickinson J, Philips T, Bonney R. Citizen science as a tool for conservation in residential ecosystems. Ecol Soc. 2007;12:1–11.
4. Schwartz MW. How conservation scientists can help develop social capital for biodiversity. Conserv Biol. 2006;20:1550–2.
5. Conrad CC, Hilchey KG. A review of citizen science and community-based environmental monitoring: issues and opportunities. Environ Monit Assess. 2010;176:273–91.
6. Greenwood JJD. Citizens, science and bird conservation. J Ornithol. 2007;148:77–124.
7. Boxshall G, Self D. UK Taxonomy and Systematics Review—2010. 2011:1–37.
8. Cohn JP. Citizen science: can volunteers do real research? Bioscience. 2008;58:192.
9. Silvertown J. A new dawn for citizen science. Trends Ecol Evol. 2009;24:467–71.
10. Fox R. Butterflies and Moths. In: Hawksworth DL, editor. The changing wildlife of GREAT Britain and Ireland. London: Taylor & Francis; 2003.
11. Science and Technology Committee. Systematics and taxonomy: Follow Up. 5th Report of Session 2007–2008 Report. House of Lords Paper 162. Stationery office books (TSO), 2008.
12. Hopkins GW, Freckleton RP. Declines in the numbers of amateur and professional taxonomists: implications for conservation. Anim Conserv. 2002;5:245–9.
13. Borrell B. Linnaeus at 300: the big name hunters. Nature. 2007;446:253–5.
14. Wilcove DS, Eisner T. The impending extinction of natural history. Chron High Educ. 2000;47:B24.
15. Pyle RM. Nature matrix: reconnecting people and nature. ORX. 2003; 37.
16. Cheesman DC. Key RS: 1 4 The extinction of experience: a threat to insect conservation? Insect Conservation Biology: Proceedings. 2007.
17. Tewksbury JJ, Anderson JGT, Bakker JD, Billo TJ, Dunwiddie PW, Groom MJ, Hampton SE, Herman SG, Levey DJ, Machnicki NJ, del Rio CM, Power ME, Rowell K, Salomon AK, Stacey L, Trombulak SC, Wheeler TA. Natural history's place in science and society. Bioscience. 2014;64:300–10.
18. Louv R. Last child in the woods. Atlantic Books Ltd; 2013.
19. Lawrence A. Taking stock of nature: participatory biodiversity assessment for policy, Planning and Practice. Cambridge: Cambridge University Press; 2010.
20. Tewksbury J, Fleischner T, Rowell K. The natural history initiative: from decline to rebirth. 2010:1–9.
21. Dayton PK. The importance of the natural sciences to conservation. Am Nat. 2003;162:1–13.
22. Novacek MJ. Engaging the public in biodiversity issues. Proc Natl Acad Sci. 2008;105:11571–8.
23. Wake DB, Vredenburg VT. Are we in the midst of the sixth mass extinction? A view from the world of amphibians. Proc Natl Acad Sci. 2008;105:11466–73.
24. Ceballos G, García A, Ehrlich PR. The sixth extinction crisis. J Cosmol. 2010;8:1821–31.
25. Dunn RR, Harris NC, Colwell RK, Koh LP, Sodhi NS. The sixth mass coextinction: are most endangered species parasites and mutualists? Proc Biol Sci. 2009;276:3037–45.
26. Rotman D, Preece J, Hammock J, Procita K, Hansen D, Parr C, Lewis D, Jacobs D. Dynamic changes in motivation in collaborative citizen-science projects. In: CSCW '12. New York, New York, USA: ACM Press; 2012:217–226.
27. Batson D, Ahmad N, Tsang J-A. Four Motives for Community Involvement. J Soc Issues. 2002;58:429–45.
28. Raddick MJ, Bracey G, Gay PL, Lintott CJ, Murray P, Schawinski K, Szalay AS, Vandenberg J. Galaxy Zoo: Exploring the Motivations of Citizen Science Volunteers. Astron Educ Rev. 2010; 9.
29. Riesch H, Potter C. Citizen science as seen by scientists: methodological, epistemological and ethical dimensions. Public Underst Sci. 2014;23:107–20.

30. Fowler A, Whyatt JD, Davies G, Ellis R. How reliable are citizen-derived scientific data? assessing the quality of contrail observations made by the general public. Trans GIS. 2013;17:488–506.
31. Tregidgo DJ, West SE, Ashmore MR. Environmental Pollution. Environ Pollut. 2013; 182(C):448–451.
32. Bone J, Archer M, Barraclough D, Eggleton P, Flight D, Head M, Jones DT, Scheib C, Voulvoulis N. Public participation in soil surveys: lessons from a pilot study in england. Environ Sci Technol. 2012;46:3687–96.
33. Ellis R, Waterton C. Environmental citizenship in the making: the participation of volunteer naturalists in UK biological recording and biodiversity policy. Sci and Pub Pol. 2004.
34. Brossard D, Lewenstein B, Bonney R. Scientific knowledge and attitude change: the impact of a citizen science project. Int J Sci Edu. 2005;27:1099–121.
35. Mackechnie C, Maskell L, Norton L, Roy D. The role of "Big Society" in monitoring the state of the natural environment. J Environ Monit. 2011;13:2687.
36. Brossard D, Lewenstein B, Bonney R. Scientific knowledge and attitude change: The impact of a citizen science project. Int J .... 2005.
37. Lawrence A. "No Personal Motive?" volunteers, biodiversity, and the false dichotomies of participation. Ethics Place Environ. 2006;9:279–98.
38. Smith FM, Timbrell H, Woolvin M, Muirhead S, Fyfe N. Enlivened geographies of volunteering: situated, embodied and emotional practices of voluntary action. Scott Geogr J. 2010;126:258–74.
39. Stebbins RA. Serious Leisure. Transaction Publishers; 2007.
40. Wentworth J. Environmental citizen science. London: Houses of Parliament Parliamentary Office of Science and Technology; 2014. p. 1–5.
41. Davies L, Fradera R, Riesch H, Lakeman Fraser P. Surveying the citizen science landscape: an exploration of the design, delivery and impact of citizen science through the lens of the Open Air Laboratories (OPAL) programme. 2016;16(s1). doi:10.1186/s12898-016-0066-z.
42. Davies L, Bell JNB, Bone J, Head M, Hill L, Howard C, Hobbs SJ, Jones DT, Power SA, Rose N, Ryder C, Seed L, Stevens G, Toumi R, Voulvoulis N, White PCL. Open Air Laboratories (OPAL): a community-driven research programme. Environ Pollut. 2011;159:2203–10.
43. Davies L, Gosling L, Bachariou C, Eastwood J, Fradera R, Manomaiudom N, Robins S (Eds). OPAL community environment report. 2013.
44. Riesch H, Potter C, Davies L. Combining citizen science and public engagement: the Open Air Laboratories Programme. J Sci Commun. 2013;12:1–19.
45. Glaser BG, Strauss AL. The Discovery of Grounded Theory. Transaction Publishers; 2009.
46. Geoghegan H. A new pattern for historical geography: working with enthusiast communities and public history. J Hist Geogr. 2014:1–3.
47. Geoghegan H. Emotional geographies of enthusiasm: belonging to the telecommunications heritage group. Area. 2012;45:40–6.

# 17

# Amphibian and reptile road-kills on tertiary roads in relation to landscape structure: using a citizen science approach with open-access land cover data

Florian Heigl[1*], Kathrin Horvath[1], Gregor Laaha[2] and Johann G. Zaller[1]

**Abstract**

**Background:** Amphibians and reptiles are among the most endangered vertebrate species worldwide. However, little is known how they are affected by road-kills on tertiary roads and whether the surrounding landscape structure can explain road-kill patterns. The aim of our study was to examine the applicability of open-access remote sensing data for a large-scale citizen science approach to describe spatial patterns of road-killed amphibians and reptiles on tertiary roads. Using a citizen science app we monitored road-kills of amphibians and reptiles along 97.5 km of tertiary roads covering agricultural, municipal and interurban roads as well as cycling paths in eastern Austria over two seasons. Surrounding landscape was assessed using open access land cover classes for the region (Coordination of Information on the Environment, CORINE). Hotspot analysis was performed using kernel density estimation (KDE+). Relations between land cover classes and amphibian and reptile road-kills were analysed with conditional probabilities and general linear models (GLM). We also estimated the potential cost-efficiency of a large scale citizen science monitoring project.

**Results:** We recorded 180 amphibian and 72 reptile road-kills comprising eight species mainly occurring on agricultural roads. KDE+ analyses revealed a significant clustering of road-killed amphibians and reptiles, which is an important information for authorities aiming to mitigate road-kills. Overall, hotspots of amphibian and reptile road-kills were next to the land cover classes arable land, suburban areas and vineyards. Conditional probabilities and GLMs identified road-kills especially next to preferred habitats of green toad, common toad and grass snake, the most often found road-killed species. A citizen science approach appeared to be more cost-efficient than monitoring by professional researchers only when more than 400 km of road are monitored.

**Conclusions:** Our findings showed that freely available remote sensing data in combination with a citizen science approach would be a cost-efficient method aiming to identify and monitor road-kill hotspots of amphibians and reptiles on a larger scale.

**Keywords:** Anurans, Kernel density estimation, Landscape ecology, Participatory science, qgis, Road mortality, Snakes

## Background

Amphibian and reptile species are endangered worldwide, suffering from numerous threats such as habitat modification and fragmentation, diseases, pollution, invasive species or climate change [1–4]. In Austria, where the current study was conducted, all 20 amphibian species and all 14 reptile species are protected by national conservation laws [5]. Focusing on habitat fragmentation, roads can have various negative effects on many vertebrate species [6–9]. The most direct negative effect of road traffic on animal populations is through fatal collisions with vehicles, i.e. road-kill [2]. Road-kill

*Correspondence: florian.heigl@boku.ac.at
[1] Institute of Zoology, University of Natural Resources and Life Sciences, Vienna, Gregor Mendel Straße 33, 1180 Vienna, Austria
Full list of author information is available at the end of the article

does not affect all taxonomic groups in the same way. Amphibians (toads, newts and salamanders) are mostly affected by road-kill when crossing roads during migration between their breeding and hibernation sites; reptiles are even attracted by the favourable microclimate on roads [4, 8, 10–12]. Amphibians and some reptile species are even more susceptible to road-kill because they get immobile in response to an approaching vehicle [13, 14]. As mentioned by Rytwinski and Fahrig [15] there are relatively few studies of the effects of roads on amphibian and reptile populations, despite the fact that amphibians and reptiles have significantly more species at risk than mammals or birds. Additionally, available data of road-kills of amphibians and reptiles for Europe is scarce and often not comparable due to different study designs [16] and because of species-specific response patterns [17, 18].

In temperate regions like Central Europe many amphibian and some reptile species require complex landscapes including wetlands for reproduction and woody areas for foraging and hibernation [19]. Hence, the composition of landscape surrounding roads is an important factor influencing the number of road-kills for both amphibian and reptile populations [1, 20, 21]. Some reptile species (e.g. European green lizard, *Lacerta viridis*, Laurenti, 1768) are more selective to a habitat than other species (e.g. Grass snake, *Natrix natrix*, Linnaeus, 1758), which migrate long distances from summer to winter habitats. Most adult toads are susceptible to road-kills when migrating to breeding ponds [1, 10], while road-kills of juvenile toads are more dispersed in space and time when moving to hibernation sites in late summer and autumn. In the northern hemisphere most road-kill studies investigate the impact of wide roads with high traffic volumes that are usually fenced off and therefore are a stronger barrier to amphibians and reptiles [6, 22, 23]. However, evidence is increasing that tertiary road networks especially affect small animal species like herpetofauna [24–27]. Nevertheless, the influences of tertiary roads on amphibian and reptile populations on a landscape level are not often studied. When monitoring herpetofaunal road-kills challenges exist because small road-killed animals disappear quickly [28], the diverse network of tertiary roads is dense and some road-killed species are difficult to identify.

The standard approach to assess the effects of road-kill on animals is to collect data on a regular basis along certain routes [29], but this method is very cost-intensive and time consuming [30]. An alternative is to use a citizen science approach, i.e. involving citizens to report road-kill sightings [29, 31–34]. However, in most citizen science projects, "presence only" data are collected, which hamper proper statistical analyses of underlying factors for road-kills [35]. The aim of the current study was to test a systematic monitoring approach that would be appropriate to engage citizens in collecting presence and absence data. Results of this pilot study would be a proof-of-concept before engaging the general public, since many challenges (e.g. motivation of citizens) exist in establishing such a citizen science monitoring approach [36]. To test this approach, we used a citizen science software for monitoring road-killed amphibians and reptiles along a fixed bicycle tour on tertiary roads. Additionally, we examine the applicability of open access remote sensing data for a large-scale citizen science approach to describe spatial patterns of road-killed amphibians and reptiles on tertiary roads. The findings will be discussed with respect to their potential for designing a cost-effective large scale road-kill monitoring system based on a citizen science approach.

## Methods

We monitored road-kills for two activity periods of amphibians and reptiles from March 2014 to October 2015 on a 97.5-km stretch of road in a rural region in eastern Austria. We use the term *tertiary road network* to summarize (I) agricultural roads (used by farmers and cyclists, speed limit 30 km $h^{-1}$), (II) cycling paths (used by cyclists only), (III) municipal roads (mainly used by residents, speed limit 50 km $h^{-1}$) and (IV) interurban roads (mainly used by residents, speed limit 100 km $h^{-1}$).

### Study area

The monitoring was conducted between the Leithagebirge and the Lake Neusiedl in Northern Burgenland, which is located in eastern Austria (Fig. 1). We chose the area for its high biodiversity in amphibian and reptile species and its relatively dense network of tertiary roads [37]. The study route consisted of 61 km of agricultural roads, 26 km of municipal roads, 8 km of interurban roads and 2.5 km of cycling paths, and is partly in the Natura 2000 sites Neusiedler See—Nordöstliches Leithagebirge (north-east of the study route) and Mattersburger Hügelland (south-west of the study route) [38]. The landscape of the study area is topographically heterogeneous with elevations ranging from 115 to 748 m. The climate of the region is considered Continental-Pannonian with 562 mm average annual precipitation and 10.7 °C annual mean temperature (Neusiedl am See; years 1981–2010) [39]. The Leithagebirge is forming the western boundary of the Natura 2000 site and at the same time framing the lake basin. The Leithagebirge exhibits a crystalline basement with accumulated reef lime stones of tertiary origin and is marked by mainly oak and mixed oak forests. In the outlying areas open farmland (vineyards interspersed with trees, bushes and grasslands

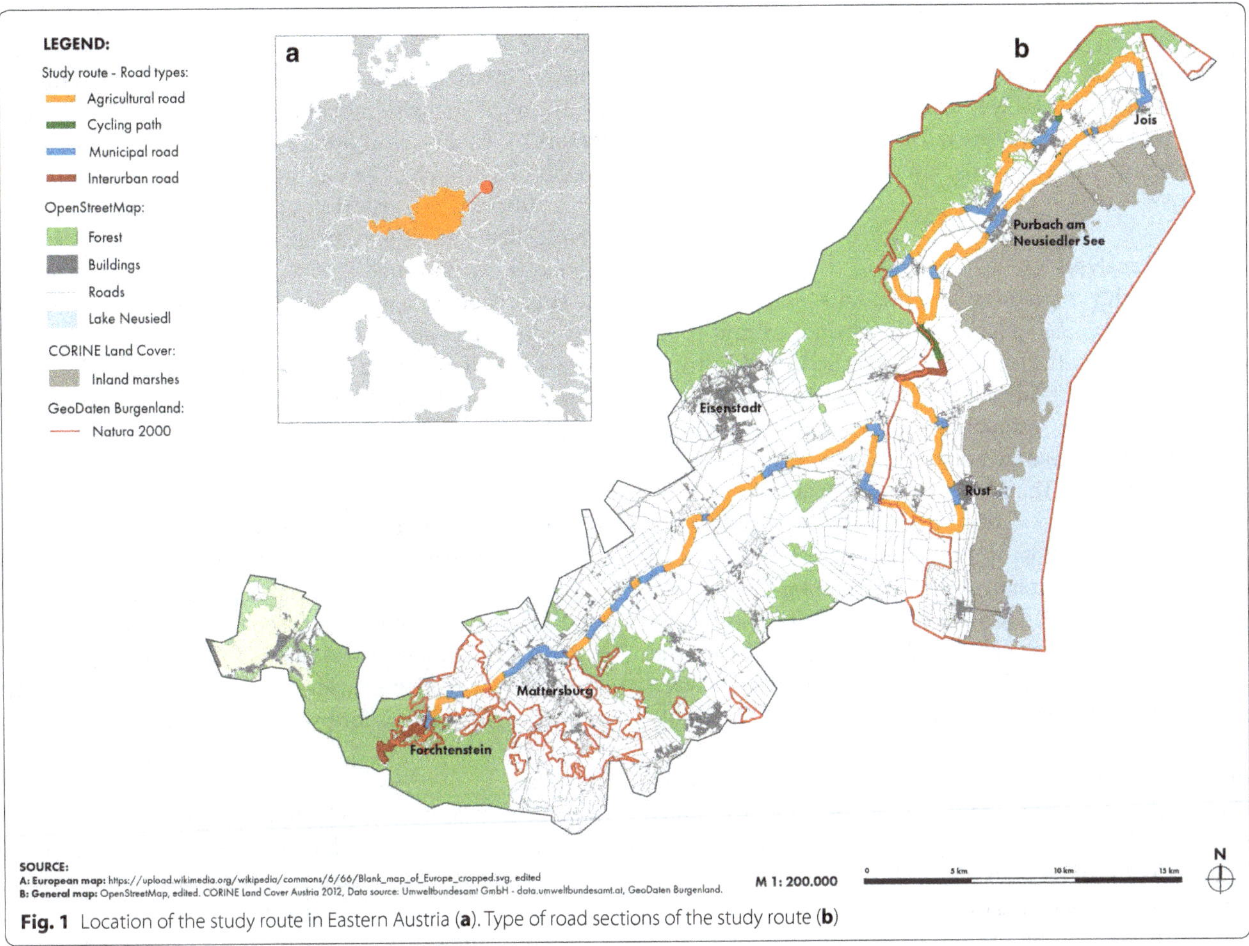


**Fig. 1** Location of the study route in Eastern Austria (**a**). Type of road sections of the study route (**b**)

residues) and dry grasslands dominate. The study area is a biodiversity hotspot in Austria as Pannonian, Alpine and Mediterranean floristic and faunal elements intermingle [38].

## Data collection

We monitored the selected route from March 2014 to October 2015 by using a smartphone app (Roadkill | SPOTTERON, NINC Media, Vienna, Austria) from the citizen science project Roadkill (http://www.roadkill.at/en). In the citizen science project volunteers collect presence only data of all vertebrate species killed on roads [35, 40]. The app was not adjusted to our approach but rather used as data collection tool in a more systematic way. In the current project, we used the Roadkill app to monitor road-killed amphibians and reptiles along a predefined route on average every 11 days. We monitored road-kills by cycling, since slower traveling speed results in higher detectability especially of smaller species [41]. We stopped at each road-killed animal and filled in the form provided by the SPOTTERON Roadkill application, which included coordinates of the spot, a picture of the animal, species name and number of individuals. All other locations where no road-kill was detected represent absence data. Our monitoring therefore results in a presence/absence dataset of locations along the route at different points in time.

## Remote sensing data

We used only open data to follow the idea of open science and to test the applicability of freely available remote sensing data in describing spatial patterns of road-killed amphibians and reptiles. We downloaded the most current CORINE (Coordination of Information on the Environment) land cover data from copernicus land cover monitoring services [42]. We divided the study route in 500 m sections and for each section assessed the surrounding landscape within a 500 m buffer on each side of the road by recording the area of each land cover class. The sum of the area of all land cover classes results in the study area. CORINE land cover (CLC) is a geographic land cover/land use database for a Pan-European region. CLC data provides information on the biophysical characteristics of the Earth's surface based on images acquired by Earth Observation satellites with a Minimal Mapping Unit (MMU) of 25 ha [43]. The CLC 2012 uses a standardized European level-3 nomenclature consisting of 37 classes [44]. We used only the 14 classes which are present in the study area. The study area consisted of a high number of vineyards, arable land and urban areas (Table 1).

### Statistical analyses

First, monthly variations of road-kills per year were analysed using Chi squared tests. Second, to find road-kill hotspots on the study route, we used the software KDE+ [45]. KDE+ identifies clusters of road-kills based on kernel density estimation and provides a measure of the significance of a hotspot based on Monte Carlo simulations [46]. Third, we divided the study route in 500 m sections to calculate conditional probabilities to assess the association of a certain land cover class with a road-kill event directly from the sample. Conditional probabilities [P(E|B)] of road-kill events (E) on each land cover class (B) were calculated to analyse which land cover classes are associated with an increased/decreased risk of road-kills, as compared to the overall probability P(E) of sections having a road-kill event. To obtain conditional probabilities, the areal fractions of land cover classes were determined for each section. The probability of a road-kill on a certain land cover class [P(E∩B)] is the total area of B of sections affected by road-kills [A(E∩B)] divided by the total area of all sections (A). [P(E∩B)] was finally divided by the overall availability (i.e., total area fraction) of this land cover class in the study area [P(B)] to obtain its conditional probability. Fourth, to analyse which land cover class is related to the number of road-kills we employed general linear models (GLM) with Poisson distributions. Possible collinearity was handled by means of stepwise model fitting and variance inflation factors (VIF). From all fitted models that do not contain predictors VIF >10, the one with the lowest value in the akaike information criterion (AIC) was chosen.

Note that conditional probabilities and general linear models are performed to 500 m road sections whereas the study route as a whole was used for the kernel density estimation. Chi squared tests and general linear models were performed using the "Rcmdr" package (R Commander Version 2.2–3) [47] in the open source program "R" (R version 3.2.4) [48].

**Table 1 CORINE land cover classes in the study area. Land cover classes in descending order of proportional area**

| Land cover class | CLC code | Area (ha) | Area (%) |
|---|---|---|---|
| Vineyards | 221 | 3025.56 | 35.13 |
| Non-irrigated arable land | 211 | 2574.66 | 29.89 |
| Discontinuous urban fabric | 112 | 1809.06 | 21.00 |
| Complex cultivation patterns | 242 | 293.52 | 3.41 |
| Land principally occupied by agriculture, with significant areas of natural vegetation | 243 | 231.77 | 2.69 |
| Broad-leaved forest | 311 | 227.79 | 2.64 |
| Pastures | 231 | 127.98 | 1.49 |
| Coniferous forest | 312 | 119.28 | 1.38 |
| Mixed forest | 313 | 96.47 | 1.12 |
| Continuous urban fabric | 111 | 52.58 | 0.61 |
| Industrial or commercial units | 121 | 24.42 | 0.28 |
| Sport and leisure facilities | 142 | 14.12 | 0.16 |
| Inland marshes | 411 | 11.89 | 0.14 |
| Transitional woodland shrub | 324 | 3.91 | 0.05 |

### Cost efficiency estimation

A rough estimation of the cost efficiency of our citizen science approach was made to compare costs of our pilot study with a classical monitoring approach. Numbers used for the calculation are based on standard staff costs in Austria, one offer provided by an Austrian engineering office and one offer provided by the software company which developed apps for project *Roadkill*.

## Results

During our monitoring of 20 months, we found 252 road-killed animals (180 amphibians and 72 reptiles) comprising eight species (Table 2). Green toad (*Bufo viridis*, Laurenti, 1768), common toad (*Bufo bufo*, Linnaeus, 1758) and grass snake (*Natrix natrix*) were the dominating species of the investigated amphibian and reptile species, respectively. Most amphibians and reptiles were killed on agricultural roads.

Road-kill reports were not equally distributed across months (amphibians: $X^2 = 136.44$, df = 7, $p < 0.001$; reptiles: $X^2 = 34.889$, df = 7, $p < 0.001$). Figure 2 shows that most amphibians were reported in July (n = 64), followed by April (n = 36) and August (n = 30), whereas most reptiles were reported in October (n = 20) and September (n = 19).

### Spatial patterns

Applying the KDE+ software to our road-kill monitoring data resulted in several hotspots including sections of 2–37 road-killed amphibians and of 2–8 road-killed reptiles, respectively (Fig. 3; Table 3). The vast majority of amphibian hotspots are next to arable land and suburban areas, whereas most reptile hotspots are located near the reed belt of the lake Neusiedl.

Green toads and common toads represented 96% of all road-killed amphibians and grass snakes represented 83% of all road-killed reptile species, therefore we focused the following analyses on these three most often found

**Table 2 Numbers of road-killed amphibians and reptiles found from March 2014–October 2015 on monitored sections of municipal roads (26 km), cycle paths (2.5 km), agricultural roads (61 km) and interurban roads (8 km)**

| Species | Municipal road | | Cycle path | | Agricultural road | | Interurban road | |
|---|---|---|---|---|---|---|---|---|
| | Rk | Rk $km^{-1}$ | Rk | Rk $km^{-1}$ | Rk | Rk $km^{-1}$ | Rk | Rk $km^{-1}$ |
| Green toad (*Bufo viridis*) | 45 | 1.73 | 1 | 0.4 | 69 | 1.13 | 4 | 0.5 |
| Common toad (*Bufo bufo*) | 4 | 0.15 | 0 | 0 | 50 | 0.82 | 1 | 0.13 |
| Agile frog (*Rana dalmatina*) | 1 | 0.04 | 0 | 0 | 3 | 0.05 | 1 | 0.13 |
| Tree frog (*Hyla arborea*) | 0 | 0 | 0 | 0 | 1 | 0.02 | 0 | 0 |
| Grass snake (*Natrix natrix*) | 7 | 0.27 | 1 | 1 | 43 | 0.7 | 9 | 1.13 |
| Lizards (*Lacertidae*) | 1 | 0.04 | 0 | 0 | 4 | 0.07 | 1 | 0.13 |
| Smooth snake (*Coronella austriaca*) | 0 | 0 | 0 | 0 | 3 | 0.05 | 0 | 0 |
| Blind worm (*Anguis fragilis*) | 0 | 0 | 0 | 0 | 2 | 0.03 | 1 | 0.13 |
| | 58 | 2.23 | 2 | 0.8 | 175 | 2.87 | 17 | 2.13 |

Numbers of road-killed animals (Rk) and road-killed animals per kilometer (RK $km^{-1}$)

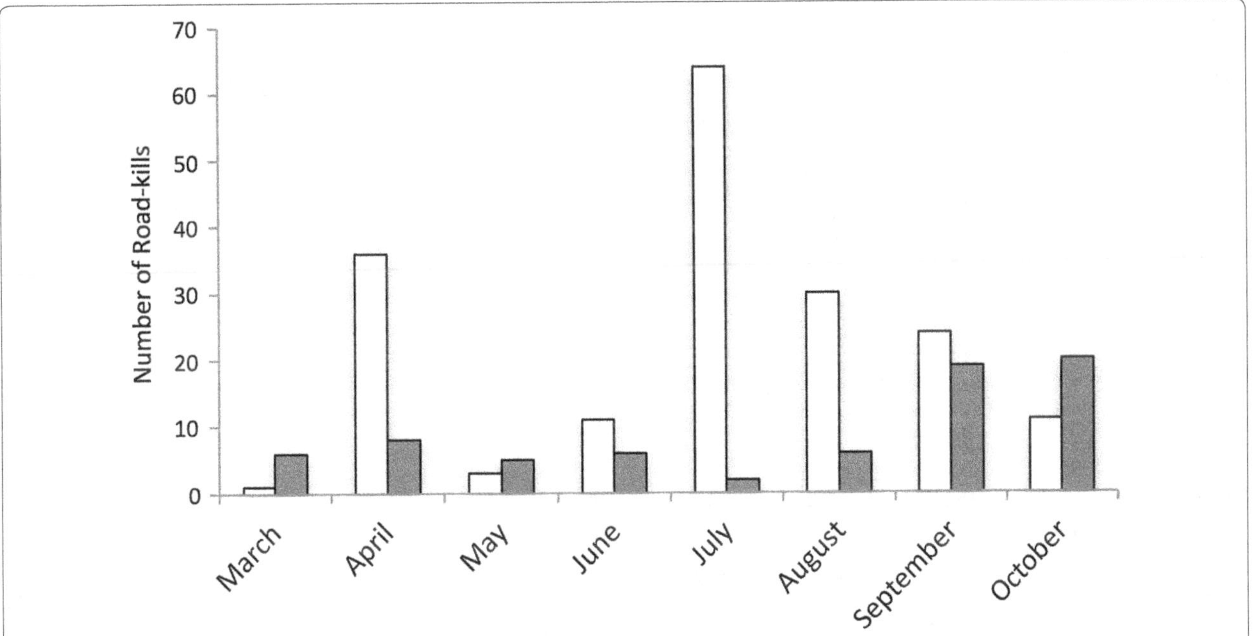


**Fig. 2** Total numbers of road-killed amphibians (*white*, n = 180) and reptiles (*grey*, n = 72) per month from March 2014–October 2015. No road-killed amphibians or reptiles were found between November 2014 and February 2015

species. Number of road-killed common toads and green toads per 500 m section varied from 0 to 35. The overall probability of a road-killed toad in the study area calculated per section was 0.36. This value was then used as a benchmark for conditional probabilities of road-kill events by land cover class (Table 4). All land cover classes with $P(E|B) > 0.36$ are associated with increased risk of road-killed common toads and green toads, whereas land cover classes with $P(E|B) < 0.36$ constrain the risk. Transitional woodland shrub (CLC 324), Sport and leisure facilities (CLC 142) and Land principally occupied by agriculture, with significant areas of natural vegetation (CLC 243) exhibit the highest conditional probabilities, whereas Industrial or commercial units (CLC 121) and Pastures (CLC 231) were associated with the lowest conditional probabilities. Using the best fitting general linear model focusing on the most often road-killed species common toad and green toad (AIC: 687.67; Table 5), a significant positive relation of Land principally occupied by agriculture, with significant areas of natural vegetation (CLC 243) and Transitional woodland shrub (CLC 324) and amphibian road-kills was seen, indicating that this land cover classes promoted amphibian road-kills. Complex cultivation patterns (CLC 242) and Vineyards (CLC

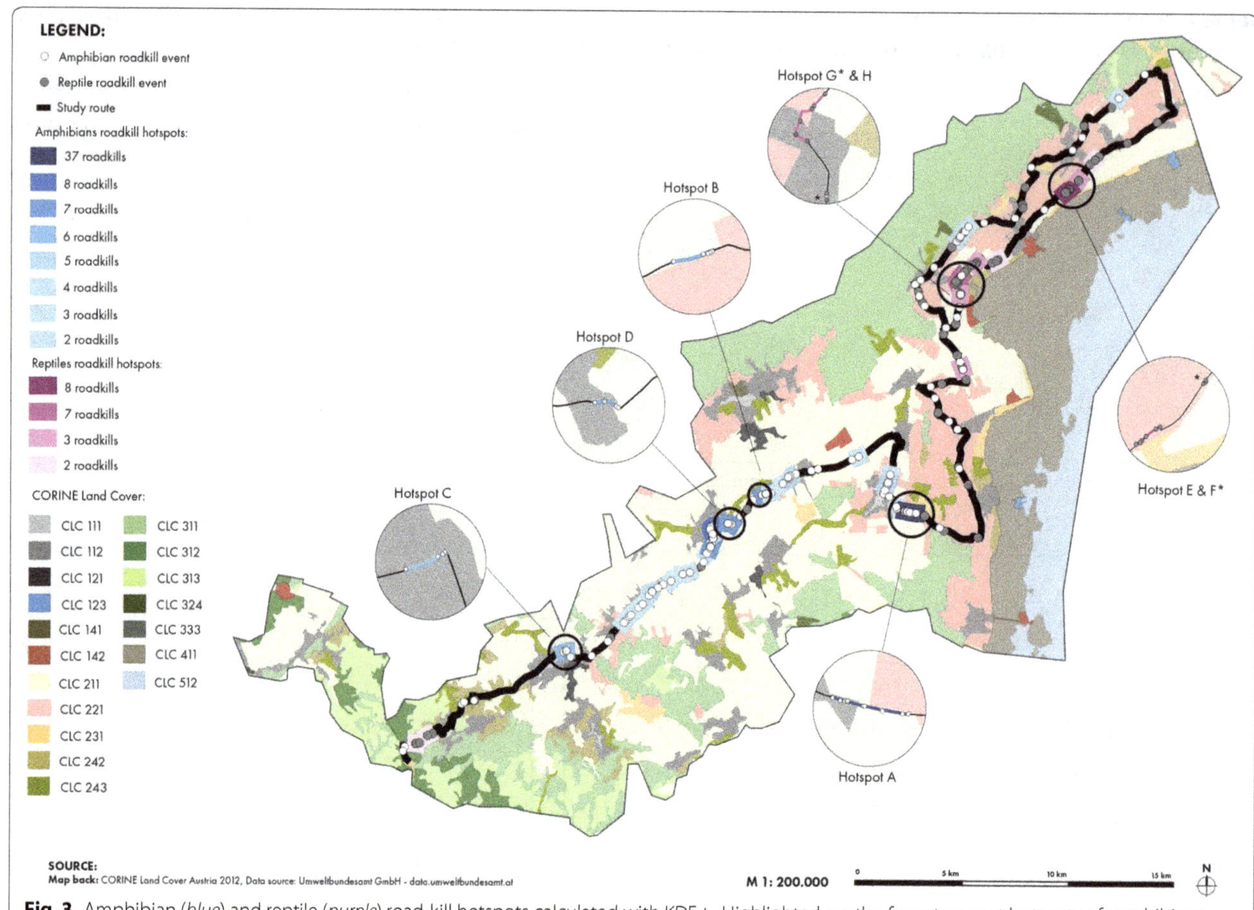


**Fig. 3** Amphibian (*blue*) and reptile (*purple*) road-kill hotspots calculated with KDE+. Highlighted are the four strongest hotspots of amphibians (*A–D*) and reptiles (*E–H*). *Asterisked letters* differ the two hotspots in one circle

**Table 3 KDE+ results of the strongest four amphibian and reptile hotspots**

| Animal group | Hotspot | Length (m) | Road-kills | Strength |
|---|---|---|---|---|
| Amphibians | A | 523.49 | 37 | 0.96 |
| | B | 202.79 | 7 | 0.82 |
| | C | 152.77 | 6 | 0.79 |
| | D | 175.54 | 7 | 0.78 |
| Reptiles | E | 249.88 | 8 | 0.80 |
| | F | 36.29 | 3 | 0.66 |
| | G | 52.25 | 3 | 0.65 |
| | H | 532.09 | 7 | 0.58 |

221) had a significantly negative relationship with the number of road-killed common toads and green toads.

Number of road-killed grass snakes per 500 m section varied from 0 to 6, with an overall probability of a road-killed grass snake per section of 0.17. Again, conditional probabilities P(E|B) were calculated for all 14 land cover classes (Table 6). Inland marshes (CLC 411) and Sport and leisure facilities (CLC 142) appeared to promote reptile road-kills, whereas Transitional woodland shrub (CLC 324), Industrial or commercial units (CLC 121) and Continuous urban fabric (CLC 111) appeared to constrain them. Using the best fitting general linear model again excluding all but the most often road-killed reptile species grass snake (AIC: 294.79; Table 7), resulted in significantly positive relations of the land cover classes Inland marshes (CLC 411), Sport and leisure facilities (CLC 142) and Complex cultivation patterns (CLC 242) with reptile road-kills. Discontinuous urban fabric (CLC 112) and Land principally occupied by agriculture, with significant areas of natural vegetation (CLC 243) were significantly negative related to the number of road-killed grass snakes.

### Cost efficiency estimation

We compared two cases for cost-efficiency estimation (Table 8), monitoring by researchers only and a citizen science approach. Based on our calculations,

**Table 4 Conditional probabilities of road-killed common toads and green toads for all land cover classes [P(E|B)]**

| CLC code | Land cover class | H(E) | H(E1) | H(E)+ H(E1) | P(E∩B) | P(B) | P(E\|B) |
|---|---|---|---|---|---|---|---|
| 324 | Transitional woodland shrub | 0.088 | 0.000 | 0.088 | 0.000 | 0.000 | 0.998 |
| 142 | Sport and leisure facilities | 0.296 | 0.023 | 0.319 | 0.002 | 0.002 | 0.924 |
| 243 | Land principally occupied by agriculture, with significant areas of natural vegetation | 3.181 | 2.081 | 5.262 | 0.016 | 0.027 | 0.606 |
| 311 | Broad-leaved forest | 2.147 | 2.998 | 5.145 | 0.011 | 0.026 | 0.416 |
| 211 | Non-irrigated arable land | 23.681 | 34.538 | 58.220 | 0.121 | 0.299 | 0.406 |
| 112 | Discontinuous urban fabric | 16.534 | 24.576 | 41.110 | 0.085 | 0.210 | 0.404 |
| 411 | Inland marshes | 0.088 | 0.180 | 0.269 | 0.000 | 0.001 | 0.327 |
| 312 | Coniferous forest | 0.862 | 1.831 | 2.694 | 0.004 | 0.014 | 0.319 |
| 221 | Vineyards | 20.752 | 47.589 | 68.340 | 0.106 | 0.351 | 0.303 |
| 111 | Continuous urban fabric | 0.334 | 0.854 | 1.188 | 0.002 | 0.006 | 0.281 |
| 242 | Complex cultivation patterns | 1.742 | 4.950 | 6.692 | 0.009 | 0.034 | 0.262 |
| 313 | Mixed forest | 0.138 | 2.094 | 2.232 | 0.001 | 0.011 | 0.063 |
| 231 | Pastures | 0.157 | 2.734 | 2.891 | 0.001 | 0.015 | 0.054 |
| 121 | Industrial or commercial units | 0.000 | 0.552 | 0.552 | 0.000 | 0.003 | 0.000 |

Proportion of each land cover class per section containing road-killed toads [H(E)] or not [H(E1)]. The probability of a road-killed toad on a certain land cover class [P(E∩B)] divided by the overall availability of this land cover class in the study area [P(B)] results in the conditional probability P(E|B). Land cover classes in descending order of P(E|B); the higher the P(E|B), the higher the probability of a road-kill on the specific land cover class

**Table 5 GLM containing land cover classes as explanatory variables that influence road-kill numbers of green toad and common toad**

| CLC code | Land cover class | VIF | Estimate | Std. error | z value | P(>\|z\|) |
|---|---|---|---|---|---|---|
| Intercept | | | 3.53E-01 | 1.51E-01 | 2.342 | 0.019 |
| 243 | Land principally occupied by agriculture, with significant areas of natural vegetation | 1.377 | 1.05E-05 | 2.26E-06 | 4.667 | 3.06E-06 |
| 242 | Complex cultivation patterns | 1.259 | −1.56E-05 | 3.57E-06 | −4.374 | 1.22E-05 |
| 221 | Vineyards | 1.276 | −2.43E-06 | 6.12E-07 | −3.964 | 7.38E-05 |
| 324 | Transitional woodland shrub | 1.086 | 5.30E-05 | 2.32E-05 | 2.284 | 0.022 |
| 231 | Pastures | 1.049 | −1.47E-05 | 8.20E-06 | −1.795 | 0.073 |
| 111 | Continuous urban fabric | 1.019 | −1.52E-05 | 8.58E-06 | −1.778 | 0.076 |
| 313 | Mixed forest | 1.14 | −2.17E-05 | 1.46E-05 | −1.486 | 0.137 |
| 311 | Broad-leaved forest | 1.176 | −2.76E-06 | 3.35E-06 | −0.825 | 0.41 |
| 142 | Sport and leisure facilities | 1.024 | −7.70E-06 | 1.25E-05 | −0.617 | 0.537 |
| 312 | Coniferous forest | 1.186 | −1.22E-06 | 2.99E-06 | −0.407 | 0.684 |
| 411 | Inland marshes | 1.041 | −1.20E-05 | 3.21E-05 | −0.372 | 0.71 |
| 112 | Discontinuous urban fabric | 1.283 | −1.16E-07 | 6.47E-07 | −0.179 | 0.858 |
| 121 | Industrial or commercial units | 1 | −2.15E-04 | 9.17E-03 | −0.023 | 0.981 |

Land cover classes in descending order of P(>|z|)

using an approach involving researchers only would result in 22,000 €. This includes monitoring road-kills by a graduate student on a study route of 100 km over two vegetation periods and recording the surrounding landscape structure and habitat descriptions by an engineering office. Alternatively, the use of citizen science in combination with open access land cover data as we supposed in our pilot study would result in 86,000 €. This includes adjusting the citizen science smartphone application of the project *Roadkill* to allow for monitoring roads, maintaining the application and professional support of the participants for a 2-year period.

## Discussion

To our knowledge, this is among the first studies testing the suitability of a citizen science approach in examining the impact of tertiary roads and the surrounding landscape on amphibian and reptile species by using freely-available remote sensing data.

**Table 6 Conditional probabilities of road-killed grass snakes for all land cover classes [P(E|B)]**

| CLC code | Land cover class | H(E) | H(E1) | H(E)+ H(E1) | P(E∩B) | P(B) | P(E\|B) |
|---|---|---|---|---|---|---|---|
| 411 | Inland marshes | 0.143 | 0.126 | 0.269 | 0.001 | 0.001 | 0.531 |
| 142 | Sport and leisure facilities | 0.147 | 0.172 | 0.319 | 0.001 | 0.002 | 0.459 |
| 242 | Complex cultivation patterns | 2.191 | 4.501 | 6.692 | 0.011 | 0.034 | 0.330 |
| 311 | Broad-leaved forest | 1.296 | 3.849 | 5.145 | 0.007 | 0.026 | 0.251 |
| 211 | Non-irrigated arable land | 11.749 | 46.471 | 58.220 | 0.060 | 0.299 | 0.202 |
| 221 | Vineyards | 12.197 | 56.143 | 68.340 | 0.063 | 0.351 | 0.178 |
| 313 | Mixed forest | 0.380 | 1.851 | 2.232 | 0.002 | 0.011 | 0.174 |
| 243 | Land principally occupied by agriculture, with significant areas of natural vegetation | 0.900 | 4.362 | 5.262 | 0.005 | 0.027 | 0.172 |
| 231 | Pastures | 0.393 | 2.497 | 2.891 | 0.002 | 0.015 | 0.136 |
| 312 | Coniferous forest | 0.286 | 2.408 | 2.694 | 0.001 | 0.014 | 0.106 |
| 112 | Discontinuous urban fabric | 4.318 | 36.792 | 41.110 | 0.022 | 0.210 | 0.105 |
| 111 | Continuous urban fabric | 0.000 | 1.188 | 1.188 | 0.000 | 0.006 | 0.000 |
| 121 | Industrial or commercial units | 0.000 | 0.552 | 0.552 | 0.000 | 0.003 | 0.000 |
| 324 | Transitional woodland shrub | 0.000 | 0.088 | 0.088 | 0.000 | 0.000 | 0.000 |

Proportion of each land cover class per section containing road-killed grass snakes [H(E)] or not [H(E1)]. The probability of a road-killed grass snake on a certain land cover class [P(E∩B)] divided by the overall availability of this land cover class in the study area [P(B)] results in the conditional probability P(E|B). Land cover classes in descending order of P(E|B); the higher the P(E|B), the higher the probability of a road-kill on the specific land cover class

**Table 7 GLM containing land cover classes as explanatory variables that influence road-kill numbers of grass snakes**

| CLC code | Land cover class | VIF | Estimate | Std. error | z value | P(>\|z\|) |
|---|---|---|---|---|---|---|
| Intercept | | | −8.82E-01 | 2.82E-01 | −3.129 | 0.002 |
| 411 | Inland marshes | 1.243 | 5.95E-05 | 1.22E-05 | 4.894 | 9.90E-07 |
| 142 | Sport and leisure facilities | 1.142 | 2.42E-05 | 1.03E-05 | 2.35 | 0.019 |
| 242 | Complex cultivation patterns | 1.579 | 5.81E-06 | 2.57E-06 | 2.262 | 0.024 |
| 243 | Land principally occupied by agriculture, with significant areas of natural vegetation | 1.483 | −1.31E-05 | 6.32E-06 | −2.08 | 0.038 |
| 112 | Discontinuous urban fabric | 1.277 | −3.08E-06 | 1.49E-06 | −2.067 | 0.039 |
| 221 | Vineyards | 1.447 | −1.21E-06 | 1.08E-06 | −1.118 | 0.264 |
| 312 | Coniferous forest | 1.099 | −5.08E-06 | 5.79E-06 | −0.878 | 0.38 |
| 231 | Pastures | 1.189 | 1.28E-06 | 3.68E-06 | 0.347 | 0.729 |
| 311 | Broad-leaved forest | 1.154 | 1.83E-07 | 4.09E-06 | 0.045 | 0.964 |
| 313 | Mixed forest | 1.113 | 2.40E-07 | 5.78E-06 | 0.042 | 0.967 |
| 111 | Continuous urban fabric | 1 | −2.81E-03 | 2.24E-01 | −0.013 | 0.989 |
| 324 | Transitional woodland shrub | 1 | −1.66E-03 | 8.79E-01 | −0.002 | 0.999 |
| 121 | Industrial or commercial units | 1 | −2.71E-04 | 1.56E-01 | −0.002 | 0.999 |

Land cover classes in descending order of P(>|z|)

Road-kills of both amphibians and reptiles were recorded during the whole vegetation period mainly on agricultural roads. The role of tertiary roads for road-kill of endangered species is generally not widely appreciated, however is perhaps more significant in regions where higher ranking roads are already equipped with efficient road-kill mitigation measures. Green toads and common toads represented 96% of all road-killed amphibians with peaks in April, August and September. These two species are also among the most abundant species in the study region [37]. From our results it seems that identifying road-kill hotspots of both species in spring would be most straightforward because of mass migration to the spawning sites during this time [49]. However, when the goal is to assess the effect of road-kill on population dynamics, surveys would need to also include surveys in late summer and autumn when individuals of both species forage in their terrestrial habitat up to 10 km away from the breeding ponds [50]. Grass snakes represented 83% of all road-killed reptile species with road-kill peaks

**Table 8 Cost-efficiency estimation for the cases researcher and citizen science**

| Cases | Costs (€) |
|---|---|
| Researcher | |
| Road-kill monitoring | 10,000 |
| Assessing surrounding land cover | 12,000 |
| Total | 22,000 |
| Citizen science | |
| App adjustment | 20,000 |
| App and website maintenance | 16,000 |
| Professional support | 50,000 |
| Total | 86,000 |

Costs are calculated in Euro for monitoring 100 km of roads over 2 years

in September and October. Grass snake is the most abundant snake species in the study region [37]. Starting in November, Grass snakes overwinter underground in areas which are not subject to freezing (e.g. compost heaps, burrows of mice) and get active again in March or April [37]. One reason for the peaks of road-killed reptiles in September and October could be that young grass snakes hatch from their nesting sites in August and sprawl into the surrounding landscape crossing our study route. Indeed, when we double-checked the photos of the road-killed reptiles it turned out that about 80% of road-killed grass snakes were juvenile. Additionally, in September and October roads are frequently used by snakes and lizards for basking during the day, but roads during these months are also more frequently used by farmers during wine harvest. Numbers of both amphibian and reptile road-kills per month suggest, that road-kill monitoring should comprise whole activity periods to get a complete overview. Here citizen science would be a suitable approach, since it would be very costly to monitor amphibians and reptiles on tertiary roads covering a wide geographic range in short time periods with a classic approach involving researchers only.

Additionally, KDE+ analyses showed significant road-kill hotspots indicating that road-kills are not randomly distributed in the landscape. This is an important information for nature conservation authorities aiming to mitigate threats for endangered amphibian and reptile species.

Conditional probabilities and general linear models applied in our study showed a positive relationship of the land cover classes Transitional woodland shrub (CLC 324) and Land principally occupied by agriculture, with significant areas of natural vegetation (CLC 243) and common toad and green toad road-kills. It was encouraging to see, that land cover classes based on a rather coarse 500-m grid matched well with the most preferred habitats of the most abundant amphibian species in the study. Green toads and common toads use water bodies only for spawning. Green toads inhabit various kinds of terrestrial sites including gravel pits, field edges, ruderal plots, dry grassland, open forests or suburban areas whereas common toads live mainly in areas with dense vegetation such as forests and scrubland areas, parks or gardens [37]. Grass snake road-kills were positive related to land cover classes Inland marshes (CLC 411), Sport and leisure facilities (CLC 142) and Complex cultivation patterns (CLC 242). Grass snakes, as the most frequently found reptile species in our study inhabit a broad range of open or semi-open habitats, including reed belts, riparian zones, forests, gardens or parks. Grass snakes can be found in the reed belt of the lake Neusiedl in high numbers matched by the CORINE land cover class Inland marshes [37]. We monitored the study route on average every 11 days and might have underestimated the number of road-kills as the persistence time of carcasses could be lower [28, 51]. Notwithstanding this limitation, our current results are in line with previous studies. This is especially important, more frequent monitorings would be difficult over a long time span with a citizen science approach. Generally, the suitability of the CORINE land cover dataset for modelling amphibian and reptile road-kills encourages us to apply our monitoring approach to a broader geographical scale.

Based on our rough cost efficiency estimation, monitoring the influence of land cover on road-kills with a citizen science approach is suitable when monitoring more than 400 km road sections. Below 400 km a conventional monitoring approach with professional researchers only seems to be more efficient. However, this calculation is just a rough estimation and should be treated with care; it is based on Austrian standard staff costs and is calculated for investigating the factor land-cover only. If other factors besides land cover are planned to be investigated, the cost efficiency could be tremendously different.

## Conclusions

Overall, our findings confirmed previous results showing that amphibian and reptile species are especially susceptible to road-kill in the vicinity of their preferred habitats [1, 52]. Nevertheless, this is interesting, as we achieved these results using freely available remote sensing data and a survey technique that could easily be adopted on a larger scale using a citizen science approach. We are confident, that the results of this pilot study can be used as basis for other citizen science projects in this field trying to enlarge their study area. A first step to the extension of our monitoring system would be to get an overview of road-killed amphibians and reptiles on a landscape scale by monitoring tertiary road networks potentially using a citizen science approach to cover this wide geographic range [35, 36, 53]. Furthermore, these data could then be used to reduce the impact of road traffic on amphibians and reptiles by installing temporal or permanent mitigation measures.

### Abbreviations

AIC: (akaike information criterion); CLC: (CORINE land cover); CORINE: (coordination of information on the environment); GLM: (general linear model); KDE+: (kernel density estimation plus); VIF: (variance inflation factors); WMTS: (web map tile service).

### Authors' contributions

FH and JGZ conceived and designed the study; FH, JGZ, KH performed the study; FH, KH and GL analysed the data; FH, JGZ and GL wrote the paper. All authors read and approved the final manuscript.


### Author details

[1] Institute of Zoology, University of Natural Resources and Life Sciences, Vienna, Gregor Mendel Straße 33, 1180 Vienna, Austria. [2] Institute of Applied Statistics and Computing, University of Natural Resources and Life Sciences, Vienna, Peter Jordan Str. 82, 1190 Vienna, Austria.



### Acknowledgements

We want to thank Philipp Hummer and Rainer Holzapfel who developed the website and apps of Project Roadkill (http://www.roadkill.at/en). Comments of two anonymous reviewers helped to improve former versions of this manuscript.


### Competing interests

The authors declare that they have no competing interests.


### Funding

This work was partly funded by the University of Natural Resources and Life Sciences Vienna, Citizen Science Network.


### References


1. Beebee TJC. Effects of Road Mortality and Mitigation Measures on Amphibian Populations. Conserv Biol. 2013. doi:10.1111/cobi.12063/abstract.
2. Rytwinski T, Fahrig L. The impacts of roads and traffic on terrestrial animal populations. Handbook road ecology. 1st ed. West Sussex: Wiley; 2015. p. 237–46.
3. Gibbons JW, Scott DE, Ryan TJ, Buhlmann KA, Tuberville TD, Metts BS, et al. The global decline of reptiles, Déjà Vu Amphibians: reptile species are declining on a global scale. Six significant threats to reptile populations are habitat loss and degradation, introduced invasive species, environmental pollution, disease, unsustainable use, and global climate change. Bioscience. 2000;50:653–66.
4. Andrews KM, Langen TA, Struijk RPJH. Reptiles: overlooked but often at risk from roads. Handbook road ecology. 1st ed. West Sussex: Wiley; 2015. p. 1–9.
5. Gollmann G. Rote Liste der in Österreich gefährdeten Lurche (Amphibia) und Kriechtiere (Reptilia). Rote Liste Gefährdeter Tiere Österr. Checklisten Gefährdungsanalysen Handl. Teil 2 Kriechtiere Lurche Fische Nachtfalter Weichtiere. Wien: Böhlau; 2007. p. 37–60.
6. Fahrig L, Rytwinski T. Effects of roads on animal abundance: an empirical review and synthesis. Ecol Soc. 2009;14(1):21. http://www.ecologyandsociety.org/vol14/iss1/art21/.
7. van der Ree R, Smith DJ, Grilo C. Handbook of road ecology. New York: Wiley; 2015.
8. Forman RTT, Sperling D, Bissonette JA. Road ecology: science and solutions. Washington, DC: Auflage: First Trade Pap; 2003.
9. Andrews KM, Nanjappa P, Riley SPD. Roads and ecological infrastructure: concepts and applications for small animals. Baltimore: Wildlife Management and Conservation; 2015.
10. Glista DJ, DeVault TL, DeWoody JA. Vertebrate road mortality predominantly impacts amphibians. Herpetol Conserv Biol. 2008;3:77–87.
11. D'Amico M, Roman J, de los Reyes L, Revilla E. Vertebrate road-kill patterns in Mediterranean habitats: who, when and where. Biol Conserv. 2015;191:234–42.
12. Teixeira FZ, Coelho IP, Esperandio IB, Rosa Oliveira N, Porto Peter F, Dornelles SS, et al. Are road-kill hotspots coincident among different vertebrate groups? Oecol Aust. 2013;17:36–47.
13. Bouchard J, Ford AT, Eigenbrod FE, Fahrig L. Behavioral responses of northern leopard frogs (*Rana pipiens*) to roads and traffic: implications for population persistence. Ecol. Soc. 2009;14:23.
14. Mazerolle MJ, Huot M, Gravel M. Behavior of amphibians on the road in response to car traffic. Herpetologica. 2005;61:380–8.
15. Rytwinski T, Fahrig L. The impacts of roads and traffic on terrestrial animal populations. In: Handbook of road ecology, 1st edn. West Sussex, UK: Wiley; 2015. p. 237–46.
16. Elzanowski A, Ciesiolkiewicz J, Kaczor M, Radwanska J, Urban R. Amphibian road mortality in Europe: a meta-analysis with new data from Poland. Eur J Wildl Res. 2009;55:33–43.
17. Kovar R, Brabec M, Vita R, Bocek R. Mortality Rate and activity patterns of an Aesculapian snake (*Zamenis longissimus*) population divided by a busy road. J. Herpetol. 2014;48:24–33.
18. Robson LE, Blouin-Demers G. Eastern Hognose snakes (*Heterodon platirhinos)* avoid crossing paved roads, but not unpaved roads. Copeia. 2013;2013:507–11.
19. Semlitsch RD. Critical elements for biologically based recovery plans of aquatic-breeding amphibians. Conserv Biol. 2002;16:619–29.
20. Marsh DM, Trenham PC. Metapopulation dynamics and amphibian conservation. Conserv Biol. 2001;15:40–9.
21. Rytwinski T, Fahrig L. Do species life history traits explain population responses to roads? A meta-analysis. Biol Conserv. 2012;147:87–98.
22. Matos C, Sillero N, Argana E. Spatial analysis of amphibian road mortality levels in northern Portugal country roads. Amphib Reptil. 2012;33:469–83.
23. Sillero N. Amphibian mortality levels on Spanish country roads: descriptive and spatial analysis. Amphib Reptil. 2008;29:337–47.
24. Marsh DM. Edge effects of gated and ungated roads on terrestrial salamanders. J Wildl Manag. 2007;71:389–94.
25. Langen TA, Ogden KM, Schwarting LL. Predicting hot spots of herpetofauna road mortality along highway networks. J Wildl Manag. 2009;73:104–14.
26. Meek R. Where do snakes cross roads? Habitat associated road crossings and mortalities in a fragmented landscape in western France. Herpetol J. 2015;25:15–9.
27. Findlay CS, Houlahan J. Anthropogenic Correlates of species richness in Southeastern Ontario Wetlands. Conserv Biol. 1997;11:1000–9.
28. Santos SM, Carvalho F, Mira A. How long do the dead survive on the road? Carcass persistence probability and implications for road-kill monitoring surveys. PLoS ONE. 2011;6:e25383.
29. Shilling FM, Perkins SE, Collinson W. Wildlife/roadkill observation and reporting systems. Handbook road ecology. 1st ed. West Sussex: Wiley; 2015. p. 492–501.
30. Costa AS, Ascensão F, Bager A. Mixed sampling protocols improve the cost-effectiveness of roadkill surveys. Biodivers Conserv. 2015;24:2953–65.
31. Vercayie D, Herremans M. Citizen science and smartphones take roadkill monitoring to the next level. Nat Conserv. 2015;11(11):29–40.
32. Shilling FM, Waetjen DP. Wildlife-vehicle collision hotspots at US highway extents: scale and data source effects. Nat. Conserv. 2015;11(11):41–60.
33. Shilling FM. Programs | Global Roadkill Network. Glob. Roadkill Netw. 2015. http://globalroadkill.net/. Accessed 9 Jan 2016.

34. Lee T, Quinn MS, Duke D. Citizen, science, highways, and wildlife: using a web-based GIS to engage citizens in collecting wildlife information. Ecol Soc. 2006;11:11.
35. Heigl F, Stretz RC, Steiner W, Suppan F, Bauer T, Laaha G, et al. comparing road-kill datasets from hunters and citizen scientists in a landscape context. Remote Sens. 2016;8:832.
36. Vercayie D, Herremans M. Citizen science and smartphones take roadkill monitoring to the next level. Nat Conserv. 2015;11:29–40.
37. Schweiger S, Grillitsch H. Die amphibien und reptilien des neusiedler see-gebiets. Naturhistorisches Museum Wien, Nationalpark Neusiedler See—Seewinkel: Wien & Illmitz; 2015.
38. European Environment Agency (EEA). Neusiedlersee - Nordöstliches Leithagebirge (AT1110137). Natura2000 standard data form for special protection areas (SPA), proposed sites for community importance (pSCI), sites of community importance (SCI) and for special areas of conservation (SAC). 2016. http://natura2000.eea.europa.eu/Natura2000/SDF.aspx?site=AT1110137. Accessed 4 Oct 2016.
39. ZAMG—Zentralanstalt für Meteorologie und Geodynamik. Klimamittel. Klimanormalperiode 1981–2010. 2012. http://www.zamg.ac.at/cms/de/klima/informationsportal-klimawandel/daten-download/klimamittel. Accessed 22 Sept 2015.
40. Heigl F, Zaller JG. Using a citizen science approach in higher education: a case study reporting roadkills in Austria. Hum Comput. 2014;1. http://hcjournal.org/ojs/index.php?journal=jhc&page=article&op=view&path%5B%5D=10. Accessed 21 Jan 2015.
41. Teixeira FZ, Pfeifer Coelho AV, Esperandio IB, Kindel A. Vertebrate road mortality estimates: effects of sampling methods and carcass removal. Biol Conserv. 2013;157:317–23.
42. CLC 2012—copernicus land monitoring services. http://land.copernicus.eu/pan-european/corine-land-cover/clc-2012. Accessed 5 July 2016.
43. Copernicus Programme. Copernicus land service—Pan-European component: CORINE land cover. 2015. http://land.copernicus.eu/user-corner/publications/clc-flyer/view. Accessed 27 May 2016.
44. Bossard M, Feranec J, Otahel J. The revised and supplemented Corine land cover nomenclature. Copenhagen: European environment agency. 2000. Report no.: 38.
45. Bíl M, Andrášik R, Svoboda T, Sedoník J. The KDE+ software: a tool for effective identification and ranking of animal-vehicle collision hotspots along networks. Landsc Ecol. 2015;31:231–7.
46. Bíl M, Andrášik R, Janoška Z. Identification of hazardous road locations of traffic accidents by means of kernel density estimation and cluster significance evaluation. Accid Anal Prev. 2013;55:265–73.
47. Fox J, The R. Commander: a basic statistics graphical user interface to R. J Stat Softw. 2005;14:1–42.
48. R Development Core Team. R: A language and environment for statistical computing. Vienna, Austria; 2008. http://www.R-project.org. Accessed 21 Sept 2016.
49. Cabela A, Grillitsch H, Tiedemann Franz. Atlas zur Verbreitung und Ökologie der Amphibien und Reptilien in Österreich. Auswertung der Herpetofaunistischen Datenbank der Herpetologischen Sammlung des Naturhistorischen Museums Wien. Wien: Umweltbundesamt. 2001.
50. Trochet A, Moulherat S, Calvez O, Stevens VM, Clobert J, Schmeller DS. A database of life-history traits of European amphibians. Biodivers Data J. 2014. http://www.ncbi.nlm.nih.gov/pmc/articles/PMC4237922/. Accessed 1 Apr 2017.
51. Ratton P, Secco H, da Rosa CA. Carcass permanency time and its implications to the roadkill data. Eur J Wildl Res. 2014;60:543–6.
52. Seo C, Thorne JH, Choi T, Kwon H, Park C-H. Disentangling roadkill: the influence of landscape and season on cumulative vertebrate mortality in South Korea. Landsc Ecol Eng. 2015;11:87–99.
53. Cosentino BJ, Marsh DM, Jones KS, Apodaca JJ, Bates C, Beach J, et al. Citizen science reveals widespread negative effects of roads on amphibian distributions. Biol Conserv. 2014;180:31–8.

# 18

# What makes a successful species? Traits facilitating survival in altered tropical forests

Mareike Hirschfeld* and Mark-Oliver Rödel

**Abstract**

**Background:** Ongoing conversion, disturbance and fragmentation of tropical forests stress this ecosystem and cause the decline or disappearance of many species. Particular traits have been identified which indicate an increasing extinction risk of a species, but traits facilitating survival in altered habitats have mostly been neglected. Here we search for traits that make a species tolerant to disturbances, thus independent of pristine forests. We identify the fauna that have an increasing effect on the ecosystem and its functioning in our human-dominated landscapes.

**Methods:** We use a unique set of published data on the occurrences of 243 frog species in pristine and altered forests throughout the tropics. We established a forest dependency index with four levels, based on these occurrence data and applied Random Forest classification and binomial Generalized Linear Models to test whether species life history traits, ecological traits or range size influence the likelihood of a species to persist in disturbed habitats.

**Results:** Our results revealed that indirect developing species exhibiting a large range size and wide elevational distribution, being independent of streams, and inhabiting the leaf litter, cope best with modifications of their natural habitats.

**Conclusion:** The traits identified in our study will likely persist in altered tropical forest systems and are comparable to those generally recognized for a low species extinction risk. Hence our findings will help to predict future frog communities in our human-dominated world.

**Keywords:** Forest degradation, Frogs, Life-history traits, Adaptation, Extinction risk, Tropics

## Background

The anthropogenic conversion of natural environments, in particular of forest habitats, is a major threat to tropical biodiversity [1]. Beside the intensive loss of forest cover [2], fragmentation of the pristine remnants further affects species [3] and limits their ability to move into adequate areas. Thus the ability to cope with altered landscapes is crucial for the persistence of a species, especially in the face of climate change.

Numerous empirical and comparative approaches on species response to environmental changes and studies relating species properties to their extinction risk were conducted on invertebrates e.g. [4–6] as well as vertebrates e.g. [7–10]. However, the general pattern which leads to the persistence of some species but the decrease or loss of other species due to forest disturbances is not fully understood. In different taxonomic groups some life-history and ecological traits show parallel patterns in their response to forest alteration, e.g. small range size [8, 10, 11] or low fecundity [12, 13] that lead to higher extinction risks. Whereas other traits, like body size exhibit a fuzzy prediction of a species' risk to decline in fragmented habitats [summary in 14]. The susceptibility of species is not determined by a single trait, but by a combination of properties which lead to a species-specific extinction risk [15–17]. So far, the majority of studies have focused on species affected by environmental changes and filter for traits increasing the extinction risk. Species not responding to habitat alterations and the traits required for their persistence in disturbed landscapes are frequently neglected. However, those species remaining are of high interest as they will make up the

*Correspondence: mail@mareikepetersen.de
Department Diversity Dynamics, Museum für Naturkunde Berlin-Leibniz Institute for Evolutionary and Biodiversity Science, Invalidenstraße 43, 10115 Berlin, Germany

majority of the fauna in our human-dominated world and thus have an increasing effect on ecosystems and their functioning [18, 19].

Frogs are strongly influenced by their environment and the degradation and conversion of natural forests is one major cause for their current global decline [20–22]. However, not all species are affected by degradation or fragmentation [23–25] and a set of life-history or ecological traits is assumed to reduce their susceptibility [8, 26].

In this study, we search for factors allowing a species to be independent of pristine areas and thus permitting their occurrence in degraded and disturbed forests, which are the dominant tropical habitats now and in future [27]. We use a unique data set comprising published records on frog species occurrences in tropical forests, forest fragments and more intense altered landscapes such as plantations or settlements. For these species we gathered life-history (e.g. body size, clutch size) and ecological traits (e.g. habitat use) as well as distribution data, which are known to affect the susceptibility of species in general [8, 26, 28] and thus might likewise influence a species response to forest degradation. We ask whether these candidate traits could predict the forest dependency of tropical frog species and whether a particular set of traits makes species less vulnerable to changes in their natural habitat and decreases their risk of extinction.

## Methods

### Data acquisition

We combined a comprehensive data set on anuran occurrences across tropical forests and human altered forest habitats with detailed information on species traits. To cover all research published on anuran distribution in pristine versus altered environments in the tropics, we did a comprehensive literature research using Google, Google Scholar, Web of Science and data bases included therein (January to August 2013). Queries using different combinations of appropriate keywords (e.g. frog, amphibian, anuran, disturbance, alteration, fragmentation, logging etc.) were applied to all data bases. Appropriate data sets covered a description of the study sites and information on the presence (and absence) of each species in the different habitat types. In addition to already published studies we added our own data on anuran occurrences from the forest zone of Cameroon (M. Hirschfeld et al. unpublished data). The survey amounted to 61 studies (see Additional file 1: anuran distribution references) covering all continents that include a tropical climate: Africa, Asia, Central- and South-America, and Australia with a total of more than 750 different anuran taxa. For our analysis we only included records with species level identifications. Species names were checked and updated if necessary according to Frost [30]. If a taxonomic name could not be unambiguously assigned to a valid species, i.e. due to cryptic species complexes, the record was not included. This resulted in a data set with 672 species.

For each valid species from the occurrence data set, its life-history and ecological traits (hereafter referred to as traits) were gathered using published literature, suitable data bases reviewed by specialists, and further web resources (see Additional file 2: anuran traits references). Additionally we included our own unpublished data, collected either in the field or from museum specimens (Museum für Naturkunde Berlin, e.g. body size, ripe eggs in female ovaries). Traits collected and used in the analysis comprised information on species distribution, morphology, biology, and ecology. We also noted the geographic (i.e. continent) and phylogenetic (family) origin of each species (see Table 1 for details). As we only considered species for our analysis where at least information on body size (either male or female) was available, the data set was reduced further to 619 species.

### Data preparation

Some of the collected trait data required processing for subsequent analysis. We used the elevational range calculated as the difference from the maximum to the minimum elevation where a species is known to occur. Regarding body size, we used the maximum body length known per species and sex or, if not available, mean values plus standard deviation. Only if maximum and/or standard deviation were not available, mean or single values were used. We supplemented the data set with sexual dimorphism, calculated as male divided by female body size. Clutch size was only available for a subset of species (345). The available data on clutch sizes were grouped objectively into ten size classes (A: 4–98, B: 100–265, C: 290–549, D: 563–905, E: 979–1652, F: 1900–3320, G: 3607–6701, H: 8357–12940, I: 17000–25000, J: 36100–40000) and species without information on clutch size were subsequently assigned to a class based on body size (see Additional file 3: clutch size classes for more details).

Studies included in our analyses focused on the comparison of anuran distribution among various landscapes. Hence, broader habitat categories were necessary to combine the results within one analysis. Based on all information available we chose three major habitat categories along a human altered degradation gradient: forest, secondary growth, and non-forest. The habitat category "forest" comprises primary forests, primary forest fragments, and selectively logged or exploited areas; "secondary growth" subsumes secondary forests, edges of primary forests, abandoned plantations (>5 years) and agricultural habitats with remaining forests (e.g. shaded coffee plantations); non-forests comprise simple structured plantations (single strata), pasture or inhabited areas such as villages.

**Table 1 Life-history and ecological traits used in the study**

| Trait | Definition | Scale | Unit/level |
|---|---|---|---|
| Range size[a] | Natural area of occurrence | Ratio | $km^2$ |
| Elevation | Min. and max. elevation in the entire area of occurrence | Ratio | m asl |
| SVL male/female | Body length, measured as snout vent length | Ratio | mm |
| Dimorphism | Calculated as male divided by female body size | Ratio | Proportion |
| Clutch size | Maximal number of total eggs deposited or maximal number of ripe eggs in the uteri of dissected females | Ratio | # |
| Clutch size class | Clutch sizes assigned to size classes | Ordinal | Ten size classes, see "Methods" for details |
| Reproduction | Development | Nominal | Direct, indirect |
| Adult habitat | Habitat where adults are usually encountered, perch height | Nominal | Aquatic, semi-aquatic, fossorial, litter (<1 m), semi-arboreal (1–3 m), arboreal (>3 m) |
| Larval habitat | Habitat where the larvae develop | Nominal | None (direct development), terrestrial, semi-aquatic, lentic, lentic and lotic, lotic, phytotelmata (plant associated water bodies, e.g. tree holes, bromeliad tank), skin[b] |
| Egg deposition | Habitat where the eggs are deposited | Nominal | Terrestrial, semi-terrestrial, aquatic, arboreal, skin[b] |
| Family | Taxonomic origin, affiliation to family | Nominal | Anuran families according to Frost [30] |
| Region of origin | Broad geographic region (i.e. continent) | Nominal | |

Given is the trait, its definition, the scale of measurement, and the unit (ratio) or levels (nominal, ordinal) of the respective trait

[a] Range size according to the IUCN Red List [29] or, if not available, for West African species to the calculated environmental niche model [70]

[b] Carried in or on adult male or female

Categorization was realized in accordance with comparative studies [4, 31, 32]. However, in consideration of the modified forest types examined in our data set, slight adaptations and a reduction of categories were necessary. We only took species into account which had information on the presence and absences in these major habitat categories. If a species was detected in several studies, its single occurrence per habitat category (although absent in other studies) was crucial to assign the species to that habitat type. Combining this reduced data set with the available trait data, the final data set amounted to 243 different species with only a few gaps for some traits. As multivariate statistics often require complete data sets, missing values in the trait data set were replaced by dummy variables. This prevents a high loss of information by excluding a trait or a species. For traits with a ratio scale we used the mean, and for traits with a nominal scale the level which occurred most often (compare Table 1). Numbers of required dummy variables in the final data set: range size = 2 (mean = 1,795,153 $km^2$), elevational range: 35 (1217.6 m), snout-vent length (SVL) males: 4 (10.5 mm), SVL females: 21 (18 mm), reproductive mode: 1 (most frequent: indirect development), adult habitat: 2 (litter); larval habitat: 11 (lentic), egg deposition site: 34 (aquatic). All analyses were conducted with the completed data set (see Additional file 4).

Based on the species occurrences in the three major habitat categories, a forest dependency index (FDI) with four levels was established (Fig. 1): dependent species solely detected in forests (D), slightly dependent species occurring in forests and habitats with secondary growth (SD), forest independent species occurring not in primary forests, i.e. only in habitats with secondary growth and/or non-forested habitats (I), and species with no response occurring in all three habitat categories or forest and non-forest habitats (NR).

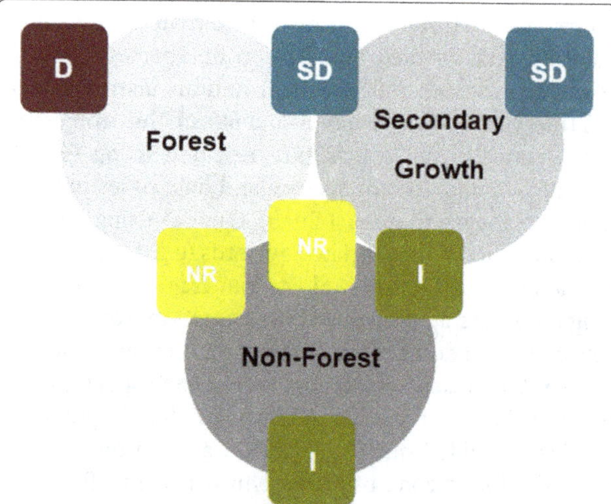


**Fig. 1** Forest dependency index. A forest dependency index (FDI) was established based on species occurrences in three major habitat categories (forest, secondary growth, non-forest); FDI: *D* dependent, solely detected in forests, *SD* slightly dependent, species occurring in forests and habitats with secondary growth, *I* forest independent species, occuring not in primary forests, i.e. only in habitats with secondary growth and/or non-forested habitats, *NR* species with no response, occurring in all three habitat categories or forest and non-forest habitats

### Statistical analysis

The distribution and trait data (ratio scale) were non-normal distributed (Shapiro–Wilk test, R package 'stats'). We thus applied the non-parametric Kruskal–Wallis test and subsequent pairwise Wilcoxon tests with false discovery rate (*fdr*) correction for parameter comparison among species with different forest dependency indices (R package 'stats'). To filter for species traits explaining the presence or absence of a species in differently degraded habitats and thus their assignment to a particular FDI we performed a Random Forest (RF) classification [33] where 1000 classification trees on bootstrap samples of the data were grown (*randomForest*, R package 'randomForest' [34]). The number of candidate variables at each node (*mtry*) was the square root of the total number of variables in the analysis (default setting). To correct for different sample sizes in the training data set, *sampsize* was adjusted according to the minimum sample size per analysis. RF was performed for the whole data set and four subsets, three comparing forest dependent species (D) with one of the other FDIs and a comparison of the groups NR and I. We incorporated all available information for a species in RF, including species distribution (range size, elevation range, region of origin) and seven traits (see Table 1). As families were evenly distributed among the different FDIs (see Additional file 5), the affiliation to a family was excluded from the analysis. Binomial Generalized Linear Models (GLM) were performed to filter for potential traits explaining the habitat dependency of a species (*glm*, R package 'stats'). Therefore, species not responding to habitat changes (NR) were defined as '0' and compared to forest dependent species (D) as well as forest independent species (I), both defined as '1'. Numerical variables (body size and sexual dimorphism) were scaled from 0 to 1. To avoid multi-co-linearity among explaining variables within one model, generalized variance inflation factors (GVIF) were calculated (*vif*, R package 'car'). Each model contained the covariates: SVL females, sexual dimorphism, clutch size class, larval habitat, adult habitat, reproductive mode, and egg deposition site. After reducing the co-linearity among the explaining variables and eliminating those with a GVIF higher than five [35], the full model only contained: SVL females, sexual dimorphism, clutch size class, and larval habitat. To test for any influence on the forest dependency of species distribution we fitted Generalized Linear Mixed Effect Models (GLMM) with range size and elevational range (both scaled from 0 to 1) as fixed and the region of origin as random factor (*glme*, R package 'lme4'). Here, a reduction of covariates due to co-linearity was not necessary. Based on the models, we predicted whether a species is either dependent on forest (non-forest) or occurs in all available habitats ($\geq$0.5 for forest, D or non-forest, I; <0.5 for habitat independent species, NR). All statistical analysis were applied using R 3.2.1 [36].

## Results

### Taxonomy

The 243 anuran species included in the analysis belonged to 26 different families. The most common families were Rhacophoridae and Hylidae, the latter representing 10–30% of the species in all forest dependency indices (FDIs). The families were equally distributed among the different FIDs (see Additional file 5), ruling out any phylogenetic influence in the data.

### Species distribution

Range sizes ranged from 6.17 to 12,217,676 km$^2$ and varied highly within each FDI (Table 1 for information on gathered traits; see Table 2; Fig. 2 for results). It differed significantly between forest dependent species (D) and species not responding to habitat alteration (NR) as well as between species slightly depending on forests (SD) and NR. All FDIs covered species with limited and wide altitudinal distribution (see Fig. 2). NR species had the broadest distribution and differed significantly from the others (see Table 2). Species in the final data set originated from Africa, Madagascar, America and Asia. The indices NR, I and SD comprised species from all four regions, only D was lacking Malagasy species (Fig. 3). The region of origin did not differ significantly between the FDIs (Pearson's $\chi^2$ test: $\chi^2 = 5.89$, df = 3, p = 0.12).

### Habitat

Overall, most species preferred litter as well as shrubs and lower tree strata (1–3 m) as adult habitat (Fig. 3). Almost 75% of the species belonging to D and I live in trees (categories semi-arboreal and arboreal); SD and NR species were mostly found on the ground. Aquatic habitats were not inhabited by SD species, while the other FDIs covered all types. The habitat use differed slightly among the FDIs (Pearson's $\chi^2$ test: $\chi^2 = 27.28$, df = 15, p = 0.03). Lentic waters constitute 35–60% of the tadpoles' habitat per FDI (Fig. 3). Lotic waters were of high importance in SD species, but less in other FDIs. All other categories were only sparsely presented, apart from no larval habitat, representing direct developing species. The larval habitat differed significantly between species assigned to different FDIs ($\chi^2 = 45.23$, df = 21, p = 0.002).

### Body size

Maximum body sizes ranged from 10 to 187 mm for males and from 18 to 287 mm for females, respectively, with a high variation for both sexes within each FDI (see

**Table 2 Distribution pattern and life history traits**

| Trait | General n = 243 | | D n = 33 | | SD n = 108 | | NR n = 83 | | I n = 19 | |
|---|---|---|---|---|---|---|---|---|---|---|
| | Mean ± SD | Range | Mean ± SD | Range | Mean ± SD | Range | Mean ± SD | Range | Mean ± SD | Range |
| a | | | | | | | | | | |
| Range size (km$^2$)[a] | 1,795,153 ± 2,784,023 | 6.17–12,217,676 | 1,511,311 ± 2,996,433 | 14.67–12,217,676 | 1,086,942 ± 2,069,589 | 6.17–10,932,823 | 2,850,720 ± 3,163,102 | 21.62–11,045,631 | 1,702,603 ± 2,983,042 | 305.09–10,419,167 |
| Elevational range (m)[a] | 1217.60 ± 591.78 | 1–3100 | 994.04 ± 582.89 | 72–3000 | 1123.0 ± 557.65 | 1–2500 | 1446.33 ± 561.84 | 400–3002 | 1144.46 ± 652.31 | 20–3100 |
| SVL males (mm)[a] | 47.13 ± 30.25 | 10.5–187 | 45.83 ± 26.96 | 18.4–146 | 45.90 ± 31.85 | 10.5–180 | 50.34 ± 31.29 | 17.0–187 | 42.44 ± 21.00 | 20.0–81 |
| SVL females (mm)[a] | 56.37 ± 34.04 | 18–287 | 63.06 ± 40.50 | 24–228.9 | 50.80 ± 26.95 | 18–185.0 | 61.99 ± 40.10 | 18–287.0 | 51.90 ± 24.22 | 23–94.0 |
| Sexual dimorphism[a] | 0.86 ± 0.26 | 0.09–3.19 | 0.79 ± 0.29 | 0.09–2.03 | 0.90 ± 0.31 | 0.25–3.19 | 0.83 ± 0.18 | 0.49–1.99 | 0.84 ± 0.19 | 0.41–1.29 |
| Clutch size[b] | 1609.62 ± 5186.19 | 4–40,000 | 1296.36 ± 3701.47 | 10–17,000 | 748.99 ± 1158.77 | 6–5018 | 2578.56 ± 7423.93 | 7–40,000 | 584.08 ± 823.73 | 4–2500 |

| Trait | Kruskal–Wallis test | | | Pairwise Wilcox test (p) | | | | | |
|---|---|---|---|---|---|---|---|---|---|
| | $\chi^2$ | df | p | D vs SD | D vs. NR | D vs I | SD vs. NR | SD vs. I | NR vs. I |
| b | | | | | | | | | |
| Range size (km$^2$)[a] | 19.72 | 3 | <0.001 | 0.29 | <0.01 | 0.32 | <0.01 | 0.50 | 0.07 |
| Elevational range (m)[a] | 21.7 | 3 | <0.0001 | 0.26 | <0.001 | 0.46 | <0.001 | 0.82 | <0.05 |
| SVL males (mm)[a] | 2.67 | 3 | 0.45 | – | – | – | – | – | – |
| SVL females (mm)[a] | 5.72 | 3 | 0.13 | – | – | – | – | – | – |
| Sexual dimorphism[a] | 8.92 | 3 | <0.05 | 0.05 | 0.38 | 0.38 | 0.11 | 0.46 | 0.71 |
| Clutch size[b] | 6.92 | 3 | 0.07 | – | – | – | – | – | – |

Given are the respective mean, standard deviation (sd), and range in general, and for each dependency index separately (a) and comparisons of traits between species of different forest dependency indices using the Kruskal–Wallis test and a pairwise Wilcox test with fdr correction as posthoc (b); forest dependency index: D = dependent (n = 33), SD = slightly dependent (n = 108), NR = non-responding (n = 83), I = forest independent (n = 19)

[a] Incorporate calculated dummy variables (see "Methods")

[b] Only measured values and therewith differing sample sizes: general = 152, D = 22, SD = 52, NR = 66, I = 12; compare Figs. 2 and 3

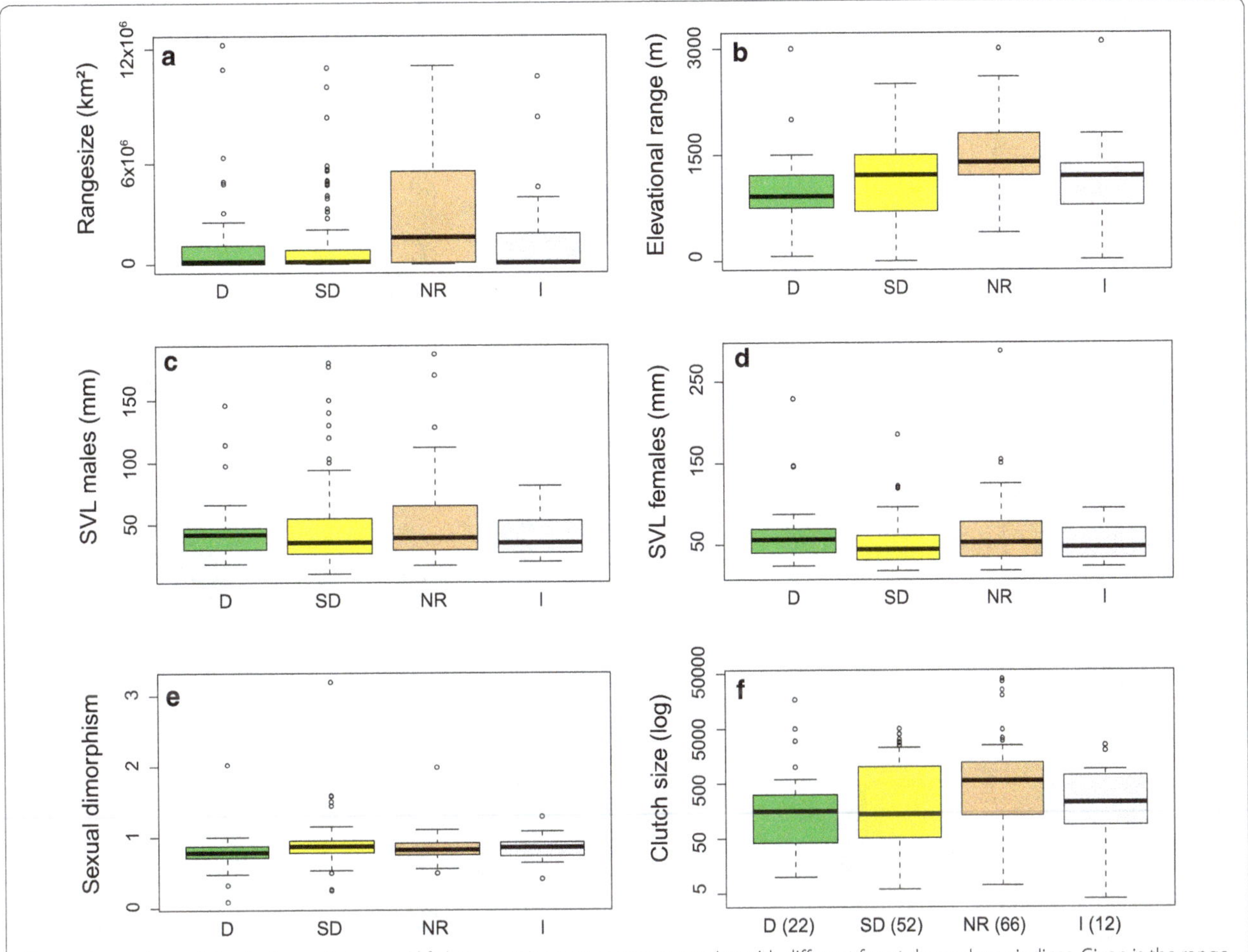

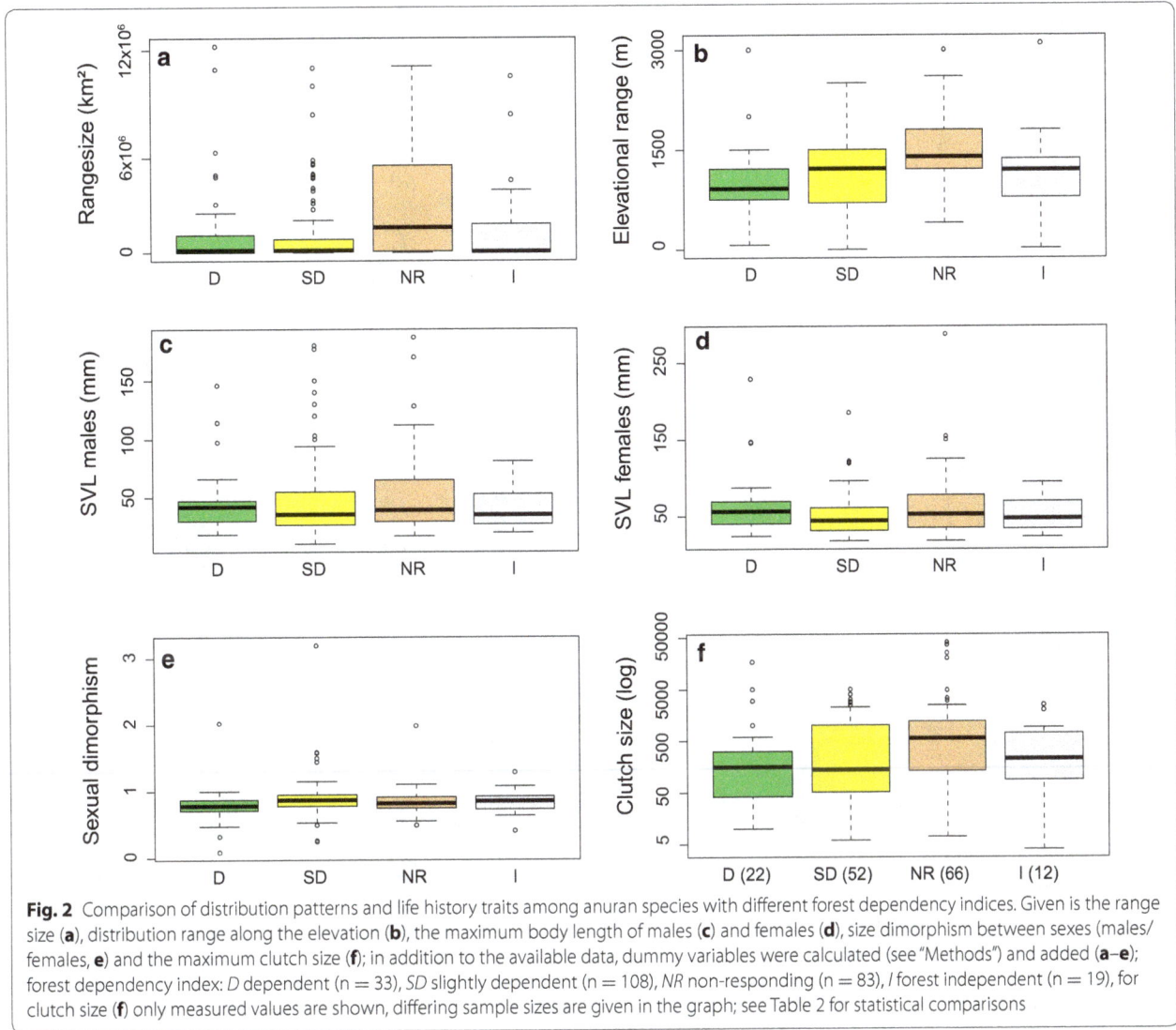

**Fig. 2** Comparison of distribution patterns and life history traits among anuran species with different forest dependency indices. Given is the range size (**a**), distribution range along the elevation (**b**), the maximum body length of males (**c**) and females (**d**), size dimorphism between sexes (males/females, **e**) and the maximum clutch size (**f**); in addition to the available data, dummy variables were calculated (see "Methods") and added (**a–e**); forest dependency index: *D* dependent (n = 33), *SD* slightly dependent (n = 108), *NR* non-responding (n = 83), *I* forest independent (n = 19), for clutch size (**f**) only measured values are shown, differing sample sizes are given in the graph; see Table 2 for statistical comparisons

Table 2; Fig. 2). It did not differ between the FDIs (see Table 2). Sexual dimorphism also did not show large differences between the indices, but the comparison between D and SD species showed a trend towards D hosting species with greater dimorphism. As female and male body size were highly correlated (Spearman Rank Correlation: $\rho = 0.88$, $p < 0.0001$, $n = 243$), we only used female body size and dimorphism in subsequent analysis.

### Reproduction

Clutch size varied between 4 and 40,000 eggs and did not differ between the FDIs (see Table 2; Fig. 2). Independent of the FDI, most clutches were in the first two size classes (4–98 and 100–265 eggs). Species belonging to I did not have clutches greater than 6700 eggs. The clutch size measured in categories likewise did not differ significantly between the FDIs (Pearson's $\chi^2$ test: $\chi^2 = 34.96$, df = 27, p = 0.14). Most species deposited their eggs in aquatic habitats (see Fig. 3). The second most common habitat was terrestrial, followed by arboreal deposition sites. There were no significant differences in egg deposition site between the FDIs ($\chi^2 = 11.52$, df = 12, p = 0.48). Almost 80% of the investigated species showed a biphasic development with free swimming tadpoles (see Fig. 3); D species had the highest proportion of direct developers (>30%). The reproductive mode did not differ significantly between the FIDs ($\chi^2 = 5.89$, df = 3, p = 0.12).

Classification by RF on the whole data set resulted in an overall error rate of 50.2%, the misclassification per FDI varied between 42.2 and 94.4% (see Table 3). Classification of subsets performed better, with an overall error rate of 20.7% (D vs. NR), 22.0% (D vs. I), 30.4% (NR vs. I), and 40.4% (D vs. I). Range size was important in

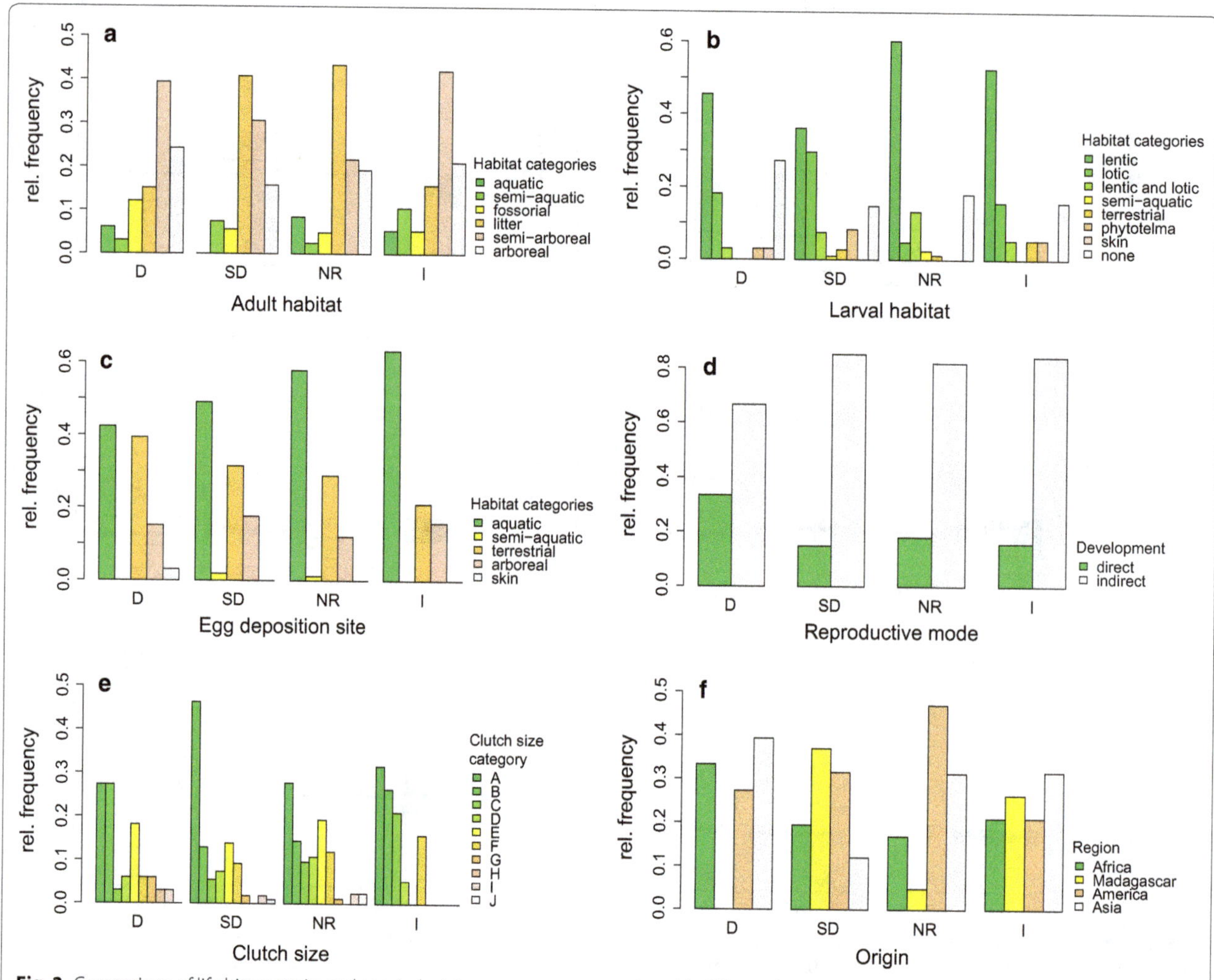

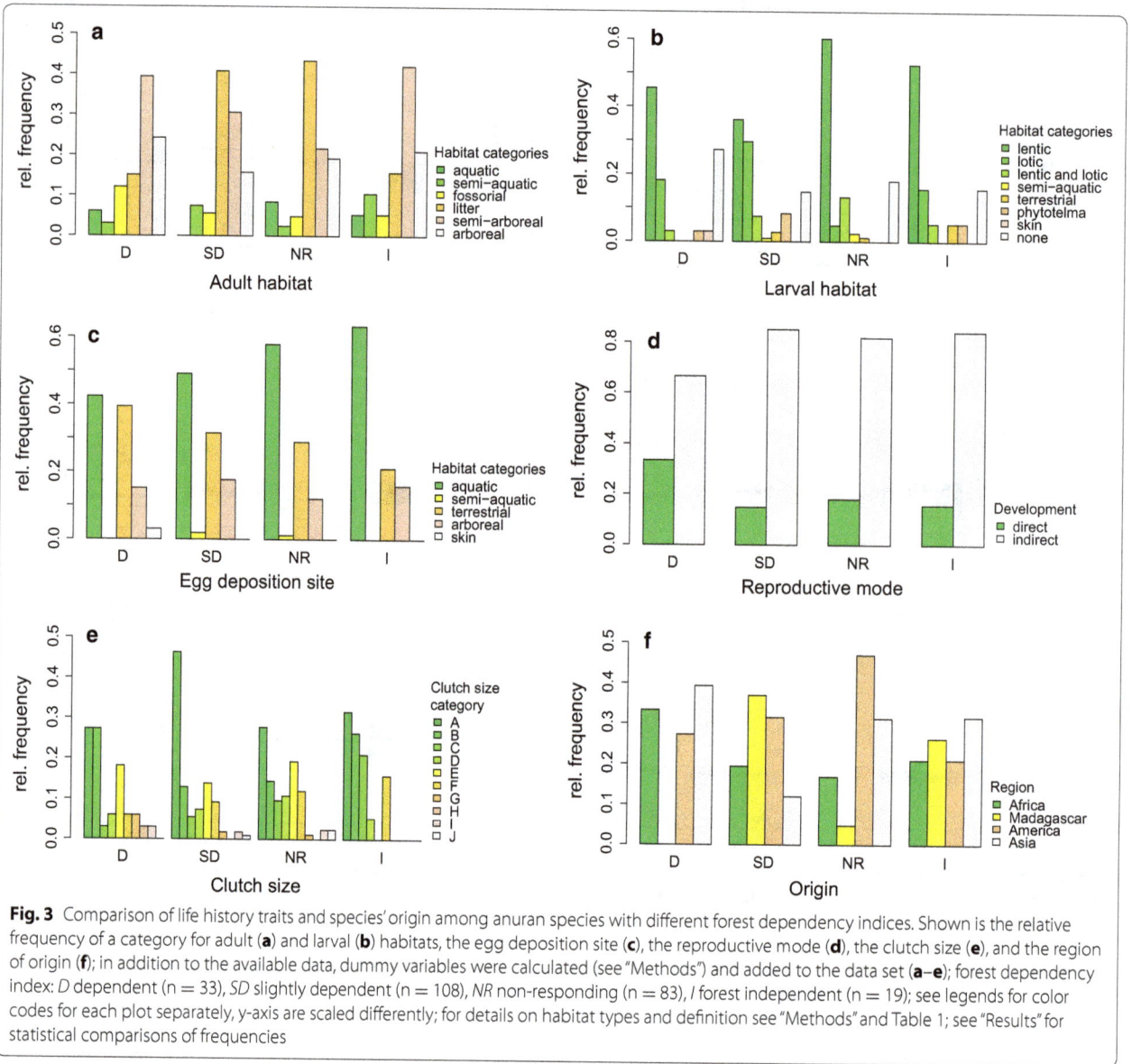

**Fig. 3** Comparison of life history traits and species' origin among anuran species with different forest dependency indices. Shown is the relative frequency of a category for adult (**a**) and larval (**b**) habitats, the egg deposition site (**c**), the reproductive mode (**d**), the clutch size (**e**), and the region of origin (**f**); in addition to the available data, dummy variables were calculated (see "Methods") and added to the data set (**a**–**e**); forest dependency index: *D* dependent (n = 33), *SD* slightly dependent (n = 108), *NR* non-responding (n = 83), *I* forest independent (n = 19); see legends for color codes for each plot separately, y-axis are scaled differently; for details on habitat types and definition see "Methods" and Table 1; see "Results" for statistical comparisons of frequencies

all, sexual dimorphism and elevational range in four, and clutch size category in three models (Table 4).

Generalized linear models (Table 5) revealed larval habitat and clutch size class as important factors explaining the dependency to forests (D vs. NR species) with the development in lotic waters being significant and clutches of class G (3607–6701 eggs) being almost significant. Based on this model, 77% of all species could be correctly assigned to the original FDIs (matches: D: 12, n = 33; NR: 77, n = 83; sample size from original data). The model for forest independent species (I vs. NR species) revealed likewise larval habitat as being important with development in lotic waters being significant. The model assigned 82% of all the species correctly to the FDI derived from field observation (I: 4, n = 19; NR: 81, n = 83). Generalized Linear Mixed Models (Table 5) fitted with species distribution revealed elevational range as being important factors when comparing both, D and NR as well as I and NR species (the latter barely non-significant). The model contrasting D and NR species assigned 78% of the species to correct FDIs (D = 13; NR = 77), based on the model comparing I and NR species 71% were correctly classified compared to the original FDIs (I = 0; NR = 83). Results of the different approaches confirm each other at least partly: RF vs. GLM: forest dependent species (D): classification overlap of 72%, matches: D: 9; NR: 74; forest independent species (I): 75%, I = 5; NR: 71; RF vs. GLMM: D: 84%, D = 16, NR = 81; I: 71%, I = 0, NR = 72; GLM vs. GLMM: D: 77%, D = 5, NR = 84; I: 94%, I = 0; NR = 96.

**Table 3 Confusion matrices of Random Forest analysis**

| | D | I | NR | SD | CE (%) | OE (%) |
|---|---|---|---|---|---|---|
| Complete data set | | | | | | 50.2 |
| D | 15 | 4 | 9 | 5 | 54.5 | |
| I | 5 | 1 | 7 | 6 | 94.4 | |
| NR | 12 | 6 | 48 | 17 | 42.2 | |
| SD | 9 | 12 | 29 | 57 | 47.2 | |
| Subset D vs. SD | | | | | | 22.0 |
| D | 17 | – | – | 16 | 48.5 | |
| SD | 15 | – | – | 93 | 12.0 | |
| Subset D vs. NR | | | | | | 20.7 |
| D | 21 | – | 12 | – | 36.4 | |
| NR | 12 | – | 71 | – | 14.5 | |
| Subset D vs. I | | | | | | 40.4 |
| D | 23 | 10 | – | – | 30.3 | |
| I | 11 | 8 | – | – | 57.9 | |
| Subset I vs. NR | | | | | | 30.4 |
| I | – | 9 | 10 | – | 52.6 | |
| NR | – | 21 | 62 | – | 25.3 | |

Confusion matrices with per class error (CE) rate and overall error (OE) rate per Random Forest analysis (complete data set and different subsets); analysis were performed with *ntree* = 1000, *mtry* = 3 and *sampsize* adjusted to the smallest sample size (R package 'randomForest'); forest dependency index: *D* dependent (n = 33), *SD* slightly dependent (n = 108), *NR* non-responding (n = 83), *I* forest independent (n = 19)

**Table 4 Importance of each variable in Random Forest analysis**

| Variable | Complete | D vs. SD | D vs. NR | D vs. I | I vs. NR |
|---|---|---|---|---|---|
| SVL females | *6.73* | 3.41 | *3.42* | 2.15 | 2.08 |
| Adult habitat | 4.70 | 3.23 | *3.64* | 0.91 | 2.37 |
| Sexual dimorphism | *9.02* | *4.00* | 2.98 | *2.63* | *2.59* |
| Clutch size category | 6.15 | *3.49* | 3.39 | *4.20* | *2.63* |
| Egg deposition site | 1.13 | 1.04 | 0.46 | 0.50 | 0.48 |
| Reproductive mode | 0.14 | 0.47 | 0.07 | 0.10 | 0.14 |
| Larval habitat | 3.46 | 2.63 | 2.10 | 0.89 | 1.12 |
| Region | 4.54 | *4.95* | 2.62 | 2.11 | 1.58 |
| Range size | *12.52* | *5.91* | *7.52* | *3.27* | *3.06* |
| Elevational range | *8.59* | 3.37 | *6.81* | *2.23* | *2.87* |

Importance of each variable per Random Forest analysis (complete data set and different subsets); the four most important variables contributing to the classification are in italics; analysis were performed with *ntree* = 1000, *mtry* = 3 and *sampsize* adjusted based on the smallest sample size for each analysis respectively (R package 'randomForest'); forest dependency index: *D* dependent (n = 33), *SD* slightly dependent (n = 108), *NR* non-responding (n = 83), *I* forest independent (n = 19)

## Discussion

Geographic range size has been identified as a vital factor predicting a species' susceptibility and extinction risk, including birds [11], mammals [10, 16], and amphibians [8, 37]. Species tolerating a wide range of abiotic factors, different habitats [38], or not responding to forest degradation (this study) likewise have the widest distribution. Here, we assign species to one of four levels of forest dependency, according to their occurrence in habitats with differently strong disturbance. Species belonging to D (forest dependent) depend on pristine forests, species assigned to the other categories (NR, I, or SD) can cope with habitat disturbances to different extents. We determined the most important traits explaining the forest dependency of a species using RF classification, GLM, and GLMM techniques. Since range size and extinction risk or habitat breadth might directly depend on each other, making it a single criterion to assess species as critically endangered in the IUCN Red List [29], we excluded it in the GLM filtering for species traits, but analyzed it separately (GLMM). Here, however, only elevational range was important for distinguishing NR from D and NR from I species. This is consistent with previous results where a wide altitudinal distribution decreases a species' vulnerability [9, 39, 40], as such species are naturally adapted to varying environmental factors (e.g. vegetation, climate) and hence might also cope better with changes of these factors caused by forest disturbances.

Body size is a central trait, usually correlated with factors such as population size, range size, clutch size or rate of exploitation, all influencing the extinction risk of a species [41–43]. It was thus typically taken into consideration when estimating a species' susceptibility. With increasing body size, studies revealed an increase (amphibians: [39], mammals: [41], birds: [44]), or, as in our data, no change in the extinction risk (amphibians: [28], birds: [40], bats: [45]). These converse results

**Table 5 Effects of species traits and distribution on habitat dependency**

| | Forest dependent species (D vs. NR) | | | | Non-forest species (I vs. NR) | | | |
|---|---|---|---|---|---|---|---|---|
| | Estimate | Std. error | z | p | Estimate | Std. error | z | p |
| GLM on species traits | | | | | | | | |
| Intercept | 0.085 | 1.47 | 0.058 | 0.95 | −2.98 | 1.37 | −2.18 | *0.03* |
| SVL females | −0.08 | 2.70 | −0.03 | 0.98 | 0.05 | 4.59 | 0.01 | 0.99 |
| Sexual dimorphism | −3.73 | 3.11 | −1.20 | 0.23 | 2.71 | 2.83 | 0.96 | 0.34 |
| Larval habitat | | | | | | | | |
| Lentic/lotic | −17.13 | 1852.28 | −0.009 | 0.99 | −1.13 | 1.30 | −0.87 | 0.38 |
| Lotic | 1.66 | 0.77 | 2.16 | *0.03* | 3.00 | 1.34 | 2.23 | *0.03* |
| None | 0.39 | 0.64 | 0.61 | 0.54 | 0.26 | 0.91 | 0.28 | 0.78 |
| Phytotelma | 18.95 | 6522.64 | 0.003 | 0.99 | 22.54 | 10,750 | 0.002 | 0.99 |
| Semi-terrestrial | −17.22 | 4611.48 | −0.004 | 0.99 | −16.31 | 6635 | −0.002 | 0.99 |
| Skin | 19.79 | 6522.64 | 0.003 | 0.99 | – | – | – | – |
| Terrestrial | −0.08 | 2.70 | −0.03 | 0.98 | 2.50 | 1.67 | 1.49 | 0.14 |
| Clutch size class | | | | | | | | |
| B | 0.67 | 0.67 | 1.01 | 0.31 | 1.14 | 0.86 | 1.34 | 0.18 |
| C | −0.57 | 1.22 | −0.47 | 0.64 | 1.73 | 1.07 | 1.61 | 0.11 |
| D | −0.19 | 1.02 | −0.19 | 0.85 | 0.20 | 1.36 | 0.15 | 0.88 |
| E | −0.02 | 0.79 | −0.03 | 0.98 | −18.48 | 2343 | −0.01 | 0.99 |
| F | −0.10 | 1.10 | −0.09 | 0.93 | 0.85 | 1.32 | 0.64 | 0.52 |
| G | 3.16 | 1.86 | 1.70 | 0.09 | −17.69 | 10,750 | −0.002 | 0.99 |
| H | 37.40 | 6780.54 | 0.01 | 0.99 | | | | |
| I | 0.59 | 1.57 | 0.38 | 0.71 | −17.37 | 7585 | −0.002 | 0.99 |
| J | −16.74 | 3995.08 | −0.004 | 0.99 | −16.46 | 7482 | −0.002 | 0.99 |
| GLMM on species distribution | | | | | | | | |
| Fixed effects | | | | | | | | |
| Intercept | 1.23 | 0.68 | 1.80 | 0.07 | −0.004 | 0.69 | −0.006 | 0.99 |
| Range size | −1.59 | 1.13 | −1.41 | 0.16 | −1.18 | 1.38 | −0.86 | 0.39 |
| Elevational range | −4.97 | 1.45 | −3.44 | *0.0006* | −3.03 | 1.56 | −1.85 | 0.06 |

Binomial models for forest dependent and non-forest species were conducted and full models (*glm*, R package 'stats'; *glme*, R packages 'lme4') after eliminating multicollinearity (*vif*, R package 'car') are presented; Generalized Linear Model (GLM): variables included: SVL females, sexual dimorphism, clutch size class, larval habitat; removed due to co-linearity: adult habitat, reproductive mode, and egg deposition site; Generalized Linear Mixed Model (GLMM): range size and elevation range as fixed and region of continent random factors (no co-linearity among explaining variables); significant effects are in italics; *D* forest dependent species (n = 33), *NR* non-responding species (n = 83), *I* forest independent species (n = 19)

emphasize the complex effects of body size and explain the variation in its influence on the vulnerability of species, differing with study systems [14] but also with the source of extinction risk [46]. According to our results, neither body size nor sexual size dimorphism seem to influence forest dependency.

Although the number of offspring explains the extinction risk in several taxa [12, 13], traits related to reproduction only had minor effects on degradation susceptibility of a frog species in our data set. Species belonging to I, however, do not deposit bigger clutches (separating I from NR species in RF). This could either be related to the larger number of I species using flowing, not stagnant, waters as larval habitat and the fact that stream breeders tend to have bigger eggs and thus smaller clutches [47], or to the absence of bigger females, depositing larger clutches (see Figure in Additional file 3) in I. A higher percentage of direct developers among forests dependent species (this study, but see [48, 49]) and an increased extinction risk of ovoviviparous anuran species in general ([8], but see [50]) can be explained by the required moist microhabitat for a direct development [51], available in pristine forests, but not necessarily in degraded or fragmented habitats [52, 53].

A species' microhabitat preferences affect its vulnerability, i.e. the availability of breeding sites, particular soil conditions or vegetation structure can be crucial for the presence of an amphibian species [e.g. 49, 54, 55]. Modified forests are accompanied by an open canopy which facilitates the growth of herbaceous strata and leads to an advantageous humid microclimate for some leaf-litter anurans. This structured understory, including downed woody debris, has been identified as an important habitat feature for amphibian populations in altered forests [56, 57] and explains the increase of ground dwelling species among degradation tolerant species [25, this study]. Degradation with an accompanying loss of canopy cover generates the most prominent microclimatic shifts in the mid-story, forming the upper strata after disturbances. The resulting decreased humidity, stronger temperature extremes, and increased solar radiation [58–60] have adverse effects on amphibians and explain the high number of semi-arboreal species in our study being forest dependent and the low number being degradation tolerant.

Forest degradation negatively impacts riparian habitats for amphibians by decreasing the amount of woody debris or leaf litter, resulting in less dissolved organic carbon [61] and by a reduction of the canopy cover, leading to higher temperatures and solar radiation [62, 63]. These unfavorable changes explain the higher number of stream breeders among species prone to degradation (this study) and the higher susceptibility of species dependent on lotic breeding sites [54] and riparian species in general [39, 50]. Although forest degradation potentially cause similar changes in lentic habitats, pond breeding amphibians might be less vulnerable or, due to different life-history strategies, even benefit from the consequences: higher temperatures for example increase the developmental rate [64, 65] and higher solar radiation favors the growth of algae [62], the primary food resource for many pond dwelling tadpoles. Compared to species not responding to habitat changes, also a higher number of non-forest species strongly depend on rivers for their tadpole development. These species might be already accustomed to open riparian habitats and thus do not suffer from the prevailing conditions like species occurring in all habitat types.

When contrasting the classification of RF, GLM, and GLMM based on the comparisons D vs. NR, ten species were always wrongly assigned. For example two species, known to occur in strongly degraded habitats [66] and to reproduce in artificial ponds [67] were assigned to D but predicted to belong to NR. Hence the models predicted the species correctly and only the incorporated information from the field was limited and did not cover the occurrences in altered habitats.

## Conclusions

Generalist species were identified as the winners in human-dominated landscapes [18, 68], but particular traits facilitating this adaptation were not yet determined. Our pan-tropical approach revealed that the dependency to forested habitats is explained by traits similar to those generally recognized for high species extinction risk. Indirect developing species exhibiting a big range size, wide elevational range, being independent of streams, and inhabiting the leaf litter are less prone to modifications of their natural habitats. As the effect of a particular trait on the vulnerability of a species might differ among threats [17, 69] and study scales (local vs. global), the generality of our results needs to be treated with caution. However, our findings point to the traits persisting in degraded habitats and thus help to identify future frog communities in our human-dominated world.

## Additional files

**Additional file 1.** Anuran distribution references. References to studies appropriate for the study. Respective data were incorporated in the primary data set on anuran occurrence in different habitat types.

**Additional file 2.** Anuran traits references. References to journal articles, books and web resources containing information on species traits for the species included in the primary data set on anuran occurrence.

**Additional file 3.** Clutch size classes. Additional methods describing the objective grouping of clutch sizes.

**Additional file 4.** Final data set. Given is the species, the traits looked at and the forest dependency index (FDI) derived from the occurrence data. Dummy variables are highlighted in grey. See Table 1 for information on the respective traits.

**Additional file 5.** Affiliation to different anuran families. Number and relative frequency of species belonging to a particular family per forest dependency index.

### Authors' contributions

MH collected data and performed statistical analysis; MH and MOR wrote the text. Both authors read and approved the final manuscript.

### Acknowledgements

M. Dahmen provided additional data on species occurrence in Cameroon. M. F. Barej, M. Dahmen, M. Emmrich, H. C. Liedtke, J. Penner, and J. C. Riemann contributed to the trait data set. F. Tillack assisted during literature research and L. Sandberger-Loua gave valuable advices analyzing the data set. A. Channing helped with a thorough language check. All supports are gratefully acknowledged.

### Competing interests

The authors declare that they have no competing interests.

### Funding

Fieldwork of MH in Cameroon was supported by scholarships from the Federal State of Berlin (Elsa-Neumann-Stipendium) and the German Academic Exchange Service (DAAD). The publication of this article was funded by the Leibniz Open Access Publishing Fund.

**References**


1. Gibson L, Lee TM, Koh LP, Brook BW, Gardner TA, Barlow J, et al. Primary forests are irreplaceable for sustaining tropical biodiversity. Nature. 2011;478(7369):378–81.
2. Hansen MC, Potapov PV, Moore R, Hancher M, Turubanova SA, Tyukavina A, et al. High-resolution global maps of 21st-century forest cover change. Science. 2013;342(6160):850–3.
3. Wright SJ. Tropical forests in a changing environment. Trends Ecol Evol. 2005;20(10):553–60.
4. Nichols E, Uriarte M, Bunker DE, Favila ME, Slade EM, Vulinec K, et al. Trait-dependent response of dung beetle populations to tropical forest conversion at local and regional scales. Ecology. 2013;94(1):180–9.
5. Williams NM, Crone EE, T'ai HR, Minckley RL, Packer L, Potts SG. Ecological and life-history traits predict bee species responses to environmental disturbances. Biol Conserv. 2010;143(10):2280–91.
6. Öckinger E, Schweiger O, Crist TO, Debinski DM, Krauss J, Kuussaari M, et al. Life-history traits predict species responses to habitat area and isolation: a cross-continental synthesis. Ecol Lett. 2010;13(8):969–79.
7. Bregman TP, Sekercioglu CH, Tobias JA. Global patterns and predictors of bird species responses to forest fragmentation: implications for ecosystem function and conservation. Biol Conserv. 2014;169:372–83.
8. Sodhi NS, Bickford D, Diesmos AC, Lee TM, Koh LP, Brook BW, et al. Measuring the meltdown: drivers of global amphibian extinction and decline. PLoS ONE. 2008;3(2):e1636.
9. Rickart EA, Balete DS, Rowe RJ, Heaney LR. Mammals of the northern Philippines: tolerance for habitat disturbance and resistance to invasive species in an endemic insular fauna. Divers Distrib. 2011;17(3):530–41.
10. Cardillo M, Mace GM, Gittleman JL, Jones KE, Bielby J, Purvis A. The predictability of extinction: biological and external correlates of decline in mammals. Proc R Soc Lond B Biol Sci. 2008;275(1641):1441–8.
11. Lee TM, Jetz W. Unravelling the structure of species extinction risk for predictive conservation science. Proc R Soc Lond B Biol Sci. 2011;278(1710):1329–38.
12. Bennett PM, Owens IP. Variation in extinction risk among birds: chance or evolutionary predisposition? Proc R Soc Lond B Biol Sci. 1997;264(1380):401–8.
13. Siliceo I, Díaz JA. A comparative study of clutch size, range size, and the conservation status of island vs. mainland lacertid lizards. Biol Conserv. 2010;143(11):2601–8.
14. Henle K, Davies KF, Kleyer M, Margules C, Settele J. Predictors of species sensitivity to fragmentation. Biodivers Conserv. 2004;13(1):207–51.
15. Davidson AD, Hamilton MJ, Boyer AG, Brown JH, Ceballos G. Multiple ecological pathways to extinction in mammals. Proc Natl Acad Sci. 2009;106(26):10702–5.
16. Peñaranda DA, Simonetti JA. Predicting and setting conservation priorities for Bolivian mammals based on biological correlates of the risk of decline. Conserv Biol. 2015;29(3):834–43.
17. Murray KA, Rosauer D, McCallum H, Skerratt LF. Integrating species traits with extrinsic threats: closing the gap between predicting and preventing species declines. Proc R Soc Lond B Biol Sci. 2011;278(1711):1515–23.
18. McKinney ML, Lockwood JL. Biotic homogenization: a few winners replacing many losers in the next mass extinction. Trends Ecol Evol. 1999;14(11):450–3.
19. Myers N. Mass extinction and evolution. Science. 1997;278(5338):597.
20. Stuart SN, Chanson JS, Cox NA, Young BE, Rodrigues ASL, Fischman DL, et al. Status and trends of amphibian declines and extinctions worldwide. Science. 2004;306(5702):1783–6.
21. Wake DB, Vredenburg VT. Are we in the midst of the sixth mass extinction? A view from the world of amphibians. Proc Natl Acad Sci. 2008;105:11466–73.
22. Catenazzi A. State of the world's amphibians. Annu Rev Environ Resour. 2015;40(1):91–119.
23. Ernst R, Linsenmair KE, Rödel M-O. Diversity erosion beyond the species level: dramatic loss of functional diversity after selective logging in two tropical amphibian communities. Biol Conserv. 2006;133(2):143–55.
24. Riemann JC, Ndriantsoa SH, Raminosoa NR, Rödel M-O, Glos J. The value of forest fragments for maintaining amphibian diversity in Madagascar. Biol Conserv. 2015;191:707–15.
25. Dixo M, Martins M. Are leaf-litter frogs and lizards affected by edge effects due to forest fragmentation in Brazilian Atlantic forest? J Trop Ecol. 2008;24(5):551–4.
26. Bielby J, Cooper N, Cunningham AA, Garner TWJ, Purvis A. Predicting susceptibility to future declines in the world's frogs. Conserv Lett. 2008;1(2):82–90.
27. Laurance WF, Sayer J, Cassman KG. Agricultural expansion and its impacts on tropical nature. Trends Ecol Evol. 2014;29(2):107–16.
28. Williams SE, Hero J-M. Rainforest frogs of the Australian wet tropics: guild classification and the ecological similarity of declining species. Proc R Soc Lond B Biol Sci. 1998;265(1396):597–602.
29. Amphibians on the IUCN Red List. http://www.iucnredlist.org/initiatives/amphibians/analysis/geographic-patterns. Accessed 3 July 2014.
30. Frost DR. Amphibian species of the world: an online reference. Version 6.0. http://research.amnh.org/herpetology/amphibia/index.html. Accessed 17 June 2014.
31. Nichols E, Larsen T, Spector S, Davis AL, Escobar F, Favila M, et al. Global dung beetle response to tropical forest modification and fragmentation: a quantitative literature review and meta-analysis. Biol Conserv. 2007;137(1):1–19.
32. Don A, Schumacher J, Freibauer A. Impact of tropical land-use change on soil organic carbon stocks—a meta-analysis. Glob Change Biol. 2011;17(4):1658–70.
33. Breiman L. Random forests. Machine learning. 2001;45(1):5–32.
34. Liaw A, Wiener M. Classification and regression by randomForest. R News. 2002;2(3):18–22.
35. Zuur AE, Ieno EN, Walker NJ, Saveliev AA, Smith GM. Mixed effects models and extensions in ecology with R. New York: Springer; 2009.
36. R Core Team. A language and environment for statistical computing. Vienna: R Foundation for Statistical Computing; 2015. http://www.R-project.org.
37. Cooper N, Bielby J, Thomas GH, Purvis A. Macroecology and extinction risk correlates of frogs. Glob Ecol Biogeogr. 2008;17(2):211–21.
38. Slatyer RA, Hirst M, Sexton JP. Niche breadth predicts geographical range size: a general ecological pattern. Ecol Lett. 2013;16(8):1104–14.
39. Lips KR, Reeve JD, Witters LR. Ecological traits predicting amphibian populations declines in Central America. Conserv Biol. 2003;17(4):1078–88.
40. White RL, Bennett PM. Elevational distribution and extinction risk in birds. PLoS ONE. 2015;10(4):e0121849.
41. Cardillo M, Mace GM, Jones KE, Bielby J, Bininda-Emonds OR, Sechrest W, et al. Multiple causes of high extinction risk in large mammal species. Science. 2005;309(5738):1239–41.
42. McKinney ML. Extinction vulnerability and selectivity: combining ecological and paleontological views. Annu Rev Ecol Syst. 1997;28:495–516.
43. Blueweiss L, Fox H, Kudzma V, Nakashima D, Peters R, Sams S. Relationships between body size and some life history parameters. Oecologia. 1978;37(2):257–72.
44. Pavlacky DC Jr, Possingham HP, Goldizen AW. Integrating life history traits and forest structure to evaluate the vulnerability of rainforest birds along gradients of deforestation and fragmentation in eastern Australia. Biol Conserv. 2015;188:89–99.
45. Jones KE, Purvis A, Gittleman JL. Biological correlates of extinction risk in bats. Am Nat. 2003;161(4):601–14.
46. Owens IPF, Bennett PM. Ecological basis of extinction risk in birds: habitat loss versus human persecution and introduced predators. Proc Natl Acad Sci. 2000;97(22):12144–8.
47. Wells KD. The ecology and behaviour of amphibians. Chicago: University of Chicago Press; 2007.
48. Ernst R, Rödel M-O. Anthropogenically induced changes of predictability in tropical anuran assemblages. Ecology. 2005;86(11):3111–8.
49. Hillers A, Veith M, Rödel M-O. Effects of forest fragmentation and habitat degradation on West African leaf-litter frogs. Conserv Biol. 2008;22(3):762–72.
50. Lips KR. Decline of a tropical montane amphibian fauna. Conserv Biol. 1998;12(1):106–17.
51. Hödl W. Reproductive diversity in Amazonian lowland frogs. In: Hanke W, editor. Biology and physiology of amphibians. Stuttgart: Gustav Fischer Verlag; 1990. p. 41–60.
52. Murcia C. Edge effects in fragmented forests: implications for conservation. Trends Ecol Evol. 1995;10(2):58–62.

53. Martius C, Höfer H, Garcia MV, Römbke J, Förster B, Hanagarth W. Microclimate in agroforestry systems in central Amazonia: does canopy closure matter to soil organisms? Agrofor Syst. 2004;60(3):291–304.
54. Ernst R, Rödel M-O, Arjoon D. On the cutting edge-the anuran fauna of the Mabura Hill Forest Reserve, central Guyana. Salamandra. 2005;41(4):179–94.
55. Urbina-Cardona JN, Olivares-Pérez M, Reynoso VH. Herpetofauna diversity and microenvironment correlates across a pasture–edge–interior ecotone in tropical rainforest fragments in the Los Tuxtlas Biosphere Reserve of Veracruz, Mexico. Biol Conserv. 2006;132(1):61–75.
56. Spear SF, Crisafulli CM, Storfer A. Genetic structure among coastal tailed frog populations at Mount St. Helens is moderated by post-disturbance management. Ecol Appl. 2012;22(3):856–69.
57. Rittenhouse TAG, Harper EB, Rehard LR, Semlitsch RD. The role of microhabitats in the desiccation and survival of anurans in recently harvested oak-hickory forest. Copeia. 2008;2008(4):807–14.
58. Chen J, Saunders SC, Crow TR, Naiman RJ, Brosofske KD, Mroz GD, et al. Microclimate in forest ecosystem and landscape ecology: variations in local climate can be used to monitor and compare the effects of different management regimes. BioScience. 1999;49(4):288–97.
59. Szarzynski J, Anhuf D. Micrometeorological conditions and canopy energy exchanges of a neotropical rain forest (Surumoni-Crane Project, Venezuela). Plant Ecol. 2001;153(1–2):231–9.
60. Turton SM, Siegenthaler DT. Immediate impacts of a severe tropical cyclone on the microclimate of a rain-forest canopy in north-east Australia. J Trop Ecol. 2004;20(5):583–6.
61. Castelle AJ, Johnson AW. Riparian vegetation effectiveness. Technical Bulletin 799: National Council for Air and Stream Improvement, Inc.; 2000.
62. Kelly DJ, Bothwell ML, Schindler DW. Effects of solar ultraviolet radiation on stream benthic communities: an intersite comparison. Ecology. 2003;84(10):2724–40.
63. Johnson SL, Jones JA. Stream temperature responses to forest harvest and debris flows in western Cascades, Oregon. Can J Fish Aquat Sci. 2000;57(S2):30–9.
64. Álvarez D, Nicieza AG. Effects of temperature and food quality on anuran larval growth and metamorphosis. Funct Ecol. 2002;16(5):640–8.
65. Atkinson D. Ectotherm life history responses to developmental temperature. In: Johnston IA, Bennett AF, editors. Animals and temperature: phenotypic and evolutionary adaptation. Cambridge: Cambridge University Press; 1996. p. 419.
66. Iskandar D, Mumpuni, Richards S. *Limnonectes grunniens*. The IUCN Red List of Threatened Species. Version 2015.2. http://www.iucnredlist.org. Accessed 19 Aug 2015.
67. Shunqing L., Datong Y, Liang F, van Dijk PP, Chanard T, Sengupta S, et al. *Rhacophorus maximus*. The IUCN Red List of Threatened Species. Version 2015.2. http://www.iucnredlist.org. Accessed 19 Aug 2015.
68. Devictor V, Julliard R, Jiguet F. Distribution of specialist and generalist species along spatial gradients of habitat disturbance and fragmentation. Oikos. 2008;117(4):507–14.
69. González-Suárez M, Gómez A, Revilla E. Which intrinsic traits predict vulnerability to extinction depends on the actual threatening processes. Ecosphere. 2013;4(6):76.
70. Penner J. Macroecology of West African amphibians. Ph. D. thesis. Berlin: Humboldt University Berlin; 2014.

# PERMISSIONS

All chapters in this book were first published in ECOLOGY, by BioMed Central; hereby published with permission under the Creative Commons Attribution License or equivalent. Every chapter published in this book has been scrutinized by our experts. Their significance has been extensively debated. The topics covered herein carry significant findings which will fuel the growth of the discipline. They may even be implemented as practical applications or may be referred to as a beginning point for another development.

The contributors of this book come from diverse backgrounds, making this book a truly international effort. This book will bring forth new frontiers with its revolutionizing research information and detailed analysis of the nascent developments around the world.

We would like to thank all the contributing authors for lending their expertise to make the book truly unique. They have played a crucial role in the development of this book. Without their invaluable contributions this book wouldn't have been possible. They have made vital efforts to compile up to date information on the varied aspects of this subject to make this book a valuable addition to the collection of many professionals and students.

This book was conceptualized with the vision of imparting up-to-date information and advanced data in this field. To ensure the same, a matchless editorial board was set up. Every individual on the board went through rigorous rounds of assessment to prove their worth. After which they invested a large part of their time researching and compiling the most relevant data for our readers.

The editorial board has been involved in producing this book since its inception. They have spent rigorous hours researching and exploring the diverse topics which have resulted in the successful publishing of this book. They have passed on their knowledge of decades through this book. To expedite this challenging task, the publisher supported the team at every step. A small team of assistant editors was also appointed to further simplify the editing procedure and attain best results for the readers.

Apart from the editorial board, the designing team has also invested a significant amount of their time in understanding the subject and creating the most relevant covers. They scrutinized every image to scout for the most suitable representation of the subject and create an appropriate cover for the book.

The publishing team has been an ardent support to the editorial, designing and production team. Their endless efforts to recruit the best for this project, has resulted in the accomplishment of this book. They are a veteran in the field of academics and their pool of knowledge is as vast as their experience in printing. Their expertise and guidance has proved useful at every step. Their uncompromising quality standards have made this book an exceptional effort. Their encouragement from time to time has been an inspiration for everyone.

The publisher and the editorial board hope that this book will prove to be a valuable piece of knowledge for researchers, students, practitioners and scholars across the globe.

# LIST OF CONTRIBUTORS

**A Townsend Peterson, Jorge Soberón and Leonard Krishtalka**
Biodiversity Institute, University of Kansas, 1345 Jayhawk Blvd., Lawrence, KS 66045, USA

**T. Ketola, K. Saarinen and L. Lindström**
Department of Biological and Environmental Science, Centre of Excellence in Biological Interactions, University of Jyvaskyla, P.O. Box 35, 40014 Jyvaskyla, Finland

**N L. Rose, S. D. Turner and B. Goldsmith**
Environmental Change Research Centre, Department of Geography, University College London, Gower St, London WC1E 6BT, UK

**L. Gosling**
Centre for Environmental Policy, Imperial College London, 13-15 Prince's Gardens, London SW7 1NA, UK

**T. A. Davidson**
Department of Bioscience, Aarhus University, Vejlsøvej 25, Silkeborg, Denmark

**Rebecca E. Hewitt**
Institute of Arctic Biology, University of Alaska Fairbanks, Fairbanks, AK 99775, USA
Center for Ecosystem Science and Society, Northern Arizona University, PO Box 5620, Flagstaff, AZ 86011, USA

**F. Stuart Chapin III**
Institute of Arctic Biology, University of Alaska Fairbanks, Fairbanks, AK 99775, USA

**Teresa N. Hollingsworth**
3 US Forest Service, Pacific Northwest Research Station, Boreal Ecology Cooperative Research Unit, Fairbanks, AK 99775, USA

**D. Lee Taylor**
Institute of Arctic Biology, University of Alaska Fairbanks, Fairbanks, AK 99775, USA
Department of Biology, University of New Mexico, Albuquerque, NM 87131, USA

**Yan-Hui Zhao**
Key Laboratory for Plant Diversity and Biogeography of East Asia, Kunming Institute of Botany, Chinese Academy of Sciences, Kunming 650201, People's Republic of China
Kunming College of Life Sciences, University of Chinese Academy of Sciences, Kunming 650201, People's Republic of China

**Zong-Xin Ren and Hong Wang**
Key Laboratory for Plant Diversity and Biogeography of East Asia, Kunming Institute of Botany, Chinese Academy of Sciences, Kunming 650201, People's Republic of China

**Amparo Lázaro**
Mediterranean Institute for Advanced Studies, c/Miquel Marquès 21, 07190 Esporles, Spain

**Peter Bernhardt**
Department of Biology, Saint Louis University, Saint Louis 63103, MO, USA

**Hai-Dong Li**
Key Laboratory for Plant Diversity and Biogeography of East Asia, Kunming Institute of Botany, Chinese Academy of Sciences, Kunming 650201, People's Republic of China
Kunming College of Life Sciences, University of Chinese Academy of Sciences, Kunming 650201, People's Republic of China

**De-Zhu Li**
Key Laboratory for Plant Diversity and Biogeography of East Asia, Kunming Institute of Botany, Chinese Academy of Sciences, Kunming 650201, People's Republic of China
Germplasm Bank of Wild Species, Kunming Institute of Botany, Chinese Academy of Sciences, Kunming 650201, People's Republic of China

**Melissa Pavez-Fox**
Instituto de Ciencias Ambientales y Evolutivas, Facultad de Ciencias, Universidad Austral de Chile, Valdivia, Chile
Magíster en Ciencias Biológicas mención Neurociencia, Facultad de Ciencias, Universidad de Valparaíso, Valparaíso 2360102, Chile

**Sergio A. Estay**
Instituto de Ciencias Ambientales y Evolutivas, Facultad de Ciencias, Universidad Austral de Chile, Valdivia, Chile
Center of Applied Ecology and Sustainability (CAPES), Facultad de Ciencias Biológicas, Pontificia Universidad Católica de Chile, Santiago 6513677, Chile

**Zhixin Wen, Deyan Ge, Lin Xia and Qisen Yang**
Key Laboratory of Zoological Systematics and Evolution, Institute of Zoology, Chinese Academy of Sciences, Beichen West Road, Beijing 100101, China

**Yi Wu**
College of Life Sciences, Guangzhou University, Guangzhou 510006, China

**Jilong Cheng and Yongbin Chang**
Key Laboratory of Zoological Systematics and Evolution, Institute of Zoology, Chinese Academy of Sciences, Beichen West Road, Beijing 100101, China
Graduate University of Chinese Academy of Sciences, Yuquan Road, Beijing 100049, China

**Zhisong Yang**
Institute of Rare Animals and Plants, China West Normal University, Nanchong 637009, China

**Kenny Helsen**
Plant Conservation and Population Biology, Department of Biology, University of Leuven, Arenbergpark 31, 3001 Heverlee, Belgium
Department of Biology, Norwegian University of Science and Technology, Høgskoleringen 5, 7034 Trondheim, Norway

**Olivier Honnay**
Plant Conservation and Population Biology, Department of Biology, University of Leuven, Arenbergpark 31, 3001 Heverlee, Belgium

**Martin Hermy**
Division Forest, Nature and Landscape Research, Department Earth and Environmental Sciences, University of Leuven, Celestijnenlaan 200E, 3001 Heverlee, Belgium

**Christoph Reisch, Sonja Schmidkonz, Katrin Meier, Quirin Schöpplein, Carina Meyer, Christian Hums and Christina Putz**
Institute of Plant Sciences, University of Regensburg, 93040 Regensburg, Germany

**Christoph Schmid**
German Research Center for Environmental Health, Research Group Comparative Microbiome Analysis, Ingolstädter Landstr. 1, 85764 Neuherberg, Germany

**Poppy Lakeman-Fraser, Laura Gosling, Roger Fradera and Linda Davies**
Centre for Environmental Policy, Imperial College London, South Kensington, London SW7 1NA, UK

**Andy J. Moffat**
Forest Research, Alice Holt Lodge, Farnham, Surrey GU10 4LH, UK

**Sarah E. West**
Stockholm Environment Institute, University of York, Heslington, York YO10 5DD, UK

**Maxwell A. Ayamba**
Department for the Natural and Built Environment, Faculty of Development and Society, Sheffield Hallam University, Sheffield S1 1WB, UK

**René van der Wal**
Aberdeen Centre for Environmental Sustainability, School of Biological Sciences, University of Aberdeen, Aberdeen AB24 3UU, UK

**Gaylord A Desurmont, Ted C J Turlings and Diane Laplanche1,**
Institute of Biology, University of Neuchâtel, Rue Emile-argand 11, 2000 Neuchâtel, Switzerland

**Florian P Schiestl**
Institute of Systematic Botany, Zollikerstrasse 107, 8008 Zurich, Switzerland

**Carsten F. Dormann**
Biometry & Environmental System Analysis, University of Freiburg, Tennenbacher Str. 4, 79106 Freiburg, Germany

**Lars von Riedmatten**
Biometry & Environmental System Analysis, University of Freiburg, Tennenbacher Str. 4, 79106 Freiburg, Germany
Computational Landscape Ecology, Helmholtz-Centre for Environmental Research, Permoser Str. 15, 04318 Leipzig, Germany

**Michael Scherer-Lorenzen**
Geobotany, Faculty of Biology, Schänzlestr. 1, 79104 Freiburg, Germany

**Linda Davies, Roger Fradera, and Poppy Lakeman-Fraser**
Centre for Environmental Policy, Imperial College London, South Kensington, London SW7 1NA, UK

**Hauke Riesch**
Department of Social Sciences, Media and Communications, Brunel University, London, Uxbridge UB8 3PH, UK

**Duarte S. Viana and Luis Santamaría**
Estación Biológica de Doñana (EBD-CSIC), C/ Américo Vespucio, s/n, 41092 Seville, Spain

**Jordi Figuerola**
Estación Biológica de Doñana (EBD-CSIC), C/ Américo Vespucio, s/n, 41092 Seville, Spain
CIBER Epidemiología y Salud Pública (CIBERESP), Seville, Spain

**Petros Damos**
Department of Environmental Conservation and Management, Faculty of Pure and Applied Sciences, Open University of Cyprus, Main OUC building: 33, Giannou Kranidioti Ave., Latsia, 2220 Nicosia, Cyprus WebScience, Mathematics Department, Faculty of Sciences, Aristotle University of Thessaloniki, University Campus, 59100 Thessaloniki, Greece
Laboratory of Applied Zoology and Parasitology, Department of Crop Production (Field Crops and Ecology, Horticulture and Viticulture and Plant Protection), Faculty of Agriculture, Forestry and Natural Environment, University Campus, 59100 Thessaloniki, Greece

**Glyn Everett**
Faculty of Environment and Technology, University of the West of England, Frenchay Campus, Coldharbour Lane, Bristol BS16 1QY, UK

**Hilary Geoghegan**
Department of Geography and Environmental Science, University of Reading, Whiteknights, Reading RG6 6DW, UK

**Florian Heigl, Kathrin Horvath, and Johann G. Zaller**
Institute of Zoology, University of Natural Resources and Life Sciences, Vienna, Gregor Mendel Straße 33, 1180 Vienna, Austria

**Gregor Laaha**
Institute of Applied Statistics and Computing, University of Natural Resources and Life Sciences, Vienna, Peter Jordan Str. 82, 1190 Vienna, Austria

**Mareike Hirschfeld and Mark-Oliver Rödel**
Department Diversity Dynamics, Museum für Naturkunde Berlin-Leibniz Institute for Evolutionary and Biodiversity Science, Invalidenstraße 43, 10115 Berlin, Germany

# Index

www.ingramcontent.com/pod-product-compliance
Lightning Source LLC
Chambersburg PA
CBHW082025190326
41458CB00010B/3275
* 9 7 8 1 6 8 2 8 6 6 5 2 8 *